Contributions to Management Science

For further volumes:
http://www.springer.com/series/1505

Alper Sönmez

Multinational Companies, Knowledge and Technology Transfer

Turkey's Automotive Industry in Focus

Springer

Alper Sönmez
Department of Economics
Middle East Technical University (METU)
Ankara
Turkey

ISSN 1431-1941
ISBN 978-3-319-02032-7 ISBN 978-3-319-02033-4 (eBook)
DOI 10.1007/978-3-319-02033-4
Springer Cham Heidelberg New York Dordrecht London

Library of Congress Control Number: 2013956627

Printed on acid-free paper

Springer is part of Springer Science+Business Media (www.springer.com)

To My Parents
Gülser & İsmail Sönmez
&
My lovely wife Özlem

Preface

This book provides evidence on how FDI leads to knowledge and technology transfer towards domestic firms by paying attention to the role of multinational companies. The author presents a comprehensive empirical research conducted at firm level in the Turkish automotive industry. Using a representative sample of face-to-face in-depth interviews with top-executives and a survey of top level managers of domestic suppliers, the research analyzes the existence, channels, intensity and determinants, and the kinds of transfers that occur at both inter- and intra-firm level in the industry. The author contends that policies aimed at attracting FDI flows should be reexamined under the findings and insights of this study since it is a necessary—although not sufficient—condition to have an efficient absorptive capacity level and/or skilled human capital stock in order to benefit from these flows. This study has policy implications for the automotive industry as well as practical recommendations for the top-executives of companies and for the public institutions in emerging countries in order to conceive and implement science and technology policies in supporting the knowledge transfers.

Ankara, Turkey
July 2013

Alper Sönmez

Abstract

This study was motivated by the importance of knowledge and technology for economic development especially in emerging countries. In this context, this book aims to provide evidence on whether FDI occurring in the Turkish economy has any effect on domestic firms, especially whether and to what extent it leads to knowledge and technology transfers (KTTs) towards domestic firms by paying attention to the role of multinationals. In order to do so, a comprehensive empirical research based on case study at firm level in the Turkish automotive industry is conducted. By this study, we aim to investigate the existence, nature, and intensity of transfers, and what—if any—kind of KTTs occur at both inter- and intra-firm level in the industry. In addition, the study aims at revealing those characteristics of domestic suppliers that influence the occurrence of such transfers, the place of the firms in the supply chain of multinationals, and the effect of the aforementioned transfers on the performance of domestic suppliers. In the course of this analysis, absorptive capacity, technological capabilities, the volume and nature of innovation, and research and development (R&D) activities of these firms will be explored as well. For these aims, two different research methods are used to collect detailed data from both suppliers and customers at the firm level. First, a detailed and original questionnaire was designed, and later a survey was conducted face-to-face with the top level managers of the 166 domestic suppliers, as the recipients of the transfers. Secondly, face-to-face in-depth interviews based on a semi-structured interview guide were carried out with the 19 top-executives of the 11 Automotive Main Manufacturers (AMMs) in Turkey, as the sources of the transfers. Our main findings in terms of KTTs at interfirm level can be summarized in this way: foreign suppliers seem to dominate local suppliers in terms of many indicators and absorptive capacity level; it is seen that the transfers occur from AMMs to their domestic suppliers mainly on *providing documentations*, *assistance for logistic management*, *quality control*, *know-how*, *R&D*, *co-design and co-development activities*, *designing*, *and cost reduction*; compared to foreign suppliers, local suppliers tend to be involved in those production-product-training-related transfers which are less knowledge-intensive and of a lesser quality; being a direct supplier of AMMs in Turkey and therefore being more close to customers in the supply chain exerts a positive effect on the number, quality, and intensity of the transfers. Moreover, performing R&D activities is found as the most important firm level factor which influences positively *the transfers, strategic collaboration activities with the AMMs, and many technology activities of the suppliers.* The technology policies on attracting more FDI flows should be reviewed under the findings and insights of this study since it is a necessary—although not sufficient—condition to have an efficient absorptive capacity level and/or skilled human capital stock in order to get benefit from these flows.

Keywords: Multinational companies, Knowledge and technology transfer, Foreign direct investment, Automotive industry, Globalization, Turkey

Acknowledgements

This book is partly based on my doctoral thesis written at Department of Economics at Middle East Technical University (METU). First of all, I would like to express my greatest gratitude to my supervisor, Assoc. Prof. Dr. Mehmet Teoman Pamukçu, for his encouragement, untiring support, comments, critiques, guidance, and resolute belief in me throughout this study. Moreover, I would like to thank him for the valuable time that he has devoted to all the periods of my study. I am equally grateful to my advisory committee members, Prof. Dr. Erol Taymaz and Prof. Dr. Ahmet Haşim Köse, for their constant supports, suggestions, insightful comments, encouragements, and constructive remarks. I should express my heart felt gratitude to Prof. Dr. Erkan Erdil, for his encouragement, providing extremely valuable comments, constructive critiques, and helpful insights for the improvement of my study. I would also like to thank Assist. Prof. Dr. İ. Semih Akçomak, for offering many helpful insights.

I am also very grateful to Assoc. Prof. Dr. Garrick Blalock who accepted me as a *visiting fellow* to the Applied Economics and Management (AEM) Department at Cornell University, the USA, for 1 year to conduct a part of my research. During that year, Dr. Blalock helped me so much to improve my work, and I am indebted to him for all he has done for me. Moreover, I would like to thank Prof. Dr. Loren Tauer, the director of AEM Department of Cornell University, for inviting me to Cornell and providing me opportunities to work on my study when I was in Cornell.

My particular gratitude goes to the participating firms without whose cooperation this study would not have been possible. At various stages of this study, I benefited greatly from the interviews with many top-executives of the companies in Turkey automotive main industry. My sincerest gratitude goes out to the people of these companies who devoted their valuable time to express and to share their ideas and experiences with me. In this context, I am especially grateful to Tülay Altay Akarsoy, Volkan Bayraktar, Ernur Mutlu, Dr. Orhan Alankuş, Hakan Özenç, Ali Göktan, Ahmet Altekin, Turgut Cankılıç, Ali Çavuşoğlu, Dr. Senem Erdem Albaş, Ferit Tam, and Önder Molva for helping me out arranging and conducting interviews. I would also like to extend my thanks to all the managers of suppliers who took time off from their busy schedules to participate and to answer my questionnaire for providing valuable information for the study.

I would like to express my acknowledgements to the institutions that provided financial supports for this study, namely, TUBITAK (Grant No: 109K587) for the project entitled *Analysis of Knowledge and Technology Transfer by Multinational Companies to Local Suppliers in the Turkish Automotive Industry*, and Faculty Development Program (OYP) of METU (BAP-08-11-DPT-2002K120510).

I want to express my thanks to Prof. Dr. Ercan Tezer (General Secretary of OSD), Celal Kaya (President of TAYSAD), and Özlem Gülşen (General Secretary of TAYSAD) for their professional supports especially for providing me letter of supports in order to conduct interviews and questionnaire survey with their member firms.

I would like to express my sincere thanks to Prof. Dr. S. Nazli Wasti Pamuksuz from METU for sharing her research experiences in the field of automotive industry. I would also like to extend my thanks to Prof. Dr. Michel Cincera for giving me the opportunity to discuss and improve my study; to Dr. Kriengkrai Techakanont for sharing his knowledge in the field of technology transfer; and to Müfit Akyos for his valuable comments and helpful insights for the improvement of my questionnaire form.

I would like to express my thanks to my colleagues of Department of Economics at METU for their assistance and friendship during my stay as a research and teaching assistant.

I wish to thank my mother, Gülser, my dad, İsmail, and my brother, Ahmet, for their support, patience, and standing behind me in this difficult endeavor. Last but not the least, I want to express my deepest gratitude to my wife Özlem, who has shared the burden of my study most, for her everlasting support and patience, encouragement and understanding, and for sharing my difficult times.

Ankara, Turkey
July 2013

Alper Sönmez

Abbreviations

AF	All Firms
AMM	Automotive Main Manufacturer
BTSO	Bursa Chamber of Commerce and Industry (Bursa Ticaret ve Sanayi Odası)
DAMM	Domestic Automotive Main Manufacturers
DO	Domestic-Oriented
DOSAB	Demirtaş Organized Industrial Zone (Demirtaş Organize Sanayi Bölgesi)
DS	Domestic Suppliers
DSF	Direct Supplier Firms
EO	Export-Oriented
EU	European Union
FA	Factor Analysis
FDI	Foreign Direct Investment
FF	Foreign Firms
FO	Fully Owned
FS	Foreign Share
GMM	Generalized Method of Moments
GOSB	Gebze Organized Industrial Zone (Gebze Organize Sanayi Bölgesi)
GPN	Global Production Networks
ICI	Istanbul Chamber of Industry (İstanbul Sanayi Odası)
ICOC	Istanbul Chamber of Commerce (İstanbul Ticaret Odası)
IMF	International Monetary Fund
ISSS	Industry and Service Statistics Survey
KOSGEB	Small and Medium Enterprises Development Organization (Küçük ve Orta Ölçekli İşletmeleri Geliştirme ve Destekleme İdaresi Başkanlığı)
KTT	Knowledge and Technology Transfer
LEVPET	Levinsohn and Petrin
LF	Local Firms
METU	Middle East Technical University

MITI	Ministry of Industry and Trade
MNC	Multinational Corporations
NACE	Nomenclature Generale des Activites Economiques dans I'Union Europeenne (General Name for Economic Activities in the European Union)
NOSAB	Nilüfer Organized Industrial Zone (Nilüfer Organize Sanayi Bölgesi)
OAMM	Overseas Automotive Main Manufacturer
OECD	Organization for Economic Cooperation and Development
OEM	Original Equipment Manufacturer
OICA	International Organization of Motor Vehicle Manufacturers
OP	Olley and Pakes
OS	Overseas Suppliers
OSD	Automotive Manufacturers Association (Otomotiv Sanayi Derneği)
OYP	Faculty Development Program (Öğretim Üyesi Yetiştirme Programı)
PC	Parent Company
PCA	Principal Components Analysis
PO	Partial Owned
R&D	Research and Development
SANTEZ	Industry Thesis Program (Sanayi Tezleri Programı)
SBSS	Structural Business Statistics Survey
SME	Small and Medium Size Enterprises
SPO	State Planning Organization
TA	Technology Agreements
TAYSAD	Association of Automotive Parts and Components Manufacturers (Taşıt Araçları Yan Sanayicileri Derneği)
TEYDEB	Technology and Innovation Support Programmes (Teknoloji ve Yenilik Destek Programları Başkanlığı)
TFP	Total Factor Productivity
TIM	Turkish Exporters Assembly (Türkiye İhracatçılar Meclisi)
TPS	Toyota Production System
TTGV	Technology Development Foundation of Turkey (Türkiye Teknoloji Geliştirme Vakfı)
TUBITAK	Scientific and Technology Research Council of Turkey (Türkiye Bilimsel ve Teknolojik Araştırma Kurumu)
TurkStat	Turkish Statistical Institute
UIB	Bursa Uludağ Exporter Union
UNCTAD	United Nations Conference on Trade and Development
UNCTC	United Nations Centre on Transnational Corporations
US	United States of America
WCM	World Class Manufacturing
WTO	World Trade Organization
YASED	International Investors Association of Turkey (Uluslararası Yatırımcılar Derneği)

Contents

1 Introduction 1
1.1 Motivation 1
1.2 Objectives of the Study and Main Research Questions 4
1.3 Data and Methodology 5
1.4 Original Contribution 6
1.5 Outline 7

2 Multinational Companies, Knowledge and Technology Transfer: Theoretical Framework 9
2.1 Definitions of Basic Concepts 9
2.1.1 Types of Technology 9
2.1.2 Externalities, Spillovers and Linkages 10
2.1.3 Horizontal and Vertical Technology Spillovers 10
2.1.4 Technology Transfer and Diffusion 11
2.1.5 Knowledge and Technology Transfer 12
2.1.6 Technological Capability and Learning 14
2.1.7 Global Production Networks and Global Supply Chain 14
2.2 Technology Diffusion Through FDI and Growth 15
2.2.1 A Picture of FDI Inflows in Turkey 18
2.3 Multinational Companies and Technology Spillovers 21
2.4 Channels of Technology Spillovers 26
2.5 Determinants of Technology Spillovers 31
2.5.1 Absorptive Capacity and Technological Gap 31
2.5.2 MNC Characteristics 33
2.5.3 Host Country Characteristics 36
2.5.4 Firm Characteristics 38
2.6 Empirical Studies on Technology Spillovers and Transfer 40
2.6.1 General Information 40
2.6.2 Evidence on Developed and Developing Countries 43

2.6.3 Evidence on Turkey 45
2.6.4 Reasons for Failure of Realizing the Technology Spillovers or for Their Negative Effect on Firm Performance 52
2.7 Theoretical Framework for Case-Study Research 53
2.7.1 General Information 53
2.7.2 Technology Transfers at Intra- and Inter-Firm Level 54
2.7.3 Forms of Technology Transfer at Intra- and Inter-Firm Level 55
2.7.4 Theoretical Framework 57
2.7.5 Studies Conducted for Foreign Countries 58
2.7.6 Studies on Turkish Automotive Industry 60
3 Research Methodology and Design 63
3.1 Introduction 63
3.2 Automotive Industry in Turkey 64
3.2.1 Basic Economic Indicators 65
3.3 Scope of the Case Study 73
3.4 The Purpose of the Study and Research Questions 74
3.5 Research Methodology 75
3.6 Data Collection Methods 77
3.6.1 Quantitative Data: Face-to-Face Questionnaire Survey with Suppliers 77
3.6.2 Qualitative Data: Face-to-Face Interviews with AMMs 99
3.7 Descriptions of Some Concepts and Differences for the Firms 106
3.7.1 Automotive Industry, Automotive Main and Supply Industry 107
3.7.2 The Relationships of Automotive Main and Supply Industry 108
3.7.3 Domestic and Overseas Automotive Main Manufacturers (DAMM and OAMM) 109
3.7.4 Domestic and Overseas Supply Industry 110
3.7.5 Domestic and Overseas Suppliers (DS and OS) 110
3.7.6 Local and Foreign Firms (LF and FF) 111
3.7.7 Parent Company (PC) 111
3.7.8 Direct Supplier Firms (DSF) 112
4 Survey Findings: Descriptive Analysis 113
4.1 General Characteristics of the Surveyed Firms 114
4.1.1 Capital Structure 114
4.1.2 Establishment Date (Age) 115
4.1.3 Number of Employees 117
4.1.4 Total Sales 117

4.1.5 Market Orientation 118
4.1.6 Technology Level of the Firms 118
4.2 KTTs at Inter-Firm Level 119
4.2.1 Transfers Related to Production Process 120
4.2.2 Transfers Related to Product 122
4.2.3 Financial Transfers 125
4.2.4 Transfers Through Training 127
4.2.5 The Reasons for Not Receiving Sufficient Amount of KTT 130
4.3 Market and Competition Structure 131
4.3.1 Major Markets 131
4.3.2 Main Competitors 133
4.3.3 Business Environment Related to Competition and Relationship 134
4.4 Technology Capabilities Related to Production and Design 135
4.4.1 Production Capabilities 135
4.4.2 International Quality Certificates 137
4.4.3 Technology Agreements 138
4.4.4 Sources of the Technologies Acquired by Firms 139
4.4.5 Design Capabilities 141
4.4.6 Co-design Activities Related to Products and Partners 142
4.5 Supply Chain 144
4.5.1 Management Structure and MNCs Among Clients 144
4.5.2 Main Customers of Surveyed Firms 145
4.5.3 Primary Sources of the Inputs 148
4.5.4 Purchasing Strategies of the Customers (Being the Sole Supplier of the Customer) 150
4.6 Cooperation 152
4.6.1 The Motivations for Cooperation Activities with Other Firms (Why Do Suppliers Cooperate with Other Firms?) 152
4.6.2 Improvements in the Production Capability as a Result of Assistance Received from the Customers 154
4.7 Direct Suppliers of the AMMs in Turkey 156
4.7.1 How the Firms Became a Direct Supplier 156
4.7.2 Cooperation Activities with AMMs 157
4.7.3 Benefits of Being a Direct Supplier 159
4.8 R&D, Innovation Activities, and Absorptive Capacity 162
4.8.1 R&D Activities (Technological Input) 162
4.8.2 Innovation Activities 164
4.8.3 Absorptive Capacity 170

4.9 Performance and Future Projections 172
4.9.1 Performance Increases: Improvements Related to Production Capability 172
4.9.2 Reasons for Performance Increases (Improvements) in Production Capability 173
4.9.3 Future Projections Related to Acquisition of New Technologies 175
4.10 Summary 177

5 Survey Findings: Econometric Analysis 183
5.1 Model 183
5.2 Explanatory Variables 185
5.3 Factor Analysis (FA) 190
5.4 Logistic Regression Analysis 191
5.4.1 Determinants of Being a Direct Supplier of AMMs in Turkey 193
5.4.2 Determinants of the Decision to Have an R&D Department 194
5.4.3 Determinants of the Innovation Activities 196
5.4.4 Determinants of the Technology Agreements (TA) 199
5.4.5 Determinants of the Sources of Technologies Acquired and/or Used 201
5.4.6 Determinants of the Co-design Activities Related to Products 204
5.5 Ordinal Logistic Regression Analysis 208
5.5.1 Determinants of the KTTs at Inter-firm Level 209
5.5.2 Determinants of the Technology Capabilities of the Suppliers 220
5.5.3 Determinants of the Inter-firm Cooperation (Why Suppliers Cooperate with Other Firms) 222
5.5.4 Determinants of the Improvements in Production Capability as a Result of Assistance Received from Customers 224
5.5.5 Determinants of the Benefits of Being a Direct Supplier 226
5.5.6 Determinants of the Cooperation Level with Partners in Innovation Activities 229
5.5.7 Determinants of the Performance Increases in Production Capability 232
5.6 Summary 235

6 Findings and Analysis of the Interviews 243
6.1 Introduction 244
6.2 General Characteristics of the AMMs Interviewed 247
6.2.1 General Information 247
6.2.2 Production Structure 248

6.2.3 R&D, Innovation, and Technology Activities 251
6.2.4 Domestic Input Rate and Major Sources of Inputs 253
6.2.5 Patent Policy 254
6.3 KTTs Through Partners in Strategic Collaborations 255
6.4 Channels of the KTTs from MNCs to AMMs Operating in Turkey at Intra-firm Level 257
6.4.1 Via Collaboration Activities with MNCs 257
6.4.2 Via Production, Management, and Quality Systems of MNCs 259
6.4.3 Via Auditing, Assessment, and Awarding Activities of MNCs 260
6.4.4 Via Training and Education Activities of MNCs 261
6.5 General Characteristics of the Direct Suppliers of the AMMs Interviewed 263
6.5.1 General Information About the Direct Suppliers 263
6.5.2 Number and Origin of the Direct Suppliers of AMMs 264
6.5.3 Capabilities of the Direct Suppliers 266
6.5.4 Expectations from the Direct Suppliers 267
6.5.5 Supplier Selection Criteria of the AMMs 267
6.5.6 Location of the Direct Suppliers 270
6.6 Channels of the KTTs from AMMs to Direct Suppliers in Turkey at Inter-firm Level 271
6.6.1 Role of Being the Sole Supplier of the AMMs 272
6.6.2 Via Joint Projects 272
6.6.3 Via Co-design Activities 273
6.6.4 Via Auditing, Assessment, and Awarding Activities 274
6.6.5 Via Global Certificates of MNCs 278
6.6.6 Via Training and Education Activities 278
6.6.7 Via Other Assistances by AMMs 280
6.7 SWOT Analysis 281
6.7.1 Strengths (Internal-Positive) 282
6.7.2 Weaknesses (Internal-Negative) 284
6.7.3 Opportunities (External-Positive) 286
6.7.4 Threats (External-Negative) 287
6.8 Summary 288

7 Conclusion 293
7.1 Introduction 293
7.2 Main Findings 295
7.3 Policy Implications and Recommendations 300
7.4 Implications for Further Research 307

Appendix A: Cover Letter for Face-to-Face Questionnaire Survey with Suppliers in Turkey Automotive Industry 309

Appendix B: Cover Letter for Interview Request from the AMMs in Turkey 313

Appendix C: Letter of Support from OSD (in Turkish) 317

Appendix D: Letter of Support from TAYSAD (in Turkish) 319

Appendix E: Fieldwork Instructions for the Questionnaire Survey 321

Appendix F: Structure of the Questionnaire Form 323

Appendix G: Questionnaire Form Applied to Suppliers in Turkey Automotive Industry 327

Appendix H: List of the Domestic Automotive Main Manufacturers (DAMM) 343

Appendix I: Semi-Structured Interview Guide used in Interviews with the AMMs in Turkey 345

Appendix J: Summary Statistics for Key Indicators in the Survey according to Firm Descriptions 349

Appendix K: Description of the Explanatory Variables used in the Econometric Analysis 351

Appendix L: Descriptive Statistics for the Explanatory Variables used in the Econometric Analysis 353

Appendix M: Correlation Matrix 355

References 357

About the Author 371

List of Figures

Fig. 2.1 Evolution of the FDI/GDP ratio in Turkey: 1999–2010 (%) 20
Fig. 2.2 Net FDI inflows in Turkey (BoP Current US$) 20
Fig. 2.3 Ranking of Turkey in the world in terms of net FDI inflows 20
Fig. 2.4 Intra- and inter-firm technology transfers 55
Fig. 3.1 World motor vehicle production by type and year 1999–2010 (million ea.) 65
Fig. 3.2 Share of countries in world motor vehicle production by year 1999–2010 66
Fig. 3.3 World motor vehicle production by country in 2010 66
Fig. 3.4 Share of Turkey's motor vehicle production in world production 67
Fig. 3.5 Motor vehicle production of Turkey by year 1999–2010 (×1,000) 68
Fig. 3.6 Export and import values of Turkey's motor vehicle industry by the years 1999–2010 (million USA dollars) 71
Fig. 3.7 Share of Turkey's motor vehicles industry in total foreign trade by the years 1999–2010 72
Fig. 3.8 Evolution of industry shares in total exports of Turkey: 1996–2010 72
Fig. 3.9 Flowchart for the questionnaire survey 78
Fig. 3.10 Distribution of sample by provinces 88
Fig. 3.11 Distribution of survey results by provinces 90
Fig. 3.12 Share of the interviewed AMMs in total sales and production in 2010 by different foreign shares 105
Fig. 3.13 The automobile production circle 106

Fig. 3.14 Illustration of the basic structure of the automotive industry and concepts used in the analyses 107
Fig. 4.1 Primary input sources of the suppliers based on main customers 150
Fig. 6.1 Share of all and interviewed AMMs in total production and sales in 2010 by different foreign ownership 246

List of Tables

Table 2.1	Empirical studies on spillovers in developing countries	44
Table 2.2	Empirical studies on spillovers for Turkey	46
Table 2.3	Forms of technology transfer at intra- and inter-firm level	56
Table 3.1	Share of AMMs in total capital, capacity, production and sales by foreign share (%)	69
Table 3.2	Distribution of the sample by union, province and corporation type	87
Table 3.3	Distribution of survey results by union and province	89
Table 3.4	Distribution of survey results by provinces and unions	90
Table 3.5	Distribution of unrealized sample by union and province	91
Table 3.6	The reasons of unrealized sample	91
Table 3.7	Basic indicators for surveyed firms and automotive supply industry	94
Table 3.8	Internal reliability of the scales	97
Table 3.9	The list of the Interviewed AMMs	104
Table 3.10	Share of the interviewed AMMs in industry level variables (%)	105
Table 4.1	Distribution of surveyed firms by foreign capital and ownership structure	115
Table 4.2	Distribution of foreign firms by origin and ownership	115
Table 4.3	General characteristics of surveyed firms	116
Table 4.4	Distribution of surveyed firms according to technological complexity of products manufactured (%)	119
Table 4.5	Types of KTTs related to the production process	121
Table 4.6	Types of KTTs related to the product	123
Table 4.7	Types of financial transfers (assistances)	126
Table 4.8	Types and modes of training activities provided by customers	128

Table 4.9 The reasons for not receiving sufficient amount of KTT 131
Table 4.10 Distribution of surveyed firms by market orientation (%) 132
Table 4.11 Distribution of surveyed firms by location of their main competitors (%) 133
Table 4.12 Perception of competition and relationship 134
Table 4.13 Distribution of surveyed firms by production capabilities 136
Table 4.14 Distribution of surveyed firms by quality certifications 137
Table 4.15 Distribution of surveyed firms by technology agreements performed 138
Table 4.16 Distribution of surveyed firms by technology sources 140
Table 4.17 Distribution of surveyed firms by design capability 141
Table 4.18 Co-design activities related to products and partners of surveyed firms 143
Table 4.19 Distribution of surveyed firms by management structure and MNCs among clients 144
Table 4.20 Distribution of surveyed firms by main customers (%) 146
Table 4.21 Main customers of surveyed firms by origin of foreign capital 147
Table 4.22 Main customers of surveyed firms by foreign ownership 147
Table 4.23 Distribution of the surveyed firms according to primary sources of inputs by importance degree (%) 149
Table 4.24 Some indicators of surveyed firms in terms of their customers 151
Table 4.25 Why do surveyed firms cooperate with other firms? 152
Table 4.26 Improvements in various fields as a result of assistance received from the customers (for the last 3-years) 155
Table 4.27 How the firms became a direct supplier 157
Table 4.28 Cooperation activities of the direct suppliers with the AMMs 158
Table 4.29 Benefits of being a direct supplier of the AMM in Turkey 160
Table 4.30 Distribution of surveyed firms by R&D activities 162
Table 4.31 Pioneers of R&D activities 164
Table 4.32 Distribution of surveyed firms by innovation activities and cooperation level with partners 165
Table 4.33 Types of innovation activities (in the last 3-years) 168
Table 4.34 Number of patents owned and applied 169
Table 4.35 Alternative indicators of absorptive capacity 170
Table 4.36 Distribution of surveyed firms by improvements in production capability (in the last 3-years) 173
Table 4.37 Reasons for performance increases (improvements) in production capability 174
Table 4.38 Future projections related to acquisition of new technologies 176
Table 5.1 Summary of the relationships between supplier characteristics and explanatory variables 190

Table 5.2 Determinants of being a direct supplier: logistic regression analysis ... 194
Table 5.3 Determinants of the decision to have an R&D department: logistic regression analysis ... 195
Table 5.4 Determinants of the innovation activities: logistic regression analysis ... 197
Table 5.5 Determinants of the product innovation activities: logistic regression analysis ... 198
Table 5.6 Factor analysis of the technology agreements (TA) ... 200
Table 5.7 Determinants of the technology agreements (TA): logistic regression analysis ... 200
Table 5.8 Factor analysis for the sources of technologies ... 202
Table 5.9 Determinants of the technology sources: logistic regression analysis ... 203
Table 5.10 Determinants of the co-product design activities: logistic regression analysis ... 205
Table 5.11 Factor analysis of the partners in cooperation for the co-product design activities ... 206
Table 5.12 Determinants of the partners in cooperation for the co-product design activities: logistic regression analysis ... 207
Table 5.13 Factor analysis of the KTTs related to the production process ... 210
Table 5.14 Determinants of the KTTs related to production process: ordinal logistic regression analysis ... 211
Table 5.15 Factor analysis of the KTTs related to the product ... 215
Table 5.16 Determinants of the KTTs related to product: ordinal logistic regression analysis ... 215
Table 5.17 Factor analysis of the KTTs through training ... 216
Table 5.18 Determinants of the KTTs through training: ordinal logistic regression analysis ... 217
Table 5.19 Factor analysis of the financial transfers ... 218
Table 5.20 Determinants of the financial transfers: ordinal logistic regression analysis ... 219
Table 5.21 Factor analysis of the technology capabilities ... 220
Table 5.22 Determinants of the technology capabilities related to production, design and R&D: ordinal logistic regression analysis ... 221
Table 5.23 Factor analysis of the inter-firm cooperation ... 223
Table 5.24 Determinants of the cooperation with other firms: ordinal logistic regression analysis ... 223
Table 5.25 Factor analysis of the improvements in production capability ... 225
Table 5.26 Determinants of the improvements in production capability: ordinal logistic regression analysis ... 225

Table 5.27 Factor analysis of the benefits pertaining to being a direct supplier 227
Table 5.28 Determinants of the benefits of being a direct supplier: ordinal logistic regression analysis 228
Table 5.29 Factor analysis of the cooperation level with partners in innovation activities 230
Table 5.30 Determinants of the cooperation level with partners in innovation activities: ordinal logistic regression analysis 231
Table 5.31 Factor analysis of the performance increases (improvements) 232
Table 5.32 Determinants of the performance increases (improvements) related to production capability: ordinal logistic regression analysis 233
Table 5.33 Recapitulative table of logistic regression analyses 236
Table 5.34 Recapitulative table of ordinal logistic regression analyses 237
Table 6.1 Number of direct suppliers of AMMs by foreign share of AMMs and origin of direct suppliers 265

Chapter 1
Introduction

Scientia est potentia

Francis Bacon

1.1 Motivation

Francis Bacon was the first person to express nearly 400 years ago that three major technological inventions—printing in literature, gunpowder in warfare and compass in navigation—had changed the world fundamentally (Bernard & Jones, 1996). The role played by the technological developments, innovations, and policies concerning science and technology in the development process of the countries has gradually gained importance with the phenomena of globalization in the twenty-first century. Although globalization and technological innovations are interrelated processes, it would not be wrong to argue that globalization in general is affected more by science, technological discoveries and innovations. It could be said that globalization has a certain effect on especially the creation and spillover of technology. The role of the technological developments and innovations in the productivity, rise in production capacity and international competitive power of the companies on the one hand, and rise in the employment rates of the countries, their industrialization levels and economic growth on the other hand have gradually increased in the twenty-first century and become a decisive factor.

Growth theories which have been developed since the time when Adam Smith suggested that the growth had been realized upon increases in the productivity as a result of division of labor and specialization, underline the fact that R&D, technology, innovation, productivity, and international trade are the keys to economic growth. Technological innovation occupies a highly important position in the modern growth theories. The importance of the concept of innovation in the development of capitalism has been pointed out by Joseph A. Schumpeter in his book *The Theory of Economic Development* (Schumpeter, 1934). The emphasis he

A. Sönmez, *Multinational Companies, Knowledge and Technology Transfer*,
Contributions to Management Science, DOI 10.1007/978-3-319-02033-4_1,

put on the role and contribution of the innovation (individual enterprise) that is in the interaction with the *creative destruction* process is important in explaining the economic growth of the countries that are considered developed today (Ghosh, 2002). According to Krugman *innovation is a process and new products are created within this process; technology transfer is also a process where old products are transformed into new ones* (Krugman, 1979). Solow (1956) suggested that the productivity of the labor would increase with the introduction of external technological innovation, which, in turn, would lead to economic growth in the long term. Since 1980s, studies concerning internal modeling of the spillover of knowledge and technological innovations, arising from the physical and human capital investments (Romer, 1986) or from the differences in the quality and variety of the inputs (Romer, 1990), have been conducted (Lee, 2008), and works have been made on the policies to best ensure the generation, transfer and use of the technology (Grosse, 1996).

With globalization process, FDI flows realized worldwide by the multinational companies (MNCs)[1] have increased significantly. In this new environment, the developing countries realized that (1) they could attain the available knowledge and technology stock by attracting FDI to their countries, (2) the technologies would be transferred to the local firms via the foreign firms operating in the country, (3) by this way their technological development level would increase in the long run, and (4) they will reach and catch up with the developed countries in the end. For all these reasons, many developing countries have started to implement a series of incentive policies in the last decade in order to attract more FDI flows.

Especially today, the most important factors determining the existence, international competitive power, and status of companies is their skills to develop new, competitive technologies—products—and production processes, to follow closely the new technologies, and to attain and adapt them. To this end, the companies today are increasingly conducting innovation and R&D activities, and establishing collaborations with various partners (such as multinationals, customers, suppliers, and universities) to obtain those technological innovations. Although innovation and technology are the major elements in the economic growth and development process of countries, the interaction between the economic development strategy and science & technology policies plays also a significant role in the long-term economic growth. Many institutions (governments, various private organizations, associations, companies, universities, etc.) come together in many developed and developing countries and devise national science & technology policies in strategic sense in order to support such activities, and the interrelationships and activities of such institutions play an important role in determining the effectiveness of those policies.

It is a common view that the direction and development of the world trade are determined by the multinationals' R&D activities, inventions, Knowledge and

[1] We will use the term "Multinational Companies" (MNCs), "Multinationals", and "Transnational Corporations" interchangeably throughout the study to refer to the same concept.

Technology Transfers (KTTs) they perform. However, there is a thin line between discovering the significance of international technology spillover and accepting creating today's technology is of global in nature. According to Keller (2004), technology today is actually not global, and there is an imbalance between the technology bases of the developed and developing countries. In our era, during which we witness global technological advancements at high-level, the developing countries still remain dependent on the developed countries to attain modern technologies. Since most of the technological innovations, advancements, and inventions in the world are made, controlled, and owned by a few multinational companies or developed countries. On the other hand, the technologies developed by the companies located in those countries could be used for the industrialization of the developing countries. Many countries that are considered developed now made use of the advanced technologies that existed during the time they were in the development phase (Tanaka, Iwaisako, & Futagami, 2007).

In sum, today, technology creates a competition advantage for the companies. It plays a major role in the success of the domestic and foreign companies, across a wide area ranging from the creation of new products to the knowledge and experience owned. Similarly, it is an important base for economic growth and development, so governments work on policies to optimize technology creation and use (Grosse, 1996). It has been recognized that economic development strategies that are effectively combined with science & technology policies serve as the key driver of long-term economic growth.

This study was motivated by the importance of knowledge and technology for economic development especially in emerging countries. It is obvious that science & technology policies, innovation, and transfer of technology are also of central importance in the process of Turkey's development, industrialization, and in the race of being an industrialized country in a developing country context. To gain competition power in international arena and to be able to race with foreign firms, Turkish industry has to do enormous enterprises about technological innovations. It does not seem possible that these enterprises can be formed by the own sources of Turkish industry without governmental supports, especially in the arena where the big international companies have been supported by their countries' technological and industrial policies. In the lights of these, first of all, Turkey has to determine the obstacles in front of the technological development and produce alternative policies to eliminate them with the cooperation of public, universities, and industry. Secondly, Turkey has to make a long term national program for technological development, innovation, and transfer of technology from developed countries in order to increase knowledge accumulation at home. In this context, there is a need for the research on whether multinationals transfer technology to domestic firms or not in emerging countries such as Turkey. In addition, it is significant and important to analyze the company, industry, and country characteristics of these technology transfers.

1.2 Objectives of the Study and Main Research Questions

In the scope of the aforementioned discussions, this study aims to provide evidence on whether FDI occurring in the Turkish economy has any effect on domestic firms, especially whether and to what extent it leads to KTTs towards domestic firms by ascribing special emphasis to the role of multinationals.[2] In this context, the main objective of the study is to investigate KTTs through multinationals in Turkey, and to find out how the domestic firms have benefited directly or indirectly from these. Also, the policies on attracting more FDI with advanced technology will be questioned. For these aims, we want to find answers for the following questions guiding this study:

- What are the effects of multinationals in Turkish manufacturing industry?
- What—if any—kind of KTTs occurs at both inter- and intra-firm levels through multinationals, and what is the nature and intensity of these transfers?
- What are the main channels of KTTs from multinationals?
- What are the determinant factors that affect KTTs from multinationals?
- What is the role of domestic firm characteristics in this process?
- Are there any differences between local and foreign firms in Turkey?
- Does technological capability (related to R&D, innovation, production, absorptive capacity, design, etc.) of the domestic firms play a role in this process?
- Do these KTTs have any effect on the performance level of the domestic firms?
- What policies should Turkish government implement to attract more FDI with advanced technology from developed countries?

In order to investigate the above questions addressed, we conducted a detailed and comprehensive empirical research based on a case study at firm-level in the Turkish automotive industry. By this case study, we investigated the transfers at both inter- and intra-firm level in the following way:

- In terms of KTTs at inter firm level, we examined and looked at our objectives from two perspectives of these interactions: the first one is from the suppliers' perspective as a recipient of these transfers by questionnaire survey method, and the second one is from the customers' perspective as a source of these transfers by in-depth interview method. Whether the FDI flows in the Turkish automotive industry have resulted in the KKTs to the local supplier firms; and if it has resulted in so, the intensity and relative importance of these transfers, as well as analyzing their effect on the performance of the firms, are the main questions

[2] From here on, *knowledge and technology transfers* (KTTs) term is used throughout the study to analyze the transfers quantitatively through backward linkages, and to see what kinds of transfers are provided from customers to suppliers. We preferred to question firms by asking *what kind of KTTs are provided from your customers*. "Knowledge transfer" term is preferred to use for the tacit transfers such as training, know-how, and assistances; on the other hand, "technology transfer" term is preferred to use for the explicit transfers such as documents, machinery, equipment, manuals, and blueprints.

that inspire the case study and that constitute the study's point of origin. Therefore, the main goal of the case study is to investigate and review what kind of the transfers are provided via the multinationals to the supplier firms operating in the Turkish automotive industry, in terms of *products* (design, joint design, joint activities, various documentation, etc.), *production processes* (various know-how, documentation, R&D, logistics, etc.), *training* and *financial transfers*. In addition, the study also aims at revealing those characteristics of the suppliers that influence the occurrence of such transfers; the place of the suppliers in the supply chain of multinationals; the effect of the aforementioned transfers on the performance of the suppliers; collaboration activities of the suppliers with their customers; their absorptive capacity, production, design and technological capabilities; the volume and nature of innovation and R&D activities of these suppliers are explored as well.
- In terms of KTTs at intra-firm level, by using the qualitative data obtained from the interviews conducted with the AMMs in Turkey, we analyze the channels of the transfers realized by the multinationals to the AMMs, the characteristics of the AMMs, their cooperation activities with the multinationals as their foreign partners (global AMMs), and their R&D and technology-related activities.

Various surveys on topics such as technology transfers, innovation, R&D, and cooperation activities of the firms are conducted by public institutions in certain countries with a well-developed automotive industry such as USA, European Union, South Korea, Thailand, Malaysia and China, where intensive collaborations among public, universities and the industry are witnessed, and vital findings are obtained for the industry to see what lies ahead. However, such studies have begun to be conducted in Turkey only recently. Therefore, another objective of the study is to fill the gap to some extent; to contribute in the ability of the industry to adapt to the sustainable competition environment; to prevent the leading role acquired in the industry from being lost to other rival countries; and to provide a contribution to guide in the country's becoming a regional production and excellence base in near future. For these purposes, we develop some policy implications and recommendations for the Turkish automotive industry.

1.3 Data and Methodology

Detailed information and data are required at the firm level for the case study investigating the transfers at both inter- and intra-firm level. For that reason, we use two different research methodologies (quantitative and qualitative) to collect primary data at firm level. Firstly, face-to-face questionnaire survey, based on original and designated questionnaire form, is conducted with the top level managers of the 166 suppliers in the Turkish automotive supply industry in order to collect quantitative data. The validity and reliability of the scales in the questionnaire are assessed by using internal consistency measured with Cronbach's alpha (α).

The *construct validity* of the scales is also tested by using Factor Analysis in Chap. 5 in order to determine correlated items in the scales and to reduce the number of related items. Secondly, in-depth semi-structured interview method is used to collect qualitative data from top-executives of the 11 AMMs operating in Turkey. The findings obtained from the interviews are used for the quantitative results on surveyed suppliers in a complementary way, and for the analysis of transfers at intra-firm level.

1.4 Original Contribution

We believe that this study would provide some valuable input for the discussions on the transfers through multinationals in emerging countries. This study also contributes to the literature as one of the first, and rare, empirical evaluations to understand the types, main channels, and determinants of the transfers at both inter- and intra-firm level, conducted with two separate field research. Because of great methodological difficulties in addition to lack of data, only a few studies investigating the KTTs empirically in some countries including Malaysia, Thailand, and South Korea appear in the literature. According to our knowledge, the interviews with the main companies and survey conducted amongst suppliers in the industry, and the econometric analysis conducted here using quantitative data collected via the questionnaire survey is one of the first studies of its kind for the industry. Here, our questionnaire form designed to collect firm-level data from the suppliers can also be regarded as an original contribution.

When it is considered that this kind of case-studies include very detailed, comprehensive and confidential information on the firms, and requires a larger period of time, research team and budget, and need very effort to be successful, it might be claimed that the study was carried out with success. To the best of our knowledge, in this context, this is one of the first empirical researches carried out successfully in national and international level on FDI-related transfers at both inter- and intra-firm level. We believe that the findings of the study would give important clues to the Turkish automotive main and supply industry firms in maintaining their current competitive status and determining the strategies they could follow in the world markets that have become increasingly competitive. At the same time, we believe that the study would contribute to the other studies conducted in this field in academic literature and that it would constitute an important reference for the public institutions as well as various private organizations and associations in Turkey (such as SPO, KOSGEB, TUBITAK, OSD, and TAYSAD) in order to construct science and technology policies in supporting the activities of the industry's innovation, R&D, and technology transfer.

In sum, this study can be seen as an original contribution and unique in terms of exploring the transfers at both inter- (between suppliers and their customers) and intra-firm level (between AMMs and their foreign partner multinationals) through two separate field research carried out with suppliers and main companies in the industry. We believe that our study could contribute to more extensive studies conducted for Turkey and other countries yet to follow, setting an example for them.

1.5 Outline

This book is organized as follows: Chap. 2 provides a review of the theoretical and empirical literature on FDI with an emphasis on the technology spillovers, multinationals, growth and transfers especially by multinationals after a brief introduction on the definitions of the basic concepts used in the study. This chapter, then, investigates the channels and determinants of spillovers, and presents the evidences of empirical studies on the spillovers and transfers for host economies and Turkey. This chapter also presents a descriptive analysis of FDI flows to Turkey. Lastly, it summarizes the analytical framework for the case-study research in Turkish automotive industry.

The research methodology and design, adopted for the case study in order to collect primary quantitative and qualitative data at firm-level in the industry, is discussed in detail in Chap. 3. This chapter presents a detailed discussion of methodologies utilized for data collection and data analysis in the research. Quantitative data are collected from the suppliers in the automotive supply industry through face-to-face questionnaire survey, and qualitative data from top-executives of the AMMs in Turkey through in-depth interviews. It also presents some statistical information on the industry. Moreover, the scope, the purpose, and main research questions of the study are discussed. Lastly, some concepts and differences for the firms constructed and used in the analyses will be described.

Chapter 4 is devoted to a broad descriptive analysis of the primary quantitative data collected through face-to-face questionnaire survey from the suppliers in the industry.

In Chap. 5, econometric analyses are conducted by using quantitative survey data in an attempt to identify the possible determinant factors impacting on different types of the transfers realized by customers to suppliers through backward linkages at inter-firm level, and on various characteristics and on activities of the suppliers. In this chapter, econometric model, estimation methods, explanatory variables and factor analysis will be discussed in detail together with the econometric findings.

Chapter 6 will analyze the qualitative data obtained from the in-depth interviews conducted with 11 AMMs operating in Turkey. It presents characteristics of the AMMs interviewed, their cooperation activities with multinationals (global AMMs) as their foreign partners; R&D and technology-related activities; and channels of the transfers realized by multinationals to the AMMs at intra-firm level. Moreover, the relations of the AMMs with their direct suppliers operating in Turkey, the channels and determinants of the transfers at inter-firm level realized by the AMMs to their direct suppliers through backward linkages will be analyzed.

Lastly, the results of SWOT analysis of the industry in terms of the transfers will be presented. The findings obtained from these analyses will be used also for the quantitative survey results attained in Chaps. 4 and 5 in a complementary way.

The last chapter summarizes what has been done in this study, the main findings and conclusions derived from them. Moreover, it is devoted to policy implications and recommendations for Turkish automotive industry, and guidelines for further researches.

Chapter 2
Multinational Companies, Knowledge and Technology Transfer: Theoretical Framework

In this chapter, we review mainly the theoretical and empirical literature on FDI with an emphasis on the technology spillovers and transfer realized by multinationals. In the first section, definitions of the basic concepts used in the study are explained briefly. Next, in the second and third sections, technology diffusion through FDI, growth and the relationship between spillovers and multinationals are presented. Then, the fourth section is devoted to the analysis of main spillover channels identified in the literature. We then examine the determinants of the spillovers in the section five. Theoretical and empirical evidence on the technology spillovers and their effects on the host economies are presented in the sixth section. Review of the related literature on spillovers for Turkey constitutes an important part of the study. The last section summarizes the theoretical framework for the case-study research on both intra- and inter-firm technology transfers in Turkish automotive industry, and reviews the literature on similar case-studies conducted on both foreign countries and Turkey.

2.1 Definitions of Basic Concepts

2.1.1 Types of Technology

We will use the term "technology" here with a broad meaning to include product and production process technology as well as knowledge and skills including management, marketing, organization, know-how, international markets, and global production networks. Some authors need to narrow the scope of the technology and grouped it into two types, namely, "hardware" and "software" technology (Techakanont, 2002, pp. 12–13). The authors call the technology embodied in physical goods as "hardware technology" (such as machinery, equipment, blueprints, and technical specifications), while they call the knowledge embodied in workers' experience and skills in terms of product and production process as

A. Sönmez, *Multinational Companies, Knowledge and Technology Transfer*,
Contributions to Management Science, DOI 10.1007/978-3-319-02033-4_2,

"software technology" (Cantwell, 1993; Kim, 2001; Teece, 1977). Moreover, some authors differentiate these technologies as "explicit" and "tacit knowledge". "Explicit knowledge" is defined as a codified and transmittable knowledge; sources (multinationals) provide this kind of knowledge to recipients (such as their affiliates or suppliers in host countries) through machinery, blueprints, technical manuals, training handouts, technical specifications, and quality control methods (Kim, 2001). On the other hand, "tacit knowledge", introduced firstly by Polanyi (1962), is defined as knowledge related to "practical knowledge", and it cannot be easily expressed, solved, transferred or declared openly to another person. It is also described as "know-how" or "embodied knowledge", and the sentence *we can know more than we can tell* best describes the notion of "tacit knowledge". In sum, the term "technology" includes both explicit and tacit knowledge here. However, in our study, we prefer to use the term "knowledge transfer" for the transfer of "tacit knowledge", and "technology" for the transfer of "explicit knowledge" in order to examine in detail both types of the technology.

2.1.2 Externalities, Spillovers and Linkages

It is necessary to point out the distinctions between "externalities", "spillovers" and "linkages" here by using the definitions of Narula and Driffield (2012). "Externalities", which are simply the positive or negative benefits from presence of multinationals or from their actions, affect local firms or host economies at no cost. On the other hand, "spillovers" are the indirect effects of multinationals on host countries. In other words, they are externalities that accrue from one firm (source) to another (recipient) through formal or informal relationships between the firms. Recipient firms must put a learning effort to benefit from such spillovers. Also, all "spillovers" are externalities, but all externalities are not spillovers. By the way, "linkages" imply transactional associations between source and recipient firms, and these may not give rise to spillovers. In other words, all spillovers are linkages, but all linkages are not spillovers. The term "spillover" has been generally used in "aggregate empirical analyses" on "production function", while the term "linkages" is preferred in studies on individual firms or multinationals as the unit of analysis (see Narula & Driffield, 2012, pp. 2–3 for further details).

2.1.3 Horizontal and Vertical Technology Spillovers

In the last few decades, FDI flows have played an important role in achieving economic growth and development especially for developing economies. FDI may affect host country economies directly through increased employment, foreign exchange earnings, capital accumulation, and also the usage of more advanced equipment and technology. However, especially in the last decade, it has been considered that the most important channel through which FDI affects developing economies is situated on the technology front/side. Indeed, the most important contribution of FDI to a developing economy consists in fostering technology transfer by bringing and diffusing new technologies, knowledge, and skills to the recipient country. Transfer of these intangible technology-related elements unintentionally from foreign to local firms is called *technology spillovers* based on FDI. In other words, spillovers are referred to the indirect effects of FDI on domestic firms in host country. These can be horizontal (intra-industry) or vertical spillovers (inter-industry) depending on whether they are disseminated within or outside the sector of activity of foreign firms that trigger these spillovers. In other words, horizontal spillovers occur from foreign firms to local firms operating in the same industry, while vertical spillovers occur from foreign firms to local firms operating in vertically linked (backward or forward) industries (see Sect. 2.4 for details). Vertical spillovers emerge from interaction of the multinationals with the suppliers and customers in the local market. The multinationals could provide technical assistance to the suppliers in the local market in order to ensure the quality of the intermediate inputs they use, could train the employees, and could help in purchasing the raw materials. We could assume that there exist the backward spillovers here. The forward spillovers, on the other hand, emerge from the local manufacturers' purchasing the intermediate or capital goods from the foreign suppliers. Since the products of the foreign suppliers are technologically more superior than the products of the local suppliers, the multinationals could provide the local firms with knowledge to ensure that these inputs are used efficiently by the manufacturers. The econometric studies analyzing the both horizontal and vertical spillovers together are very limited, and they have generally revealed significant positive vertical spillovers than horizontal ones (Pamukçu & Taymaz, 2009, p. 9).

2.1.4 Technology Transfer and Diffusion

What is meant by the "technology transfer" here is the process through which technology is intentionally transmitted between countries or firms. In other words, we define "technology transfer" as the direct type of the spillovers that occurs voluntarily from technology source (such as multinationals or foreign firms) to recipient (such as local firms, affiliate of multinational or suppliers in host country) by way of embodied in the equipment supplied (such as machinery, manual, and equipment) or disembodied in the forms of software, patents, knowledge, or know-how and skills provided by training and education activities (see Bennett, 2002; Radosevic, 1999 for further details on *technology* and *technology transfer*). In this process, there are three factors that play an important role in technology transfer; the source, the recipient, and "the technology itself" (Techakanont, 2002). On the other hand, "technology diffusion" is generally used to refer to the unintentional transfer of technology that may occur as a result of reverse engineering or imitation (Rath, 1994). This type of technology transfer is also called "technology spillover" as a positive externality effect in the literature and occurs through horizontal and vertical linkages by means of various channels (see Sect. 2.4).

Technology transfers via FDI have been through multinationals in a host country. MNCs operating in host country may transfer some of its technology in terms of product, production process, and organization to recipient firms, which are generally affiliates or the local suppliers of multinationals. However, today attracting multinational is not the only way to obtain technology transfers; they can also be transferred in two ways. The first one is contractual agreements between source and recipient for the transfer of specific technologies that include licensing and management arrangements, technological assistance agreements, purchase of machinery-equipment-assembly apparatus, recruitment of the foreign specialists, patents, and brands. The second one is the transfer of skills, experiences, and knowledge through technical assistance, know-how agreements, labor mobility, turnkey projects, service, and engineering agreements (UNCTC, 1994).

2.1.5 Knowledge and Technology Transfer

"Knowledge transfer" and "technology transfer" concepts are used interchangeably in innovation and development literature. Nevertheless, while "technology transfer" involves rather the transfer of capital goods such as machinery and equipment, "knowledge transfer" includes mostly transfer of tacit knowledge such as know-how, management, and technical skills. Knowledge transfers seem to be more important than technology transfers since they ensure learning of new information, technical and organizational skills (UNCTAD, 1997). Knowledge transfers are realized from one company to another; and its potential benefits depend on the long-term relationships between the companies concerned, the level of the knowledge transferred, and the abilities of the recipient company. The technology transfer process—this term will hereinafter be used in a way to cover the transfer of both the explicit and tacit knowledge—involves acquisition of the technology, absorption of the technology, adaptation of it to local conditions, improving and developing it, its spillover to other companies (Eden, Levitas, & Martinez, 1996; Tung, 1994). Business literature generally refers to three kinds of technology transfer (Grosse, 1996): Product (knowledge used in order to produce any product), process (knowledge used in order to organize the inputs and operate the machinery during production), and transfer of the management skills (knowledge used in order to operate an enterprise). Techakanont and Terdudomtham (2004) clearly explain the concept of technology transfer and its process in the following excerpt:

> The term of technology transfer in its broadest sense defines the process of creation of the knowledge, skills and experience by the recipient of the technology, as a direct result of the contributions of the sources from where the technology is provided. It could be said that the transferring process mentioned above has been completed upon understanding the transferred technology well, absorbing it, adapting it to the local conditions, ensuring its maintenance, sustainment and effective use, by the recipient of the technology. In other

> words, the technology transfer process is the process of internalization or learning of the transferred technology-related knowledge and experience by the recipient. Effective realization of the transfer process depends on two important factors: (i) that the technology recipient has a certain basis of knowledge, and (ii) the intensity of the effort to develop the existing knowledge base. These factors are vitally important in that they determine how fast and how successfully the transferred technologies are internalized by the local suppliers. Especially intensity of effort to develop the existing knowledge base is more important than the existing knowledge base itself because the former creates that latter, but not vice versa (Techakanont & Terdudomtham, 2004, pp. 156–160).

The "efforts" emphasized in the last sentence of the citation above ensure that the recipient company understand, implement, absorb and evaluate the new technologies better, and develop the accumulation of knowledge about them. Therefore, such efforts ensure success of the technology transfer. A successful technology transfer would result in advancement of the technological capabilities of the recipient company personnel playing a role in the transfer, and also in increased productivity of the company's production process.

Recipients can assimilate the transferred "explicit knowledge" into products and production processes by using their "tacit knowledge" (see Kim, 1997; Lall, 1996; McKelvey, 1998; Nelson & Winter, 1982 on discussions about tacit knowledge). Therefore, it is necessary for the recipients to have a certain level of "knowledge base" (tacit knowledge) to benefit from the transferred technology. In other words, acquiring of the external knowledge by recipients requires experienced and skilled human capital. Developing of the "knowledge base" can be possible through specific ways such as extensive personnel contacts and regular interactions with the personnel of foreign firms in host or home countries, mutual trust between source and recipient firms, social networks, regular visits to the plants of foreign customers in order to get on-the-job or off-the-job training activities provided by foreign firms, various assistances of foreign firms in terms of quality control systems, production process, and distribution. For instance, in certain periods, MNCs invite production and management personnel of the local firms in host countries to visit their plants or headquarters in home or host country. They show them how production systems, lines, and organization work by off-the-job training activities, and give them various on-the-job training activities. Moreover, MNCs can send their own expert personnel to assist their local suppliers in solving the technical problems encountered in the application of MNC's production and engineering systems (a kind of tacit knowledge). Consequently, "knowledge base" is based on experience and know-how of the personnel, and this can be accumulated and acquired in many years through training, research, and practice (see Ernst & Kim, 2002 for further details). The creation of new technology and knowledge could also be possible by interaction of the explicit and tacit knowledge categories (Collins, 2001; Goffin & Koners, 2011; Wagner & Sternberg, 1985).

2.1.6 Technological Capability and Learning

According to Kim (2001), *technological capability* can be defined as an ability to make efficient use of technological knowledge, which is gained from either internal or external sources, in manufacturing, production process, R&D, design, engineering, and innovation. It is expected that firms with higher technological capability can acquire, assimilate, and improve new technologies more easily, and also they can deal with creating new technologies. On the other hand, *technological learning* is the process through which technological capability is improved, accumulated and formed. Such ability may be developed through "learning by research" in developed countries, or through imitative "learning by doing" process in developing countries (Kim, 2001, p. 297).

Technological capability of a firm can be considered as a continuous process to acquire, absorb, internalize the knowledge and adapt into local conditions, and improve it. It is also determined by accumulation of skills and knowledge, and by efforts to improve this. The firm-level capabilities can be categorized as "investment capabilities" (the skills needed to obtain new technologies for design, construction, equipment), "process and production capabilities" (quality control, operation, monitoring, controlling, maintenance, research, design, and innovation), and "linkage capabilities" (the skills needed to transmit knowledge and technology to and receive them from various institutions, suppliers, customers) (Lall, 1992, p. 168). The complexity degree of these capabilities can be assessed by activities from which these capabilities arise such as "experience", "search", and "research". Firms must improve and develop their capabilities in order to internalize and get a new technology into production. The factors that have impact on the activities of the firms in developing their capabilities can be summarized as the size of the firm; ability of the technologies from the market; dealing with R&D and innovation activities; cooperation activities with other firms in terms of design, product development, etc.; high-quality human capital; "organizational and managerial skills"; "ability to change structures to absorb new technologies"; "access to appropriate embodied technology"; and "access to external technical knowledge" from various sources (such as foreign firms, MNCs, local firms, special engineering and consultancy firms, laboratories, testing facilities, and standards institutions) (see Lall, 1992 for further details).

2.1.7 Global Production Networks and Global Supply Chain

Globalization process has changed all world economies. One of the most important effects of globalization has occurred in the choices of firms in their location choices for production, R&D, and innovation activities through "international foreign trade", development of "information and communication technologies", and liberalization policies of the host countries to attract high FDI flows (Narula & Dunning,

2000). In this process, the global production networks (GPN) have emerged due to increase in international competition and the need for "the organization of international trade". These networks have been established by multinationals to integrate their supply, technology, R&D, innovation, and production centers into GPN. They consist of hierarchical layers (such as "independent suppliers and subcontractors", "joint-ventures", "affiliates", "R&D alliances", and "subsidiaries"), and they have increased the knowledge diffusion among such actors. Local suppliers in host countries of multinationals (lower-tier suppliers) are at the bottom of these hierarchical layers. "Global supply chain" is also established as a result of these networks in order to provide integrated intermediate goods and services to the firms in GPN (based on Narula & Dunning, 2000).

These networks have played an important role in fostering the advanced technology and knowledge diffusion across countries and firms in two ways:

- By increasing their capabilities; multinationals demand high quality products at low cost and to be delivered on-time. In other words, their working standards and technical specifications are very high and these may put high pressures on local suppliers, especially in developing countries. If local suppliers could not fulfill the demands of multinationals in terms of quality, cost, delivery, durability, and reliability, they may face the threat of expulsion from the production and supply networks of multinationals. Therefore multinationals are endeavoring to upgrade their suppliers' capabilities and productivity.
- Multinationals as the final-user of the products supplied have provided new opportunities for local firms to benefit from the knowledge diffusion. Multinationals may provide financial and technical assistance, knowledge and technology transfers (including engineering, product and production process, on-the-job and/or off-the-job training activities, various blueprints, machinery, raw material, managerial systems, etc.) to the local suppliers in the GPN in order to raise their technical capabilities in terms of production and process because multinationals must be sure about the quality of the products to be supplied (see Ernst & Kim, 2002; Narula & Dunning, 2000 for further details).

2.2 Technology Diffusion Through FDI and Growth

Developing countries consider that FDI by multinationals give rise to technology diffusion and it is a major channel for gaining access to advanced technologies and knowledge of developed countries (Borensztein, Gregorio, & Lee, 1998). In their study, Borensztein et al. (1998) and Xu (2000) verify this by showing that FDI is the most important source of economic growth for developing countries, and they reveal that FDI flows from developed countries are an important channel through which technology transfers have been provided to local firms of developing countries. Moreover, they show that technology transfer through FDI has a more positive effect on economic growth than domestic investment when host countries have a sufficient level of human capital. They suggest that the stock of human capital plays an important role in the absorptive capacity by emphasizing the role of the education level in order to get benefit from FDI. In a similar study, Blomström and Kokko (1997) also show that there is a positive relationship between the benefits of FDI and technological capability level of domestic firms. In addition, Romer (1986, 1990) in his "endogenous growth theory" emphasized the role of human capital in order to acquire the foreign technologies and to benefit from FDI in host countries by suggesting that human capital would lead to economic growth by using new technologies and efficient production techniques.

The simple models in the neoclassical economic theory assume that technology is freely available across, and within countries, and among all firms. This theory basically assumes that technological knowledge is a public good and non-rivalry, when it is created every person that is willing to pay the price of technology may benefit from it, in other words, it is non-excludable. This view expresses itself best in the motto *free international flows of capital and technology*. According to this view, all firms are considered to be operating on the same production function and they decide how much labor and capital will be used in accordance with their relative factor prices. In this process, developing countries can transfer advanced technologies from developed countries, and adaptation and assimilating of these technologies are not required, and technical efforts by firms in the learning and absorbing of these technologies are not seen as important factors (see Lall, 1992). In other words, knowledge, capability, cost, and tacit nature of technologies are not seen important; it is assumed that they can be easily acquired by developing countries, and there is no need for adaptation of such technologies to local conditions (Bell & Pavitt, 1997). In line with the neoclassical growth theory, Wang (1990) also conducted a model by assuming that increase in "knowledge" used in the production is determined as a function of FDI (Borensztein et al., 1998). Even though neo-classical approach suggest that there is not any obstacle to restrict the spillovers from multinationals to local firms, and that spillovers occur efficiently, findings and evidence on developing countries in the literature show that there are very limited benefits from multinationals, or they have revealed negative spillovers from multinationals (Narula & Marin, 2003).[1]

On the other hand, "unconventional" approaches, inspired by "evolutionary theory"—developed by Nelson and Winter (1982) and explained in Dosi (1988)—propose that "indigenous technological efforts" of the firms in developing countries in adapting, learning, and improving of new technologies from developed ones play a deterministic role by implying that markets are inefficient in the diffusion of technology (Lall, 1992). This approach is mainly based on these assumptions: (i) knowledge and technology are not easily available and fully transferrable across firms since they are created as a result of long R&D and

[1] For instance, Haddad and Harrison (1993) for Morocco; Aitken and Harrison (1999) for Venezuela, Konings (2001) for Bulgaria and Romania, Djankov and Hoekman (2000) for Czech Republic have found negative evidences.

innovation activities at high costs; (ii) they are protected by creators in order to prevent the diffusion of these technologies to other rival firms; (iii) the adaptation of these technologies requires high technological capability and capacity, skilled and experienced human capital, effort, investment, and learning process by the recipient due to tacit character of these technologies. These also explain why there are differences among firms in terms of their technology levels, production processes, and capabilities (based on Lall, 1992; Pack & Saggi, 1997). Therefore, the approach of "the evolutionary theory" seems more realistic than the neoclassical approach.

In the traditional growth theory, the role of the technology was left as an unexplained residual, whereas recent literature on growth revealed that state of the domestic technology relative to that of the rest of the world plays an important role in the process of economic development (Barro & Sala-i-Martin, 1995). According to this theory, economic development of a backward country depends on its absorptive capacity, technological capabilities, infrastructure, and human capital stock. In other words, economic development, and growth are explained by a "catch-up" process in the level of technology and it depends on the extent of absorption, adaptation, acquisition and implementation of new technologies, and knowledge that are already in use in developed countries (Borensztein et al., 1998). Therefore, in modern economic growth theory, stock of knowledge, its intensity and generation play a very important role and determine a country's economic growth (Kuznets, 1966).

Today, technology is seen as a major determinant for industrial development and economic growth. According to Pack and Westphal (1986), *industrial development* is a process through which accumulation of knowledge and technology capabilities are conducted especially by learning and translating them into new product and production process innovations. The studies related to the developed countries revealed that 50% of the long-term economic growth stems from technological developments that improve R&D, innovation, productivity or lead to new products and processes (Grossman & Helpman, 1991). One of the important examples is South Korea, which has achieved to transform its economy from agrarian to industrialized one during the past 4 decades through technology and industrial development. Therefore, many governments in developing countries nowadays have been studying and examining the South Korea experience as a case study to be successful in their own countries by applying science and technology policies or strategies (see Kim, 1998).

Kim (1998) postulates that industrial development process of developing countries consists of three stages: "acquisition", "assimilation", and "improvement" of foreign technologies from developed countries through technology transfers rather than relying entirely on foreign sources. In the first stage, developing countries acquire technologies from advanced countries through "assembly processes", "product specifications", "production know-how", "technical personnel", and/or "components and parts" because of insufficient technological capabilities of local firms. In other words, firms in the developing countries try to imitate foreign technologies and products through reverse engineering; therefore, this stage consists of "assembly production" for "standard products" and requires only an "engineering" capability. In the second stage of industrialization, "production" and "product design" technologies are quickly improved by "assimilation of foreign technologies". In this stage, firms undertake "creative imitation" and try to produce differentiated products by using their technical capabilities on "engineering" and "development". In the last stage, after the successful assimilation of foreign technology, local firms shift to "original innovation" and they can undertake research, development, and engineering activities by their increased scientific, technological, and engineering capability in order to improve and develop new technologies and products. In these stages, if they are successful, countries accumulate knowledge and technology, and they increase their technological capabilities, and in the end, they may be a candidate to be an advanced country (see Kim, 1998, 2001 for further details).

Developing countries benefit from globalization process through liberalization of capital flows and export oriented policies. These policies also provide them with access to advanced technologies and sophisticated products invented in developed countries, which trigger growth rates, and thus they may have a chance to catch up with the developed countries (Keller, 1996). However, when a small part of the technology can be obtained by these processes (blueprints, manuals, specifications, etc.), they are not sufficient to bring those countries to the level of developed countries successfully without adopting, assimilating, and understanding the knowledge embodied in these technologies. Therefore, it is also necessary to have an efficient absorptive capacity level and a skilled and experienced human capital stock (engineers, workers, and managers) in order to get benefit from these technologies (see Evenson & Westphal, 1995; Grossman & Helpman, 1991; Keller, 1994; Pack, 1992). For instance, in their studies, Borensztein et al. (1998) and Xu (2000) reveal that economic growth is positively affected by FDI in only developing countries with a certain threshold level of absorptive capacity.

In sum, technology spillovers from FDI and the accumulation of knowledge are seen as a key determinant factor for productivity of local firms in and economic growth of developing countries. These can be increased by R&D and innovation activities, or by knowledge diffusion from external sources. It is considered that FDI give rise to inflow of advanced knowledge and technologies into the host countries; hence, today knowledge diffusion from FDI is a very important factor especially for the developing countries to access the latest technologies. FDI can bring technology embodied in goods and services, and knowledge as intangible assets; however, to benefit from them is not an automatic process, and it requires that recipient must have some capabilities to absorb, assimilate, and adopt such technologies (see Kinoshita, 2001 for details).

2.2.1 A Picture of FDI Inflows in Turkey

Although the first law on foreign capital was enacted in 1954, Turkey was a relatively closed market to foreign investment until 1980. After pursuing inward-oriented economic policies based on an import-substitution development strategy implemented through Five-Year Development Plans since the 1960s, Turkey switched to outward-oriented policies following a severe balance of payment crisis in the early 1980s. In 1989, liberalization of international capital flows occurred. Signing the Customs Union agreement with the European Union in 1995 further contributed to the liberalization of its economy. The policies focused on attracting foreign investment, promoting export activities, and minimizing state intervention. Note that efforts to open the Turkish economy to international markets were not enough to attract more FDI. For instance, from 1950 to 1980 the cumulative authorized FDI had only reached US$229M (Önis, 1994). After the government initiated a stabilization program in 1980 that paved the way to an open economy, the legislative background was also reorganized to eliminate favoritism among foreign investors, local content requirements, minimum export requirements, and restrictions on transfer of capital and profits (Akpınar, 2001; Erdilek, 1986). In addition to changes in the regulatory framework, privatization of state economic enterprises, liberalization of the financial system, elimination of restrictions on foreign exchange, establishment of a stock exchange and heavy investment in telecommunications technology all contributed to the development of a favorable environment for FDI throughout the 1980s. However, in the following decade, two major economic crises in 1994 and 1999 as well as reliance on short term capital flows resulted in a relatively poor FDI performance.[2]

When we look at the 2000s, we see a much more favorable environment for foreign investors with a strongly regulated financial system, a low inflation rate and the establishment of a Coordination Council for the Improvement of the Investment Climate. Following the enactment of the new foreign capital law in June 2003, minimum capital requirements and permits were eliminated; the ownership of property by foreigners without any restrictions, the right to international arbitration and employment of expatriates were granted. Partly as a result of these measures a sharp rise occurred in FDI from 0.56% of GDP in 2003 to 3.8% in 2006 which was followed by a fall after 2006 (see Fig. 2.1).

Policies implemented after the early 1980s to open up Turkish economy were not enough to increase the FDI inflows to Turkey. As can be seen in Fig. 2.2, until the year 2001, annual FDI flows to Turkey were rather low (below US$1B) compared to other emerging economies (see UNCTAD, 2005). Total cumulative net FDI inflows were nearly US$10.7B between 1974 and 2000, corresponding to an annual average of US$400M. Moreover, the place of Turkey was in the 35th rank by US$982M in 2000 in terms of FDI inflows among all countries in the world (see Fig. 2.3). From 2001 onwards there has been an important increase in the FDI flows: total cumulative net FDI inflows reached nearly US$100B between 2001 and 2010

[2] See Kepenek and Yentürk (2003), Önis (1994), Erdilek (1986; 2003), Akpınar (2001), Alıcı and Ucal (2003), Yılmaz (2006), Yılmaz and Barbaros (2005), Taymaz and Yılmaz (2007), Pamukçu and Taymaz (2009) for detailed information on FDI flows in Turkey in terms of historical background, obstacles, structure, discussions and various statistical data.

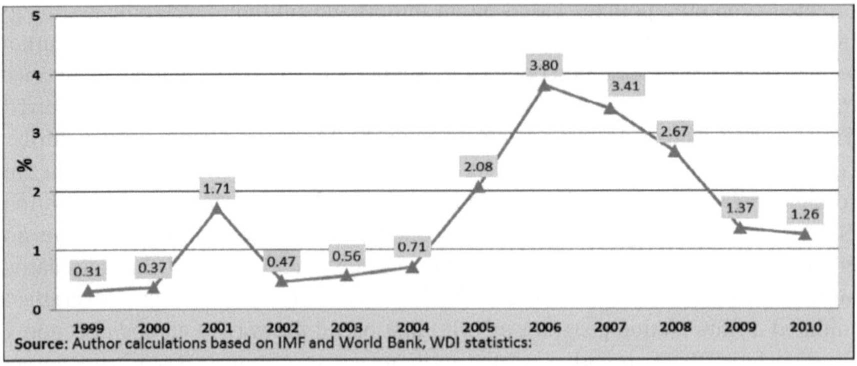

Fig. 2.1 Evolution of the FDI/GDP ratio in Turkey: 1999–2010 (%)

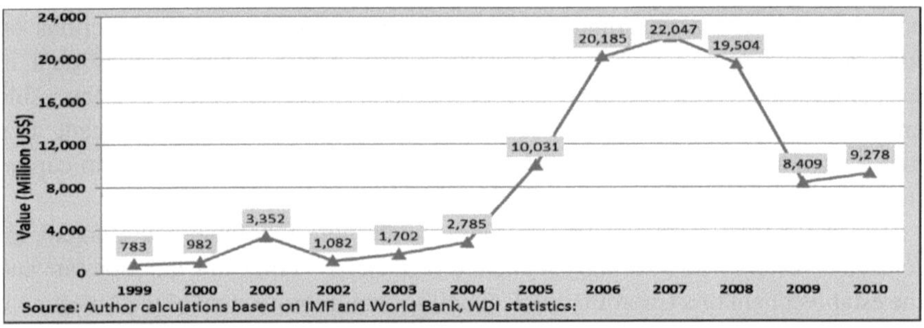

Fig. 2.2 Net FDI inflows in Turkey (BoP Current US$)

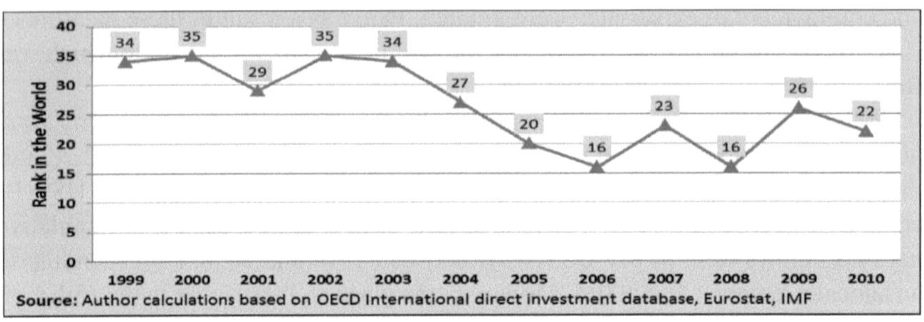

Fig. 2.3 Ranking of Turkey in the world in terms of net FDI inflows

(annual average of US$10B) especially after the Turkish government has started to liberalize its investment policy in 2004 (last stage of privatization program). As illustrated in Fig. 2.2, FDI inflows peaked in 2007 with US$22B. Moreover, as seen from Fig. 2.3, the ranking of Turkey in the world in terms of FDI flows for the period 1999–2010 increased from 35 (in 2000) to 16 (in both 2006 and 2008) for the first time within all countries that attract most FDI flows in the world.

2.3 Multinational Companies and Technology Spillovers

It is known that multinationals are the major players in global production, FDI flows, R&D, and innovation; they are the technology producers of the world. A significant part of the R&D investments in the world are made by these companies (Borensztein et al., 1998). While producing technology, these companies also control the worldwide technology (Eden et al., 1996). Significant information and statistics on the production and technology activities of those companies can be found in various publications and reports, published by OECD and UNCTAD annually (such as main science and technology indicators, and world investment reports). For instance, according to UNCTAD (2011) statistics, the value added generated by multinationals worldwide in 2010 was nearly US$16 trillion, about a quarter of global GDP. Also, total sales/(value added) generated by foreign affiliates of multinationals in the world accounted for US$33 trillion/(US$7 trillion) in 2010. They also conducted exports that are worth more than US$6 trillion, about one-third of global exports. The trade volume of the multinationals' headquarters and their affiliates located in other countries is nearly one-third of the total world trade. Eighty percent of the R&D activities in the OECD countries are performed by the multinational companies who have more than 10,000 employees. Again, 75–80% of the private R&D expenses worldwide are made by the same companies. However, the majority of the technology production and basic R&D activities are realized in the home country of these multinationals. For example, USA-based multinationals conduct only 13% of their R&D research in other countries (UNCTAD, 1997, 2003). More than 65% of the global R&D research total is conducted only in five countries. These countries are the USA (33%), Japan (13%), China (9%), Germany (6%), and France (4%) (OECD, 2009). When we add the next five countries—South Korea, England, Russia, Canada and Italy—, this ratio increases to 80%. This situation reflects that four-fifths of the world's R&D is conducted by just ten countries (OECD, 2009). All these statistics show that multinationals have become the predominant decision-makers in global economic system, technology, innovation, and R&D. For these reasons, multinationals are seen as the creators, major sources, and diffusers of new technologies for the host countries.[3] Therefore, all studies that investigate the technology spillovers have focused on these companies.

Multinational companies invest substantially in R&D, innovation, test facilities, design, and advanced and sophisticated new technological fields, and their products are recognized internationally. These activities are much costly and risky, and require high budgets. Therefore, multinationals owe their dominant position to these activities as the major producers and sources of advanced technologies.

---

[3] See Lenger (2004) on the discussion of MNCs and for further details.

Moreover, they have established their production, distribution and marketing networks worldwide, by which they can operate in many locations of the world and they can protect their strategic technologies through many ways (intellectual property rights, patents, and other informal ways) (UNCTC, 1994).

In the last decades, together with globalization process, FDI accompanied by multinationals is considered a major factor for economic growth and development strategy. Hence, almost all governments, particularly developing countries which do not want to reinvent the wheel, compete with each other in order to attract more FDI flows by liberalizing their FDI regulations, by creating favorable conditions, and by offering incentives in a wide range for multinationals[4] (see Narula & Driffield, 2012; Narula & Dunning, 2000; Narula & Marin, 2003; UN, 1999). Moreover, FDI is also being promoted by international institutions (WTO, OECD, IMF, EU, UN, UNCTAD, Washington consensus,[5] etc.) that help countries to attract more multinationals by giving them advice, training, and various assistances (Narula & Driffield, 2012). Because of these reasons, the relationships of the governments with MNCs have gained impetus, and the role of the MNCs in developing countries has significantly increased in their economies (UNCTAD, 1997, 1998). Accordingly, FDI flows by MNCs have increased significantly and it is seen as the most important external source of technology for developing countries (UNCTAD, 2000). In sum, developing countries rely on MNCs for their technological development and for transferring of the advanced technologies.

According to the general view, it is considered that MNCs associated with FDI will bring with them superior technologies (which includes modern and advanced technologies in terms of product, production process, marketing, and distribution), know-how, skills (in terms of organization/management/marketing), export contacts, well established and designed systems for relationships with suppliers and customers on top of the inflow of physical capital (Aitken & Harrison, 1999; Blomström & Kokko, 1998; Caves, 1971; Hymer, 1960; Kokko, 1996). Host countries hope that they will have a chance to gain access to the technology by inviting foreign investment, and they hope that they will acquire these in the end. It is also considered that FDI will result in benefits for national income, employment, capital formation, productivity, foreign trade, and technology structure of the host country. All of these will increase the technological capability and productivity level of the host country firms through their interactions with MNCs. That means they rely on FDI as a source of technology and capital (Lall, 1997; UNCTAD, 1998). These kinds of benefits arise through externalities discussed above with the

---

[4] These incentives are usually in the form of special tax concessions, lower income taxes, exemptions from import duties, extension of tax holidays, subsidies for infrastructure and direct subsidies.

[5] The term was firstly used by the economist John Williamson in 1989. It refers to ten recommendations of three institutions (the World Bank, the IMF and the US Treasury department) for developing countries based on neo-classical economic theory to establish a more market oriented economy.

terms *FDI productivity spillovers* (external effects) or *technology spillovers*,[6] and it is considered as the most important benefit from FDI (see Blomström & Kokko, 1998 for further details).

In the literature, technology spillovers are associated with the direct or indirect benefits of MNCs, and they generally occur when the entry of MNCs lead to productivity increases in the local firms of host country. Direct benefits can be summarized as increases in physical capital, employment, and usage of advanced machinery and equipment. On the other hand, indirect benefits are the various technology spillovers from MNCs to local firms. According to this view, technologies of MNCs is in the nature of public good to some extent and generate positive externalities because they cannot be fully internalized by MNCs and local firms of host country may benefit from these externalities. In other words, technology spillovers are thus a matter of externalities being provided by foreign firms in host country to local ones (see Blomström, 1986). These are considered as one of the most significant channels for the diffusion of modern technology across countries, rather than formal technology transfer arrangements (Romer, 1990).

Spillovers from MNCs impact also positively on human capital level in host country by direct and indirect ways. Entry of a multinational into domestic market is expected to directly increase the employment level and capability level of the domestic human capital by providing on-the-job or off-the-job training and education, by "learning by doing" activities, and by transferring their "technological knowledge" to their domestic workers (engineers, R&D, production, and management personnel). These education and training activities are expected to increase the quality and tacit knowledge level of the workers in host country. In addition, local firms increase their own employment because of "increased economic activity", and it is expected that MNCs provide training and technical assistance to their direct suppliers in host country. Direct suppliers of MNCs gain access to more productive human capital stock: their staff are employed and trained by MNCs, or they hire workers who already have the knowledge and technology of MNCs (see Narula & Marin, 2003 for further details).

MNCs invest generally in overseas countries in the expectation of acquiring higher profits by using their firm-specific advantages. These advantages of MNCs are associated with technology (including economies of scale, patent, capital intensity, human capital, innovation, R&D, etc.), "managerial ability", "organizational skills", "the ability to coordinate resources or supply chains", international distribution and production networks, "access to finance", and "the knowledge of markets" (Dunning, 1979; Narula & Driffield, 2012). Dunning (1993) explains potential sources of these advantages within "OLI model" (which stands for ownership, location, and internalization advantages) that may have impact on a firm's decision to become a MNC. According to this, "ownership advantages" (O) of MNCs allow them to compete successfully in host countries due to their firm-specific advantages such as better marketing techniques, export contacts,

[6] We will use the terms "productivity spillovers" and "technology spillovers" interchangeably throughout the study to refer to the same concept.

reputation, superior managerial, production, and technological capabilities, and so on. Dunning uses "location advantages" (L) term in order to refer to the benefits offered by host country to attract the MNC (investment and tax incentives, special tariffs, etc.) and the access to good infrastructure and production factors of host country at low cost. Lastly, "internalization advantages" (I) refers to advantages of MNCs to invest in a country rather than exporting, licensing, or joint venture. In short, according to Dunning, determinants of MNCs' activity abroad are influenced by these three factors.

MNCs generally establish an affiliate in overseas countries in order to produce and sell goods. These affiliates use local inputs of capital, labor, and intermediate products. Moreover, these affiliates provide new technologies to host countries by both development through R&D activities in host country and importing from its parent company in home country (Hines, 1995). According to the 'industrial organization' approach, affiliates of MNCs can compete with local firms in host country because of their firm-specific intangible assets[7] (see Caves, 1996). These assets are in the form of explicit technology (machinery, tools, equipment) and tacit knowledge (capabilities, experience, know-how) about "organizational governance" based on intra-firm hierarchies and on "the advantages of common governance" (Lall & Narula, 2004). Due to less costly and easily transferable/transferred specifications of these assets, MNCs prefer direct investment in host country rather than licensing these assets to local firms, and they protect their advantages by this way (Aitken, Harrison, & Lipsey, 1996). On the other hand, some studies that analyzed why MNCs invest in the host countries rather than directly exporting or giving license of their products/technologies revealed that one of the reasons is to protect these advantages due to market failures and non-existence of proprietary rights (see Caves, 1996; Görg & Greenaway, 2004; Markusen, 1995). Therefore, even if explicit technology can be acquired by local firms, tacit knowledge cannot be easily gained by such firms. The only way that local firms can profit from external benefits of MNCs is the technology transfers from MNCs, but MNCs will not easily transfer the source of their advantages to the local firms in host country.

Due to globalization process, MNCs have a dominant role in the technology transfers, and they play an important role in the transfer of advanced technologies across countries. Technology transfer from developed countries through MNCs may provide increases in knowledge base and technological capabilities, and may serve as a vehicle for technological change and learning; however, experiences reveal that relying on fully-owned or joint-venture MNCs are not necessarily an effective way to acquire foreign technologies. These firms may transfer production technology, but they do not transfer their engineering, innovation capabilities, and

[7] Such as trademarks, technology, knowledge, organizational, marketing and managing skills, export experiences and contacts worldwide, production methods and marketing advantages, coordinated strategic relationships with suppliers and/or customers, distribution networks established worldwide, and reputation.

their tacit knowledge when parent company used FDI to exploit the local market. Therefore, these kinds of FDI as a means of technology transfer may lead to foreign dependency (see Kim, 1998 for further details).

Technology spillovers depend on many factors. One of them is "the scope and competence" of the MNC affiliate. According to this view, MNC affiliates with complex production systems and advanced technology need high level of local capabilities, skilled human capital, strong clusters, service firms, government incentives, substitutions, and institutions. That is one of the important reasons of why MNCs generally prefer to establish and to conduct their basic R&D and design activities in their home countries. If countries can provide these, then MNCs will establish high-tech affiliates in host country, and high quality of spillovers from these affiliates can be expected (Lall & Narula, 2004). In their study, Wang and Blomström (1992) developed a model to analyze the international technology transfer, and they revealed that these were resulting from the strategic interactions between affiliates of MNCs and local firms in host country. They also point out that learning efforts of local firms have positive impact on such technology transfers provided by MNCs. On the other hand, establishment structure of the affiliates also determines the quality of spillovers. For instance, spillovers from an affiliate in the structure of sales office are limited than that of manufacturing plants.

Consequently, in the last decades, there has been a significant change in developing countries in terms of openness and attracting more FDI flows. The main reasons for this change can be summarized as globalized production networks, easy access to international markets and production networks of MNCs, and need of acquiring advanced technologies of MNCs. In this process, the role of MNCs as the creators of high technologies has grown in all countries and they have become the leader of technology transfers. They are seen as the only source of the technology creation and as dominant actors in international technology diffusion. However, the MNCs that use high-technology generally prefer to invest in developing countries with high level capabilities in terms of human capital, better infrastructure, and production. That is, liberalizing FDI policies is seen as a necessary condition for developing countries, but not sufficient to attract more MNC with high technology and to benefit from it. Besides, it requires strong local technological capabilities that MNCs need (Lall, 1997; Lall & Narula, 2004). In other words, there is no any reason to believe that FDI flows through MNCs lead to an increase in exports, competition, capacity, and productivity of local firms and result in technology spillovers to local firms, which ultimately determine economic growth in the long run. Conversely, it is necessary to have a certain "knowledge base", "intensity of effort", and "technological capacity" in order to benefit from spillovers acquired through MNC activities (Lall & Narula, 2004).

2.4 Channels of Technology Spillovers

This section discusses in detail the major channels that earlier literature identified through which the spillovers from foreign firms to local firms can occur. There are a number of spillover channels identified in the literature, but we categorized these as FDI, horizontal and vertical (backward and forward) linkages with foreign firms, "demonstration/imitation", "labor turnover" (skill acquisition), "competition", and "exports" (see Blomström & Kokko, 1998; Görg & Greenaway, 2004; Kinoshita, 1999; Kokko, 1992; Wang & Blomström, 1992). In the literature, three of these channels—demonstration, labor turnover, and competition—are referred as horizontal ones.

Via FDI: It is known that FDI is generally the most important channel through which superior technologies (equipment, machines, skills, knowledge, etc.) can be transferred from developed countries to developing ones. Developing countries, where local technologies and capabilities are very low, hope that MNCs, by establishing plants in host country, may provide technology transfers to their suppliers and sub-contractors, and they may also allow these firms to access their international networks. By this way, these countries may upgrade their technologies and conduct high-tech activities, and local firms may enhance their productivity (Glass & Saggi, 2002). Hence, FDI is seen as the largest source of external financing of the development and growth for less developed countries. For this aim, many governments have followed policies to ease restrictions on FDI and offered special incentives to MNCs such as tax exemptions, areas for establishing plants free of charge, and subsidies for infrastructure to attract foreign investment. The basic idea behind these policies as discussed before is to benefit from the spillovers occurring from FDI through technology spillover (Aitken & Harrison, 1999).

Via Vertical Linkages: Another important channel through which technology spillovers occur is vertical linkages (backward and forward), the relationship between foreign and local firms. The importance of this channel for host countries to benefit from FDI was introduced to the literature by Lall (1980) and Mead (1984), and developed further by Markusen and Venables (1999), Blomström and Kokko (1999), Pack and Saggi (2001). The spillovers that occur through these linkages are also referred as vertical spillovers (or inter-industry spillovers). These spillovers occur through forward and backward linkages of MNCs with local firms who become their suppliers and customers. According to general view, spillovers associated with vertical linkages with foreign firms give rise to more advanced, better quality and specific transfers to local firms, and more improvement in the productivity since MNCs voluntarily share their firm-specific technologies only with their closer local suppliers (backward linkages) and customers (forward linkages) for strategic reasons. Moreover, foreign firms may need to assist their local suppliers or customers to raise the quality of products supplied, and by this way local firms also improve their productivity.

- ***Backward Linkages***: In the case of backward linkages (selling output to foreign customers), the entry of MNCs may impact local suppliers positively by increasing the demand for local inputs. Moreover, local firms as suppliers or sub-contractors of MNCs may be forced to produce output with higher quality standards and technical specifications, and by using more efficient technologies. Therefore, MNCs may provide technical assistance or introduce new technologies for their suppliers in order to raise the quality of the intermediate products manufactured according to technical specifications enforced by MNCs (supply side) (Moran, 2001). Furthermore, MNCs may provide on-the-job and/or off-the-job training activities to both management and production personnel of their suppliers to ensure the organizational and management structure, delivery times, and production standards of the goods to be procured. MNCs may help their local suppliers to set up production facilities by providing technical assistance. In addition, MNCs may insist on using the high quality standards and procedures of their own on the production processes, or they may help their suppliers by giving materials-tools, or by providing some incentives or financial assistance to upgrade their technologies (demand-side) (Blomström & Kokko, 1999; Lall, 1980).

The spillovers and technology transfers provided by MNCs to their suppliers through backward linkages discussed in the preceding paragraph are well known practices especially in the global automotive industry. Global automotive manufacturers provide various technologies to their overseas affiliates or joint-ventures in host country in order to develop their production technology, to increase efficiency, and to provide firm-specific benefits. Foreign affiliates as the final users of the products being supplied by local firms assist their local suppliers in various ways as summarized in the preceding paragraph. The strategic relationships between buyers and suppliers in this industry are very strong and depend on mutual trust established in many years. Moreover, there are very close communication and strategic relationships between such firms in every step of the production, and supplier firms have actively participated in the production and design processes of products. These are also the most important reasons why we analyze the Turkish automotive industry as a case study in this book (see Sect. 3.3).

- ***Forward Linkages***: In the case of forward linkages (local firms' purchasing intermediate inputs from foreign suppliers), the most important benefit from MNCs is the supply of higher quality inputs at lower price to local firms (Markusen & Venables, 1999). Furthermore, MNCs may provide various assistances (such as technical services, manuals, training activities, seminars, and courses) to their local customers about using the products supplied by MNCs more efficiently (Görg & Greenaway, 2004). It must be kept in mind that if increases in production quality lead to increases in prices, and if absorptive capacity of the local firms is insufficient to benefit from this upgrade of quality, they will be negatively affected by increased costs (Javorcik, 2004). This situation is called *negative forward spillovers* and it may also help to explain the negative findings on such spillovers (see Sect. 2.6.4 for other reasons of negative spillovers).

Because of the reasons discussed above, vertical linkages are considered as one of the most important channels to increase the technological capabilities of local firms (see Lall, 1980; Lin & Saggi, 2004; Markusen & Venables, 1999; Navaretti & Venables, 2004; Rodriquez-Clare, 1996). In addition, vertical spillovers are seen as more important than horizontal ones (intra-industry spillovers) in terms of technology spillovers because of direct and close strategic relationships between foreign and local firms (Moran, 2001).

Empirical studies for vertical spillovers have produced mixed results for each country. For instance, Kugler (2001) finds positive horizontal (intra-industry) spillovers but no evidence for vertical spillovers (inter-industry) in Colombia manufacturing industry. Javorcik (2004) in her study for Lithuania and Blalock and Gertler (2003) for Indonesia find positive productivity spillovers through backward linkages but no evidence for forward and horizontal linkages. Moreover, the estimations of Driffield, Munday, and Roberts (2002) for UK show that there are positive spillovers through forward linkages but not significant spillovers through backward linkages. In other studies for UK, Harris and Robinson (2004) find that horizontal spillovers are more important than vertical ones, and there are negative spillovers in many sectors. Girma, Görg, and Pisu (2003) in their study for UK also find that export oriented domestic firms benefit more from vertical linkages than domestic orientated ones.

In sum, firm-specific technology of the foreign firms may spillover to local firms in host country through vertical linkages. With these linkages, local firms are exposed to new products, production and marketing techniques, and to more rigorous quality and working conditions, and they may receive direct technical support and training activities from upstream or downstream foreign firms. Moreover, foreign firms may also act as a stable customer of the local firms for their input demands, and this gives rise to maintaining long-term and mutual trust that is based on strategic relationships with foreign firms. In all these cases, foreign firms would raise the productivity gains for local firms through vertical linkages (see Aitken & Harrison, 1999 for further details).

Via Demonstration/Imitation: *Demonstration* by MNCs or *imitation* by local firms is probably the most evident spillover channel. Foreign technology embodied in FDI can be acquired by observing foreign firms or by imitating some technologies used by MNCs. Local firms, by these ways, can improve their productivity and efficiency. For instance, when a new MNC enters a market, it introduces new technologies in terms of production, process, quality control, distribution systems, and organization into market, and local firms may improve their capabilities and production methods by adopting and copying these through imitation or reverse engineering. These externalities are also called as "contagion" effects (see Blomström & Kokko, 1998; Cantwell, 1989; Das, 1987; Kokko, 1996; Wang & Blomström, 1992 for details). These are generally observed among firms operating in the same industries, so these are also referred as horizontal spillovers. A number of case studies have identified the presence of spillovers through demonstration in various countries (see Blomström & Kokko, 1997).

Via Labor Turnover (Skill Acquisition): One of the channels through which spillovers occur to local firms is labor turnover. Training of the employees at the MNCs and recruitment of this trained workforce by the local firms could enable the local firms, though indirectly, to access the production and management skills of the MNCs. According to this, workers employed by MNCs acquire knowledge of its technology by training activities and experiences, and this knowledge may move to local firms by labor turnovers. Role of labor turnover as a channel of technology spillover was examined by Glass and Saggi (2002). It is expected that labor turnover mostly occur among firms within same industries, so it is also referred as horizontal spillovers.

It is agreed that, as a result of having firm-specific intangible assets, MNCs demand highly skilled and experienced workers from labor market. Moreover, when a worker joins the firm, they invest a lot in regular training activities, and they train their employees continuously in order to provide the latest technologies. Therefore, skills and knowledge of the domestic workers employed and trained by MNCs may spillover to local firms when they set up their own firms or when they are hired by local firms in host country (Kokko, 1996). Moreover, hiring workers who have acquired knowledge and advanced technologies of MNCs is seen as one of the important channels for the transfer of tacit knowledge in terms of management, quality control, distribution, and marketing systems. On the other hand, it should be noted that this channel also has a negative impact on local firms as MNCs may attract the skilled workers away from domestic firms by offering higher salaries (Girma & Wakelin, 2001). For theoretical discussion for the role of labor turnover in the technology transfers from MNCs see Glass and Saggi (2002).

It is very difficult to measure and analyze the spillovers that occur through this channel because it is not possible to track employees, worked for MNCs and hired by local firms, and their effects on local firms (Glass & Saggi, 2002). One of the methods to measure the spillovers through labor turnover is "wage spillovers". As a result of higher training investment expenditures, it is expected that MNCs should put upward pressure on wages and often pay higher wages to prevent the labor turnover, because marginal productivity of workers in the plants are very high (Lipsey & Sjöholm, 2004). On the other hand, if local firms want to attract skilled workers who have tacit knowledge of MNCs have to pay also higher wages and give more incentives than MNCs. As a result of these, if they compete on the same labor market, higher wages in foreign firms spillover to local firms, and this could be an indicator for technology spillovers associated with higher "wage spillovers" to local firms. Therefore, some studies which examine the impact of FDI on domestic wages hypothesized that rises in equilibrium wages in response to increases in FDI may represent an evidence for technology spillovers through labor turnover. In other words, high wage differentials may present the lack of technology spillovers between foreign and local firms (see Aitken et al., 1996; Girma, Greenaway, & Wakelin, 2001; Görg, Strobl, & Walsh, 2003; Lipsey & Sjöholm, 2001 for further details and testing of wage spillovers). The studies that examine wage spillovers are very limited[8] and they generally have adopted a regression framework as in the case of productivity spillovers: wages paid in local firms are used as the dependent variable and some proxies for foreign presence at both firm- or sector-level are added. Moreover, firm-specific (age, firm size, exports and capital intensity, share of skilled employees in the workforce, R&D expenditures, and so on) and sector-specific control variables are added as explanatory variables (see Pamukçu & Taymaz, 2009 for further details).

Via Competition: Presence of MNCs increases competition in host countries and this is seen as one of the major channels of spillovers. Entries of MNCs into domestic market should bring newer and advanced technologies; these force domestic firms (receivers) to use their available technology more efficiently, to improve or upgrade their technology in order to be competitive and to get much benefit from these technologies. Moreover, this situation also forces other local firms to use their existing technologies and sources more efficiently in order to cope with the "competition", to survive and to protect their market shares. In addition, they have to seek new technologies to use in the production process and to increase their productivity levels. As a result of competition effect, acquiring and adaptation process of new technologies will also be accelerated (see Aitken & Harrison, 1997; Blomström & Sjöholm, 1999; Dunning, 1993; Glass & Saggi, 1998, 2002; Markusen & Venables, 1999; Wang & Blomström, 1992). Due to the nature of competition within intra- and inter-industries, this channel affects both horizontal and vertical spillovers. However, it should be noted that this channel may also affect the productivity of local firms negatively. If foreign firms enter into domestic manufacturing market at a lower marginal cost than local firms, they may steal some market share from local firms, and they may force them to manufacture on a less efficient scale. These give rise to higher average costs for local firms and their productivity will be lower (Aitken & Harrison, 1999, p. 607).

Via Exports: The last important channel through which spillovers occur is the exports (export access). In other words, the spillovers occur when local firms have gain to access to foreign markets because of the presence of export oriented MNCs (Aitken, Hanson, & Harrison, 1997; Blomström & Kokko, 1998; Greenaway, Sousa, & Wakelin, 2004). Local firms can benefit from the MNCs' knowledge on global markets by observing their production, management, and exporting activities and techniques (*demonstration effects*). MNCs may share their information on foreign markets and may enable local suppliers to use their distribution networks in foreign countries. In other words, local firms can increase their export shares by learning from export oriented MNCs and using their distribution networks since MNCs have superior experience and knowledge on entering world markets, international marketing, production, exporting, and they have strong networks in various countries. Therefore, the collaboration between local firms and export oriented MNCs may provide international marketing channels for local firms. Furthermore, export competition with MNCs at host country and in foreign markets may lead to

[8] See, Görg and Strobl (2001, 2003) for details.

productivity increases for local firms. In sum, as a result of these effects, local firms may improve their export capabilities and may be successful exporters, and they can compete more successfully in export markets (Görg & Greenaway, 2004, p. 194).

Several studies have revealed the positive impact of MNCs on the export capacity of local firms (Aitken et al., 1997; Kokko, Zejan, & Tansini, 2001; Rhee, 1990). In addition, the local firms that have linkages with MNCs may easily access foreign markets or may reduce the entry costs into foreign market by benefiting from the knowledge and distribution networks of MNCs (through imitation or collaboration). For instance, the backward linkages with MNCs may provide knowledge about product and process technologies, foreign market conditions and arrangements, and tastes of foreign consumers in terms of design, product quality, and packaging (see Aitken et al., 1997; Barrios, Görg, & Strobl, 2003; Blomström & Kokko, 1998; Greenaway et al., 2004). For the empirical studies about the effects of exporting access on spillovers, see Aitken et al. (1997); Kokko et al. (2001); Greenaway et al. (2004); Banga (2003); Barrios et al. (2003).

Consequently, all of the above channels are seen important because they are complex and interconnected (Kinoshita, 2001). However, it is clear that how much the local firms can benefit from these spillovers is dependent on their efforts and technology capabilities.

2.5 Determinants of Technology Spillovers

This section briefly discusses the determinants of technology spillovers that occur MNCs to local firms in host country. In the light of literature, key determinants can be categorized under four headings: Absorptive capacity and technological gap, MNC, host country, and firm characteristics.

2.5.1 Absorptive Capacity and Technological Gap

One of the most important factors that affect technology spillovers is the absorptive capacity of both host country and local firms. Although absorptive capacity is evaluated as firm specific, it can be also associated with a country or location. For instance, the success of a firm to absorb and to internalize efficiently transferred technology depends also on "the institutional and organizational framework", "infrastructure", "the processes that create and distribute scientific knowledge", "cultural, political and financial structure of the host country" (Lorentzen, 2005; Narula & Driffield, 2012).

Absorptive capacity is defined as an ability to internalize technology created by foreign firms and "modifying it to fit their own specific applications, processes, and routines" (Narula & Marin, 2003, p. 23). The results of the studies show that technology diffusion does not automatically occur; it requires the recipient to have a certain "knowledge base" and "technological capability" to collaborate successfully with source so that it can absorb and adopt such technologies (Kinoshita, 2001; Wang & Blomström, 1992). In the literature, it is agreed that "existing knowledge base" and "intensity of effort" are two important elements of absorptive capacity in order to internalize and to assimilate new knowledge. Today, firms learn new technologies by their "existing knowledge base", and this also influences their learning processes and their knowledge in the future. "Intensity of effort", on the other hand, refers to the activities of local firms to solve technical problems encountered in production, and represents a firm's physical energy and intellectual entrepreneurship to internalize and to convert foreign technology (Cohen & Levinthal, 1990). Moreover, absorptive capacity of the local firms determines the speed and duration of the internalization process of the transferred technology. In the literature, "knowledge base" is considered less important than the "the intensity of effort" because the latter creates the former, but not vice versa (Techakanont & Terdudomtham, 2004; Ullrich, 1998).

"Technology gap" term is used to refer to the absorptive capacity and it is simply defined as the distance between source (home country or foreign firms) and recipient (host country or local firms) in terms of technological competence and development level (Borensztein, et al., 1998; Xu, 2000). It can be measured in many ways, such as the difference between the productivity of local firms and the average productivity of foreign firms in the sector, or the ratio between the labor productivity of a firm and average productivity of the sector. Furthermore, it signals to MNCs about the recipient's absorptive capacity related to human capital, institutions, infrastructure, R&D capacity, distribution networks, and technology capabilities.

In the literature, there are two contradictory views about the effects of technology gap on spillovers. Findlay (1978), in his pioneering contribution, introduced the role of *relative backwardness* on technology spillovers. He proposed that the greater the technology gap between two countries/(two firms) in terms of technological capability or development level, the greater the opportunities for less developed country/(less developed firm). He also suggested that higher technological gap accelerates the technology spillovers due to greater pressure for change and more rapid adaptation of new technologies (see Castellani & Zanfei, 2002; Sjöholm, 1999b). In other words, technology gap will facilitate the technology transfer processes through relationships between those who have the advanced technology and knowledge (such as developed country or MNCs) and those who wish to acquire, assimilate, and absorb them (such as host country or local firms) (Narula & Driffield, 2012). However, it is necessary to point out that Findlay does not completely ignore the role of technological capability to benefit from spillovers by suggesting that the gap must not be too wide, or a threshold level of capability in terms of scientific and technical knowledge is necessary in order to use and acquire technologies. In contrast to this view, some studies (Cantwell, 1989; Glass & Saggi, 1998; Kathuria, 1996; Kokko, 1994; Kokko, Tansini, & Zejan, 1996; Liu, Siler, Wang, & Wei, 2000; Perez, 1997; Wang & Blomström, 1992) reveal that there is a negative relationship between technology gap and technology spillovers; the greater technology gap means lower qualities of technology transferred and lower potential gains for spillovers. This line of thought assumes that it is necessary to have only a reasonable technological gap with MNCs for the recipient to benefit from technology spillovers associated with MNCs. It is suggested that if the technology gap is higher for the local firms with low-tech level, technology spillovers may be lower for these firms because of their lack of "absorptive capacity" (Kokko, 1996). In other words, a large technology gap is a signal to MNCs about low absorptive capacity of domestic firms, and it decreases the gains from spillovers by domestic firms.

The findings of the studies support the second claim above. For instance, in their study for Tanzania, Portelli and Narula (2004) find that larger technology gaps between local and foreign firms give rise to fewer backward linkages and to lower quality technological spillovers. Narula (2004) also shows that countries with high absorptive capacity receive FDI flows that include advanced and high level technologies. In another study, Borensztein et al. (1998) revealed that higher productivity growth from FDI depends on the level of absorptive capacity. In a similar vein, Narula and Marin (2003) confirm that only firms with high absorptive capacity are to benefit from FDI spillovers. Moreover, Xu (2000) shows that skilled human capital stock of the host country plays an important role in reaping benefits from technology transfers. Therefore, the second claim may be more rational than Findlay's claim because one of the conditions to benefit from spillovers is that local firms must have certain technological capabilities relative to their foreign counterparts. Meanwhile, it can be claimed that the absorptive capacity of host country firms is more important than being able to access the technologies of MNCs, and this capacity determines whether or not they benefit from FDI (Görg & Strobl, 2003). Therefore, backwardness is not seen as an advantage for the countries; actually, it could be a disadvantage (Bell & Pavitt, 1997).

In sum, absorptive capacity is seen by researchers as an important factor in order to internalize the benefits resulted from MNCs at both the firm and the host country level. Moreover, according to FDI assisted development view, local firms as collaborators and suppliers of MNCs must have a certain level of technological capability to usefully internalize spillovers, and to learn from MNCs, and to undertake specific duties assigned by MNCs. Therefore, a specific threshold level of absorptive capacity required to benefit from spillovers is seen as a key determinant on spillovers (Narula & Driffield, 2012).

2.5.2 MNC Characteristics

This section deals with the characteristics of the MNCs that impact on spillovers.

Motives of MNCs: According to Narula and Dunning (2000), the main motives of FDI through MNCs whether to invest or not in a country (especially in developing countries) are summarized in four ways: "resource seeking", "market seeking", "efficiency seeking (to restructure existing foreign production through rationalization)" and "strategic asset seeking" motives. They define these motives as follows:

- "Resource seeking" motive is the motive to obtain resources and production factors at lower costs.[9]
- Motive for the "market seeking" is to invest in a country to supply goods to host country or other neighbor countries. In other words, MNCs may invest in a country in order to access host country market,[10] to protect and to expand their market shares by expanding their production bases.
- "Efficiency seeking" motive for the MNCs is to invest in a country in order to increase efficiency by using different production factors, market structures, experienced and skilled domestic workers in host country, policies, customer preferences, etc.
- Lastly, "strategic asset seeking" is the motive to acquire and to benefit from the strategic assets of host country such as "marketing", "technology", "marketing", "infrastructure", investment policies, and human capital stock (see Narula & Dunning, 2000, pp. 150–152 for further details).

It is stated that, MNCs aim to generate economic rent by using their advanced production techniques in the first three motives, while they move and invest in a country to acquire new assets that protect or enhance existing assets in the last motive. In general, MNCs have generally moved with these four motives discussed above to make an investment decision in a country: using natural resources in host country; using cheap labor stock; entering new markets and expanding their market; improving production technologies; developing new technologies; and obtaining "new strategic assets" (Narula & Dunning, 2000, pp. 150–152). Until a few decades ago, the majority of the MNCs engaged in the first two motives and invested in host countries; therefore, developing countries received most of their FDI owing to these motives. In other words, developing countries have generally attracted MNCs which try to generate economic rent by using their "firm-specific intangible assets" (Lall & Narula, 2004, p. 451). Today, however, we see that MNCs have changed their motives, and the majority of them have engaged in the last two motives to manage and integrate their activities across borders, and to maximize their cross-border efficiencies. These motives also have a prominent role in the transfers and spillovers since they are also interrelated with strategies of MNCs (Narula & Driffield, 2012; Narula & Marin, 2003). Henceforth, the main motives of MNCs may affect the extent and quality of spillovers, and linkages with local firms.

The Entry Mode of MNCs: The other important factor that affects the occurrence of spillovers is the entry mode of MNCs into host country. MNCs may enter into domestic market either through a "merger or acquisition" (M&A), entering into

---

[9] These could be in the form of natural or human resource-seeking.

[10] Especially they invest in host countries where import substitution policies have restricted import from the foreign markets in developed countries.

market by purchasing an existing domestic firm or by joint venture with one, or through "greenfield investment", entering a market by building a new firm. It is suggested that new technologies have been introduced to the market slowly in the case of M&A, while they are instantaneous in greenfield investment mode. According to this view, MNCs established in the form of M&A integrate into the domestic market and sources more locally than those established through greenfield investment. M&A entry mode creates more linkages with local firms, and thus it may give rise to more spillovers. On the other hand, in greenfield investment mode, because MNCs rely on their own technologies and systems that differ from host country, and they import such inputs heavily from their home country, the spillovers may be restricted (based on Crespo & Fontoura, 2007, p. 414). Moreover, affiliates of MNCs established in the form of greenfield investment are expected to have weak linkages with local firms than those established in the form of M&A investment (Narula & Driffield, 2012; UNCTAD, 2000). Therefore, it is expected that MNCs established through M&A will result in greater vertical spillovers than those established through greenfield investment.

Foreign Ownership Structure: Although evidences from studies have produced mixed results, another factor that may affect spillovers is "the ownership structure of the MNC affiliates" established in host country. There are two contradictory views about the role of foreign ownership in the literature (based on Blomström & Sjöholm, 1999). According to one view, MNC affiliates with minority foreign ownership reduces the spillovers to local firms because parent company may hesitate to transfer more advanced technology, know-how, and systems to its affiliate in host country due to its reduced control over the affiliate. In other words, there is a positive relationship between technology spillovers and the degree of foreign ownership. When the MNCs have the majority of the shares, it would bring along its much newer technologies and its managerial skill along with it, since they would have a control over the profit and the firm. Newer and more technology could mean more spillover (Ramachandran, 1993). The contrary view suggests that minority ownership MNC affiliates may have more positive impact on benefiting from spillovers than those with majority ownership because it is expected that MNC affiliates with minority foreign ownership need to create more linkages with their partners and local firms in host country by transferring more advanced technologies in order to increase their advantages over domestic market and to get high profits. These are expected to lead to more knowledge and technology diffusion to local firms in host country. In other words, larger domestic ownership in MNC affiliates is better to access foreign technology because it may create more linkages with the local firms and the parent firm cannot prevent leakage of technologies. For this reason, some of the developing countries press on the foreign investors to make them establish joint ventures with the local firms (Blomström & Sjöholm, 1999).

The Market Orientation Strategy of MNCs: It is one of the most important factors that matter in spillovers. Export-oriented MNCs are expected to conduct less frequent linkages with local firms since they purchase less locally than domestic oriented ones in order to produce goods with higher quality requirements and technical specifications for export markets. Moreover, if intermediate products are used intensively, and if costs of communication are high between parent and affiliate, and if intermediate goods that are produced in home and host country are similar, it is expected that more linkages are created by MNCs (Görg & Strobl, 2003; Lall & Narula, 2004; Narula & Driffield, 2012).

The Origin of MNC: It is another factor that may also generate different spillovers to local firms because of their different nationalities, cultures, technologies, different modes of transfers, working conditions, and systems (see Crespo & Fontoura, 2007 for details).

2.5.3 Host Country Characteristics

The literature on the determinants of technology spillovers reveals that the host country characteristics play an important role in the decisions of MNCs about whether to invest or not in that country, and impact on both the technology spillovers and the adaptation of new technologies by local firms. The major characteristics identified by econometric analyses and case studies in the literature are summarized under four headings: (i) economic development level, (ii) knowledge base or available skilled human capital stock, (iii) infrastructure, and (iv) technological capability level of the host country. These also determine the composition and quality of technology spillovers and the level of benefiting from it. Furthermore, foreign trade policy, foreign investment policy, intellectual property rights, market size, competition level, investment incentives, policy stability, or overall macroeconomic outlook of the host country are other host country characteristics that may determine the linkages with MNCs (based on Blomström & Kokko, 1997; Görg & Greenaway, 2004). For instance, countries with larger market size, with larger capacity, with better skilled human capital stock, and higher technological capability are expected to attract MNCs with advanced production technologies and knowledge, often with R&D and design departments. Moreover, it is agreed that there is a positive relationship between the quality of linkages created with MNCs and average technology level of the local firms in host country.

Economic Development Level: Empirical findings related to developed and developing countries show that there is a positive correlation between development level of host country and benefiting from spillovers (Blomström, Lipsey, & Zejan, 1994). The studies on developed countries provide consistent results on positive spillovers, while studies on developing countries produce mixed results (positive, negative, or insignificant results). These studies support above argument by suggesting that developed countries have better infrastructure, advanced financial institutions, advanced technological equipment, high competitive sectors, local firms which have high technological capability, skilled human capital, better inter and intra-sectoral linkages, and these give rise to more positive spillovers and to more benefits from these.

Supply and Quality of the Human Capital Stock: It is the other most important factor and used as a proxy for knowledge base because advanced technologies need skilled and educated workers. MNCs may decide to invest at a location if there is enough educated and skilled human capital. Findings show that MNCs with high technology invest only in countries with higher skilled human capital stock. Therefore, it is necessary to have certain qualified human capital stock in order to get benefit from advanced technologies and to host advanced foreign technology (see Keller, 1996, 2004). The quality of human capital may be generated by host countries through formal education, formal training, on-the-job and off-the-job training activities.

Infrastructure and Incentives: Spillovers may also depend on the infrastructure of the host country (such as communication, and logistics) and incentives given by host country for R&D, innovation, and investment. In a similar way to human capital stock, MNCs with highly-intensive technological manufacturing need high-tech and developed infrastructure. Incentives by government to attract more FDI, to develop infrastructure, and to increase absorptive capacity of the local firms play an important role in promoting the spillovers. Although individual firms are responsible to benefit from spillovers, it is clear that entire economic system and infrastructure are very crucial to successfully absorb technologies, and they positively affect individual firms (see Narula & Marin, 2003).

Technological Capability Level: The possibility of attracting MNCs with high-tech production systems and advanced technologies, in order to acquire skill and technology from them, is higher for the countries with higher technological capabilities. However, these MNCs seek incentive public policies and special investment promotions conducted by host countries, so these are the other factors that positively impact spillovers (Lall & Narula, 2004). The major factors that impact the host country capabilities can be grouped as "physical and human capital", "infrastructure", "technological effort", FDI policies, incentives for FDI, trade regimes, institutions, and financial markets (Lall, 1992).

Foreign Trade Policy: According to this argument, trade regime of the host country has an impact on spillovers; however, the direction of the relationship is not certain and varies according to countries and host country policies. Yet, the general view suggests that export-oriented strategy is to attract more FDI flows than import substituting strategy. According to this view, in import substituting strategy, the size of domestic market is so small for foreign firms that they do not need to use their best technologies, marketing, and international distribution networks to produce only for the domestic market (Kokko et al., 2001). It is observed that MNCs prefer to transfer new and advanced technologies to their affiliates in developing countries with export-oriented markets, and these create higher technology diffusion to domestic firms (Dutz, Us, & Yılmaz, 2005). Empirical findings on different countries have produced mixed evidences on this argument. A number of studies find positive spillovers for countries with export-oriented policies, while others find positive evidences for countries with domestic-oriented policy (Kokko et al., 2001).

Intellectual Property Rights (IPR): The policies of the host country in terms of IPR are also affecting the decisions of MNCs whether to invest or not in the host country, thus the occurrence of spillovers. If there is not any protection about the IPR, MNCs generally prefer not to invest in that country, or they prefer to bring with them only products and production process technologies with low-technological level.

The Foreign Investment Policy: This argument suggests that foreign trade-related investment measures and government policies, such as foreign ownership restrictions, incentives, substitutions, local content requirements, minimum export measures, and duties, impact on FDI inflows, and thus the technology spillovers accompany those inflows (Blomström, Globerman, & Kokko, 2001). With the accelerating globalization wave, government incentives for foreign investment have become much more important to attract FDI and it is considered as one of the important determinants of spillovers (Kokko, 2003).

The Structure of the Financial Sector: It is another key determinant factor. It is expected that host countries with highly developed financial market facilitate the spillovers to local firms. In this case, foreign or local firms who are in need of capital can be established by borrowing from financial markets at relatively lower costs, and firms can focus on engaging in their production activities (Alvaro, Chanda, Kalemli-Ozcan, & Sayek, 2004).

Competition Level: According to the results of some studies, competition level of the host country may be a factor that affects spillovers from MNCs. However, findings could not reach consistent evidences. Some of the authors argue that competition level may negatively affect spillovers due to high costs in order to obtain advanced technologies (Barry, Görg, & Strobl, 2005; Blomström et al., 2001), while others suggest that high competition level encourages MNCs to transfer more advanced technologies and this enhances the spillovers to local firms (Kokko, 1996; Sjöholm, 1999a; Wang & Blomström, 1992).

Location Advantage: Location advantages of host country where investment will be made also determine the decision of a MNC whether to enter a given market or not, and its interaction level with host economy and local firms. Moreover, quality of the FDI flows and firm specific assets that will be brought by MNC depend on these advantages (Narula & Marin, 2003). MNCs generally prefer to invest in countries where there is the opportunity to export to neighboring countries; in other words, they invest in the host country in order to use it as a production base for neighbor countries.

2.5.4 Firm Characteristics

Another important factor that affects the occurrence of spillovers is characteristics of the firms in host country such as export intensity, size, age, human capital, ownership status, R&D capability, and so on.

Export Orientation: It is argued that export-oriented firms benefit less from spillovers associated with foreign firms in host country because they already produce for foreign markets and face significant competition in these markets.

Therefore, it is expected that their capabilities are very high and they do not need to create extensive linkages with MNCs in host country (Blomström & Sjöholm, 1999). According to this view, non-exporting local firms will benefit more from technology spillovers.

Size: It is expected that firms with larger size (in terms of production, turnover, and employment) benefit more from spillovers because small firms do not have enough capacity to compete with MNCs and to imitate technologies and systems used by MNCs. Aitken and Harrison (1999) in their study for Venezuela confirmed this hypothesis. However, in the literature there are studies that find the opposite results. For instance, Girma and Wakelin (2001) and Sinani and Meyer (2004) find that small-sized firms benefit more from FDI spillovers.

Human Capital and R&D Capability: The other factors are the human capital and R&D capability of the local firms. The studies point out those firms with highly educated employees or higher R&D capability benefit more from the presence of foreign firms (see Blalock & Gertler, 2004; Kathuria, 2000; Kinoshita, 2001).

Technological Capability: As proposed in the literature, assimilating knowledge and technology and benefiting from technology spillovers from MNCs depend on both "the complexity of the technology transferred" and on the technological capability of recipient firms, which is measured by the absorptive capacity. Evidence for the absorptive capacity of the firms can be interpreted and measured in many ways. Some of the proxies used to measure "the technology gap" or "absorptive capacity" of the recipient firms can be summarized in this way: the difference between the firm's labor productivity and the average labor productivity of foreign firms in the sector; R&D expenditures; payments on patents; capital intensity; export intensity; number of patents; number of engineers, educated and skilled workers.

The studies on various countries have produced supporting results on positive relationship between technological capability and spillovers. For instance, Kokko (1994), in his study for Mexico, concludes that technology spillovers occur only in sectors in which technology gaps between foreign and local firms are not too large. In another study for Uruguay, Kokko et al. (1996) found that local firms with low technology gap with MNCs could benefit and absorb the knowledge transfers from MNCs. Similarly, Girma, Greenaway, and Wakelin (2001), using firm-level panel data for UK, found evidence for spillovers to local firms with a small gap between their productivity level and the industry productivity level. In another study for Spain, Barrios and Strobl (2002) showed that there were positive spillovers from foreign presence to export-oriented local firms but not to domestic-oriented local firms. In a similar study, Blomström and Sjöholm (1999) also found that technology spillovers were restricted to non-exporting local firms. They proposed that export-oriented firms are using more advanced technologies, and also they are exposed to international competition. Therefore, their technological capabilities are high, and as a result of these, they benefit more from positive spillovers than non-exporters do. Kinoshita (2001), in his study for the Czech Republic, also found evidence of positive spillovers for local firms with high R&D capability. Furthermore, Keller and Yeaple (2003) in their study for the USA showed the evidence of positive FDI spillovers for the firms operating in high-tech sectors and conducting R&D activities. The studies once more reveal that absorptive capacity matters for spillovers. Therefore, it can be hypothesized that local firms, which have high technological capabilities and skilled human capital, with lower technology gap with MNCs may absorb and assimilate the spillovers more easily from MNCs (for more details, see Görg & Greenaway, 2004).

Geographic Location: Another characteristic that affects the technology spillovers may be the geographic location of the firms. Several studies have investigated the role of geographic dimension in order to benefit from spillovers. These studies hypothesize that spillovers from foreign firms would be received firstly by their neighbor local firms, and then they diffuse to more distant local firms, thus benefiting from spillovers is higher for the firms located nearby MNCs (Aitken & Harrison, 1991, 1999; Audretsch, 1998; Audretsch & Feldman, 1996). In other words, technology spillovers decrease with distance. Some of these studies failed to find evidence of spillovers at the regional level (Sjöholm, 1999a for Indonesia), some found negative spillovers for the same sector in any region of the country (Aitken & Harrison, 1999 for Venezuela), and some found evidence for positive spillovers in the same region and sector (Aitken & Harrison, 1991; Girma & Wakelin, 2001 for UK).

As can be seen from the studies in the literature, the determinants of technology spillovers have produced contrary results, and these make it difficult to reach a definite conclusion about most of the determinant factors. However, the common result among them is the importance of the absorptive capacity of both recipient firms and host country to benefit positively from spillovers (Crespo & Fontoura, 2007, p. 420).

2.6 Empirical Studies on Technology Spillovers and Transfer

2.6.1 General Information

In the literature, a number of surveys review theoretical and empirical studies on spillovers through FDI.[11] These identify potential sources of spillovers, then present findings of available econometric studies and discuss reasons that may explain the positive or negative effects of spillovers on the productivity of local firms in host economies.

In the theoretical literature, the studies on the effects of FDI on the host countries of MNCs date back to the early 1960s. The first authors that systematically

[11] See Blomström and Kokko (1998), Görg and Greenaway (2001; 2004), Saggi (2002), Haskel et al. (2002), Crespo and Fontoura (2007) for detailed literature review and theoretical discussions on technology spillovers.

discussed technology spillovers as a consequence of FDI were MacDougall (1960), Corden (1967), Caves (1971, 1974) for Australia, Globerman (1979) for Canada, Blomström and Persson (1983) for Mexico. In these studies, the authors generally tried to understand the costs and benefits of FDI on host countries by impacting on foreign trade, taxes, or balance of payments, and they analyzed whether the foreign presence has any significant impact on the productivity of domestic firms by using cross-section or panel data. Caves (1974) showed that there is a positive relationship between the foreign share and productivity level of domestic firms by using the share of foreign firms in employment at manufacturing sector level. Globerman (1979) also found positive evidence of spillovers on domestic firms in Canada manufacturing industry by using labor productivity as a dependent variable. In their study, Blomström and Persson (1983), using labor productivity as a measure of efficiency and relating this to capital intensity, labor quality, scale of production, foreign participation, and using different degree of concentration indices (Herfindahl index), provided also support for positive spillovers in Mexican industries by using cross-section data from the 1970 census. In these studies, the authors defined the dependent variable as the total value added per employee in domestic firms, and assumed the key independent variable as foreign share in total employment or value added in order to use as a proxy for spillovers together with other independent variables.

The importance of technology spillovers has emerged with the results of empirical studies rather than theoretical studies.[12] The empirical studies on technology spillovers from FDI that provide important evidence on the presence and pattern of spillover effects can be summarized under three types[13] (Blomström & Kokko, 1998).

- The first type is the studies at micro level by analyzing the impact of spillovers on firm productivity.
- The studies in second type are at the macro level and analyze the effects of spillovers on the economic growth of host country.
- The third type studies are in the form of case studies specific to an industry and these studies are conducted to obtain information and to reveal unknown details of the issues that cannot be captured by empirical studies (see Temenggung, 2006).[14]

The first two type of studies discussed above are generally in the form of econometrical analysis about the relation between foreign presence and productivity of local firms, or about the effects of foreign presence on the host country. These

[12] For theoretical studies, see Findlay (1978), Das (1987), Wang (1990), Wang and Blomström (1992). For some early empirical studies, see Dunning (1958), Safarian (1966), and Gabriel (1967).

[13] See Görg and Strobl (2001) and Blomström and Kokko (1998) for the details on this literature.

[14] These studies are generally employed by using interviews and mailing questionnaires. For instance, see Hobday (1995) and Kim (1997).

studies generally estimate a production function and focus on a productivity dependent variable such as labor productivity or value-added of local firms and regress this on a range of explanatory variables (concentration ratio, factor inputs, skilled human capital, scale of the firm, foreign ownership, etc.) by using cross-section or panel data while controlling for other potentially important factors. In the literature, the studies that use panel data are seen more appropriate, reliable, and informative than those using cross-section data due to firm- and sector-specific factors, and time effects, and because it is expected that spillovers can be detected in the long-term due to its dynamic nature (Aitken & Harrison, 1999; Görg & Strobl, 2001; Haddad & Harrison, 1993). In other words, it is difficult to control for industry- and firm-specific factors in studies "using cross-section data where the time dimension is absent". For example, MNCs may invest in most productive sectors to benefit from high productivity levels, and this may produce positive productivity spillover effect although "foreign presence is the result, not the cause of high productivity levels observed in the sector". Therefore, the findings of the econometric studies using cross-section data are hardly reliable (Pamukçu & Taymaz, 2009, p. 8). In addition, a set of proxy variables are calculated to analyze the effects of spillovers from MNCs such as share of foreign firms in total output, in total sales, or in employment in a given sector. The evidences from literature also show that factors affecting the spillovers depend on the characteristics of host country, sector, and firm.

As can be seen in Table 2.1 and in Sect. 2.6.2, according to the literature available, the empirical studies on the importance of technology spillovers through FDI between MNCs and local firms have produced mixed results. Some studies found positive spillovers, which mean that foreign presence impacts positively on the productivity of local firms, while some reveal that spillover effects on domestic productivity from FDI did not produce significant results or found negative spillovers. For instance, Haddad and Harrison (1993) for Morocco is insignificant; Aitken and Harrison (1999) for Venezuela is negative; Djankov and Hoekman (1998, 2000) for Czech Republic is negative; Okamoto (1999) for Japan is ambiguous; Kathuria (2000) for India is ambiguous; Konings (2001) for Bulgaria and Romania is negative and for Poland is ambiguous; Li, Liu, and Parker (2001) for China is ambiguous.

The last type is the case studies focusing directly on spillovers that analyze linkages between MNCs and their local suppliers by focusing on technology transfers (see Hanson, 2001; Larrain, Lopez-Calva, & Rodriquez-Clare, 2000). The first case studies that produced evidence on technology spillovers to local firms are Germidis (1977), Mansfield and Romeo (1980), Rhee and Belot (1989), and Mody, Sanders, Suri, Rao, and Contreras (1991). Case studies show that presence of MNCs may give rise to *technology diffusion* to local firms by direct knowledge and technology transfers. For instance, according to Blomström and Kokko (1998):

- MNCs may transfer their techniques, quality systems, standardization procedures to their local suppliers in order to upgrade their capabilities and to be sure about the quality of the products

- They force local suppliers to increase their managerial capabilities and to adopt marketing techniques used by MNCs
- They may provide know-how by demonstrating new technologies and by training workers of local suppliers
- They increase the efficiency and productivity of local firms by forcing them in a competition with other potential suppliers and by demanding lower prices for the products supplied

These case studies have given some important clues about spillovers from MNCs; however, they say little about how these spillovers occur, what their intensity is; their channels and determinants; and how important they are in general. To analyze these questions, an ideal study of technology spillovers from MNCs would require a large number of detailed micro data at firm level, both quantitative and qualitative, and would have to cover several years since spillovers do not occur instantaneously so that it would be possible to make significant conclusions (Blomström & Kokko, 1998). To the best of our knowledge, a few comprehensive analyses of such studies have been made (see, Keller, 2004; Larrain et al., 2000; Moran, 2001). Therefore, we focus on this type of study in this book and we believe that future research should focus on this type of studies.

2.6.2 Evidence on Developed and Developing Countries

It is seen that the empirical studies on developed countries generally have provided consistent results that foreign presence positively affects the productivity of local firms; in other words, they find positive spillovers (for instance, Caves (1974) for Australia; Globerman (1979) for Canada; Nadiri (1991) for France, UK, Japan, and Germany; Griffith (1999), Liu et al. (2000), Driffield (2001), Haskel, Pereira, and Slaughter (2002), Harris and Robinson (2003) for UK; Barrios and Strobl (2002) for Spain; Görg and Strobl (2003) for Ireland; Keller and Yeaple (2003) for USA). On the other hand, the studies on developing countries have not produced consistent results; while a number of studies find positive spillovers, some find negative or insignificant (ambiguous) results on spillovers (see Görg & Greenaway, 2004 for further details on productivity spillovers in developing and developed economies as well as the ones in transition).

Table 2.1 based on information provided in Görg and Greenaway (2004) displays the results of some empirical studies on spillovers conducted by various authors at firm or industry level in manufacturing industries of developing countries in terms of country, period covered, type of dataset used (cross-section or panel data), aggregation level (industry or micro-level data), and the results obtained. For instance, Haddad and Harrison (1993) could not find any evidence of technology spillovers for Morocco at micro or industry level by using 5-year panel data for the period 1985–1989, while Blomström and Sjöholm (1999) find positive spillovers at micro level for Indonesia by using cross-sectional data on 1991. Djankov and

Table 2.1 Empirical studies on spillovers in developing countries

Author	Country	Period	Data	Aggregation level	Result
Blomström and Persson (1983)	Mexico	1970	Cross-sectional	Industry	+
Blomström (1986)	Mexico	1970–1975	Cross-sectional	Industry	+
Haddad and Harrison (1993)	Morocco	1985/1989	Panel	Micro and industry	?
Blomström and Wolff (1994)	Mexico	1970–1975	Cross-sectional	Industry	+
Kokko (1994)	Mexico	1970	Cross-sectional	Industry	+
Kokko (1996)	Mexico	1970	Cross-sectional	Industry	+
Kokko et al. (1996)	Uruguay	1990	Cross-sectional	Micro	?
Blomström and Sjöholm (1999)	Indonesia	1991	Cross-sectional	Micro	+
Sjöholm (1999a)	Indonesia	1980–1991	Cross-sectional	Micro	+
Sjöholm (1999b)	Indonesia	1980–1991	Cross-sectional	Micro	+
Chuang and Lin (1999)	Taiwan	1991	Cross-sectional	Micro	+
Aitken and Harrison (1999)	Venezuela	1976/1989	Panel	Micro	–
Djankov and Hoekman (2000)	Czech Republic	1993/1996	Panel	Micro	–
Kathuria (2000)	India	1976/1989	Panel	Micro	?
Kokko et al. (2001)	Uruguay	1988	Cross-sectional	Micro	?
Zukowska-Gagelmann (2000)	Poland	1993/1997	Panel	Micro	–
Kugler (2001)	Colombia	1974/1998	Panel	Industry	?
Kinoshita (2001)	Czech Republic	1995/1998	Panel	Micro	?
Bosco (2001)	Hungary	1993/1997	Panel	Micro	?
Konings (2001)	Bulgaria	1993/1997	Panel	Micro	–
	Poland	1994/1997			?
	Romania	1993/1997			–
Damijan, Majcen, Knell, and Rojec (2001)	Bulgaria Czech Republic Hungary, Poland Romania, Estonia Slovakia, Slovenia	1994/1998	Panel	Micro	? Or – + only for Romania

(continued)

Table 2.1 (continued)

Author	Country	Period	Data	Aggregation level	Result
Li et al. (2001)	China	1995	Cross-sectional	Industry	+
Lopez-Cordova (2002)	Mexico	1993/1999	panel	Micro	−, ?
Görg and Strobl (2002)	Ghana	1991/1997	Panel	Micro	+
Javorcik (2004)	Lithuania	1996/2000	Panel	Micro	?

Source: Extracted from Görg and Greenaway (2004, pp. 177–178)
Note: Micro data may be firm-, establishment- or plant-level data

Hoekman (1998, 2000) find negative spillovers at micro level for Czech Republic by using 4-year panel data for the period 1993–1996. Moreover, Aitken and Harrison (1999) show that there are negative effects of foreign investment at micro level on domestically owned plants in the Venezuelan manufacturing sector by using 14-year panel data for the period 1976–1989. Kokko (1994, 1996) studied Mexico at industry level by using cross-sectional manufacturing data for the period 1970. He finds evidence for positive spillovers and suggests that the effects of technology spillovers on local firms are conditional on the technology level of local firms. Kokko created three independent variables in order to test spillovers, which are average patent payments per employee, average capital intensity of foreign firms, and labor productivity gap between local and foreign firms at industry level.

As can be seen in Table 2.1, the effects of spillovers among countries are different. In these studies, the same econometric models and estimation techniques have been used; especially a proxy variable of productivity as a dependent variable—labor, gross output, or total factor productivity—is regressed on an indicator of foreign presence at the sector level (foreign share in total employment, in output, in sales, etc.), and a number of control variables (share, Herfindahl index, scale, exports, size, age, and so on) are supposed to be correlated with the dependent variable.

2.6.3 *Evidence on Turkey*

This section reviews some of the empirical studies on spillovers for Turkish manufacturing industry in detail. Empirical studies on spillovers for Turkey which started especially after 2000s are very limited and they all concern the pre-2001 period where annual FDI flows to Turkey were rather low (see Table 2.2). There are various important reasons for the scope and limited number of studies on this issue. One of the important reasons is the unavailability of appropriate data at the sector- or firm-level in order to study spillovers. This situation is still valid. The other reason may be due to fact that FDI flows to Turkey were very low till the end of the 1980s. According to statistics, total FDI inflows to Turkey were nearly US $10.7B between 1974 and 2000, corresponding to an annual average of US$400M

Table 2.2 Empirical studies on spillovers for Turkey

Author	Period	Data	Data source	Aggregation level	Estimation method	Result
Aslanoğlu (2000)	1993	Cross-sectional	ICOC	Micro	OLS	?
Alıcı and Ucal (2003)	1987/2002	Series	TurkStat	Industry	Granger non-causality analysis	No significant effect
Taymaz and Lenger (2004)	1983/2000	Panel	TurkStat	Industry	Arellano-bond type of GMM estimation	–
Ayvaz et al. (2006)	2001	Cross-sectional	ICI	Micro	Regression analysis	+
Lenger and Taymaz (2006)	1995/2000	Panel	TurkStat	Micro	OLS (binary choice model)	No significant effect
Bertinelli et al. (2006)	1983/1994	Panel	TurkStat	Micro	OLS and FE	–
Taymaz and Yılmaz (2008a)	1990/1996	Panel	TurkStat	Micro	Olley and Pakes/OLS	+ Backward spillovers
Aksoy (2008)	1983/2001	Panel	TurkStat	Industry	OLS and FE	+ Horizontal spillovers
Pamukçu and Taymaz (2009)	1983/2001	Panel	TurkStat	Micro	FE and GMM	–
Çetin (2009)	2003–2004	Cross-sectional	TurkStat	Micro	Probit	Firm size, skill level, export, and capital intensity have positive effect on technology transfer
Köymen and Sayek (2010)	1990–2001	Panel	TurkStat	Micro	Levinsohn-Petrin	+ Backward spillovers

Legend: Istanbul Chamber of Commerce (ICOC); Istanbul Chamber of Industry (ICI)

for 26 years. The average annual FDI inflows were nearly US$800M in 1990s, in contrast to US$200M in 1980s. However, FDI flows into the country increased steadily after 2001; total cumulative net FDI inflows reached nearly US$100B for the years 2001–2010 (see Sect. 2.2.1). The last reason may be the globalization polices in 1980s that encouraged all countries to implement liberal policies. Therefore, Turkey also started to liberalize its economy in 1980s by changing its policy to export-promoting industrialization and by opening its doors to foreign investment (see Aslanoğlu, 2000 for details).

Aslanoğlu (2000) is considered as a pioneer in doing the first econometric study analyzing the effects of FDI on Turkish manufacturing industry. He tried to analyze the effects of spillovers by using the cross sectional data for the year 1993 from the survey data of Istanbul Chamber of Commerce (ICOC) on the largest 500 industrial firms of Turkey.[15] He constructed five single equation econometric models. The aim of the first two models is to estimate the effect of foreign presence[16] on the productivity and competitiveness of domestic firms. The results show that foreign presence increases competition in domestic industries, while it has no any significant impact on the productivity of domestic firms. Other two models are constructed to analyze the impact of technology gap between domestic and foreign firms on the productivity and market growth of domestic firms. However, he could not find significant relation between those variables. In final model, the impact of the initial technology gap on the change in technology gap is analyzed, and a significant correlation is found between such variables.

In their study, Alıcı and Ucal (2003) examine the effect of liberalization process on economic growth of Turkey by emphasizing the export led growth (ELG) strategy. The authors aim to find a causal link between exports, FDI, and output by conducting unit root and the Granger non-causality tests from 1987-I to 2002-IV. For this aim, three variables are utilized: export price index, industrial production index, and FDI flows into Turkey. The data for the first two variables come from TurkStat, and those for FDI variable from the Undersecretariat of Treasury. The results neither produced significant positive spillovers from FDI to export nor a kind of FDI-led export growth linkage. In other words, results do not confirm the existence of FDI-growth relationship. They suggest that policies should focus on attracting more FDI in order to gain the spillovers to output and FDI-led export growth.

In their empirical study, Taymaz and Lenger (2004) provide evidence in favor of negative spillovers in Turkish manufacturing industry. The authors examine the productivity spillovers from MNCs by using panel data from TurkStat at industry level over the period 1983–2000. The data consist of 28 industries in three-digit

---

[15] He uses the 1993 data for the entire analysis; however, the data of the year 1988 is used for creating some independent variables due to lack of data.

[16] Various proxies are used in these models for foreign presence such as the share of foreign firms in total employment, total sales, total value added, and total net assets of an industry (Aslanoğlu, 2000, p. 1122).

level in various categories such as public firms, and private small, medium, and large sized firms in Turkey (Taymaz & Lenger, 2004, p. 2). The study investigates the role of the size of the recipient firms, the intensity of the R&D activities, the ownership structure, and whether spillovers change by time or not. Although the results show that spillovers from MNCs differ with respect to size of the recipient firms and by time, the study reveals the negative spillovers for the domestic sectors.

Lenger and Taymaz (2006) also examine empirically the innovation and technology transfer activities of local and foreign firms in Turkish manufacturing industry. The authors construct a model for estimating the impacts of horizontal, vertical, and labor spillovers on technological activities of firms by taking into account two dummy variables: *in-house innovative activities* and *technology transfer from abroad*. They also test whether foreign firms are more innovative and/or whether they transfer technology mainly from abroad, and whether they have any impact on the technology transfer decisions of local firms. The data used for innovativeness dummy variable come from the two innovation surveys conducted by TurkStat and cover the periods 1995–1997 and 1998–2000. On the other hand, the data for technology transfer dummy variable come from the "Annual Structural Business Statistics" survey conducted by TurkStat for the same periods. In the analysis, several proxies are used for horizontal, vertical, and labor spillovers. The three proxies for horizontal spillovers are the market share of foreign firms, the ratio of foreign firms' R&D expenditures to total output in the province in which the firm operates, and at the industry level. The two variables used as proxies for vertical spillovers are the proportion of firm's inputs produced by foreign firms and the proportion of firm's output used by foreign firms. These proxies measure the weighted average of foreign market share in supplier and user industries, respectively. In the calculation of these proxies, the input (one sector's share in inputs used by another sector) and output (the share of one sector in total consumption of another sector's output) coefficients of the sectors are also calculated by utilizing the 1996 Input-Output Table of Turkish manufacturing industry. The last proxy used for labor turnover is the ratio of the number of separations from foreign firms to the total number of employees in a given sector. The authors find that foreign firms are more innovative than their local counterparts, and transfer technology mostly from their parent firms in overseas. In addition, they show that horizontal spillovers are insignificant, while vertical spillovers are ambiguous. The major finding of the study is that labor turnover is the main channel of spillovers by emphasizing the importance of tacit knowledge. Also, the findings confirm that technology cannot easily be transferred through passive mechanisms such as demonstration effects, and reverse engineering (Lenger & Taymaz, 2006, p. 152).

In another study, Ayvaz, Baldemir, and Ürüt (2006) analyze the effects of FDI on both labor productivity and the relationship between capital efficiency and labor productivity by using dummy variables regression model in Turkish manufacturing industry. The data used in the study are obtained from "the largest 500 industrial firms of Turkey" report conducted in 2001 by the Istanbul Chamber of Industry (ICI). The sample consists of 199 firms in total, specifically 19 foreign firms with full ownership, 159 local firm with no foreign ownership, and 21 public firms. The authors find significant positive spillover of foreign capital on labor productivity, and suggest that local firms should increase their capital and labor resources to compete successfully with foreign firms. However, the study is seen very problematic since it covers a very small sample to conduct this kind of analysis and also actually does not examine spillover effects; only the difference between foreign and local firms in terms of productivity efficiency is tested. Therefore, the findings of this study are not reliable enough (Aksoy, 2008, p. 49).

Another empirical study is the one by Taymaz and Yılmaz (2008a), which analyze the direct and indirect effects of foreign ownership on productivity by using firm-level unbalanced panel data for the period 1990–1996 in the Turkish manufacturing industry. The authors use the Olley-Pakes[17] production function estimates for productivity and OLS regression analyses to test the horizontal and vertical linkages. The sample consists of firms with 25 or more employees due to lack of key variables needed for the study in the 10–24 employee size group, and the sample is limited only to private establishments. In the calculation of linkages, the authors benefit from the sectoral output shares of foreign firms and Input-Output matrix in 1990 to identify linkages across plants. Based on Olley-Pakes approach using total factor productivity, they find that foreign firms are more productive than their local counterparts. Also, majority-owned foreign firms are more productive than minority-owned foreign firms, and fully foreign owned firms are more productive compared to majority foreign owned ones (Taymaz & Yılmaz, 2008a, p. 31). In addition, when industry based measures of linkages are used, they find that productivity spillovers from foreign firms to local firms occur through horizontal and vertical linkages. However, the authors argue that there is a very high correlation between horizontal and vertical linkage measures, so the findings may generate false results with industry based measures. Therefore, they use product-based measures of linkages in order to overcome multicollinearity problem. In this case, the results mostly lose their economic and statistical significance compared to industry based measures of linkages, and they find positive significant spillovers that occur through backward linkages only.

In their paper, Bertinelli, Pamukçu, and Strobl (2006) examine the impact of horizontal FDI spillovers on firms' productivity in Turkish manufacturing industry. Also, the authors test the effect of technology gap, which is measured by the distance between local firms' productivity and that of the industry leader at the four-digit level (relative productivity). The study uses firm-level panel data obtained from the "Annual Surveys of Manufacturing Industry" of TurkStat and covers the period 1983–1994. The authors estimate a productivity equation by using OLS as well as FE methods. An index of total factor productivity in local firms is used as the dependent variable in the productivity equation. Moreover, to analyze productivity spillovers, four different proxies for spillovers based on the extent of foreign presence at the sector level are constructed: the share of foreign enterprises in the number of employees or in gross output at the four-digit sector level, and the

[17] For details, see Olley and Pakes (1996).

indicators of foreign presence for enterprises with foreign equity that is at least 10% of their capital and for those with majority-owned foreign firms (at least 50%). Besides control variables, three firm-specific variables are constructed in the study: the share of skilled labor in total employment (skill level); the ratio of firm sales to total sales of its four-digit sector (market share); the ratio of a firm's gross output to that of sector-level firm average (scale). The authors also introduce two sector level variables in the productivity equation to measure different aspects of the degree of competition: "the share of imports in the domestic demand at the four-digit sector level to domestic demand (import penetration)" and the variables that account for the market structure prevailing at the four-digit sector of the firm by using two alternative indicators, which are the share of the four largest firms in sales at the four-digit sector level (CR4) and the Herfindahl index of concentration of sales (Herfindahl index). The findings show that foreign firms do not generate any spillover that positively impact local firms' productivity levels. Conversely, all four proxies for spillovers produce evidence of negative spillovers on local firms' productivity levels.

Another attempt to investigate the horizontal technology spillovers through FDI in the Turkish manufacturing industry is made by Aksoy (2008). The analysis on spillovers is conducted over the period 1983–2001 at sector level by using the dataset of TurkStat including sectoral level determinants of 89 different sectors. In this analysis, sectoral market shares of the foreign firms are used as proxies for horizontal spillovers. The author also divides the sample into high-tech and low-tech industries in order to test spillovers. In the study, a production function in the form of Cobb-Douglas is estimated by OLS and FE methods. The findings of the study have produced positive evidences on horizontal spillovers at sectoral level.

In a similar study, Pamukçu and Taymaz (2009) examine empirically the existence of FDI-related productivity and wage spillovers in the Turkish manufacturing industry by using the panel dataset of TurkStat over the period 1983–2001. The study is carried out at firm-level, and a production function approach in the form of Cobb-Douglas is adopted in order to assess the impact of horizontal spillovers on firm-level total factor productivity. Firm-level output measured by its gross output is used as dependent variable, while firm's capital stock, labor, raw materials, and energy inputs are used as explanatory variables. Market share of foreign firms at sector level and at the regional level are used as two proxies for horizontal spillovers. Also, firm- and sector-specific variables are added to the equation in order to control for the influence of other determinants of firm-level total factor productivity and to avoid an omitted variable bias. In the estimation of productivity spillovers, FE estimation method is preferred to use in order not to suffer from an omitted variable bias due to unobserved time-invariant firm-specific factors. On the other hand, a dynamic wage bargaining model based on McDonald and Solow (1981) is used in the econometric analysis of wage spillovers. This wage model is empirically tested by using both FE method and Generalized Method of Moments (GMM) a la Arellano-Bond (see Pamukçu & Taymaz, 2009, p. 57). The findings of the study show that FDI-related productivity spillovers exert a negative effect on the productivity of domestic firms and a significant and positive impact on the wages paid by domestic firms.

Çetin (2009) also investigates the determinants of technology transfer decisions of firms in Turkish manufacturing industries by using probit model. The dataset used in the study is based on the merged data set of both 2003 and 2004 "Structural Business Statistics" survey conducted by TurkStat at enterprise level. The dependent variable is a dummy variable as the firm's decision to transfer technology, and the author uses the "expenditure on licenses, patents, and trademarks" as a proxy for such variable. If the expenditure of a firm on licenses, patents, and trademarks is positive, then it is assumed that technology transfer decision equals to one. The firm-level factors tested are size, skilled labor, capital intensity, expenditure on R&D, and dummy variables for export and import intensity and foreign ownership of the firms. Also, sector level variables are import penetration, Herfindahl index, sectoral license expenditure, share of foreign-owned firms' output, and sector dummies classified according to technology levels. The findings of the study show that firm size, general skill level, export behavior, and capital intensity have significant effect of technology transfer decision of the firm, while foreign ownership does not. The effects of sectoral characteristics are also statistically significant.

The effect of human capital on the productivity spillovers from FDI to domestic firms in Turkish manufacturing industry is investigated by Köymen and Sayek (2010) for the period 1990–2001. In their paper, the role of human capital of local firms is analyzed in order to reveal through which channel of linkages (horizontal, backward or forward) it enhances the local firms' total productivities and acts as an absorptive capacity. In the study, total factor productivity (TFP) of the firms is estimated by using Levinsohn-Petrin (Lev-Pet)[18] methodology based on firm-level unbalanced panel dataset obtained from TurkStat. Then, TFP is regressed on three industry-based linkages (backward, forward, and horizontal) to test the spillovers from FDI. The findings of the study produced evidence on positive productivity spillovers through backward linkages, while there was no evidence on horizontal and forward linkages that play a role in contributing to the productivities of local firms. One of the important result of the study, although it does not take into account the role of the firm characteristics on spillovers, is that such characteristics are important determinants of spillovers from FDI and they should be taken into consideration in the spillover analysis.

Table 2.2 above summarizes the empirical studies on spillovers for Turkish manufacturing industry discussed before. The table provides a comparison of the studies in terms of period covered, data used, aggregation level, estimation method, and the result obtained. As can be seen from the table, the findings of previous studies on FDI-spillovers are pertaining to the pre-2001 period.

Consequently, empirical studies on whether the MNC-based technology spillovers have an effect on the performance of the local firms, and if yes, the direction and dimension of the effect, yield controversial results differing from each other;

[18] For details, see Levinsohn and Petrin (2003).

some provide negative results, while some produce positive results either significant or insignificant. Therefore, we can say that there is no consensus on the effects of spillovers from MNCs. The results vary depending on the different econometric estimation methods adopted, type of the datasets used (cross sectional or panel), aggregation levels at sector or firm level, various methodological problems, different proxies used for spillovers, different foreign share variables, unidentified factors, country concerned, sector specific factors, different time periods analyzed, etc. in the empirical studies. For a comprehensive evaluation of the econometric studies conducted on the effect of the MNC-based technology spillovers on the performance of the firms, see Görg and Greenaway (2004) and Smeets (2008). Although these empirical studies based on published formal data have produced very important results in terms of the positive effects of FDI on host country and existence of positive benefits from FDI, they are not able to provide sufficient information about how these benefits take place, in what forms, and by which channels. Therefore, it is necessary to conduct more detailed case study researches as conducted in this book in order to question and understand the real situations at firm level and in order to clarify the determinants, channels, and dynamics of the KTTs that take place.

2.6.4 Reasons for Failure of Realizing the Technology Spillovers or for Their Negative Effect on Firm Performance

The reasons why there is no technology spillover from the MNCs to the local firms in host country or why they are not adequate could be linked to many factors. The main reasons based on Aitken and Harrison (1999) could be summarized as follows:

- The workforce circulation (labor turnover) among the local and foreign firms is limited (especially due to high salaries in the MNCs)
- There are only a few local firms to render sub-contracting/supply services from MNCs (the MNCs' bringing along their own suppliers with them)
- The local firms do not perform any R&D activities, or their R&D activities are inadequate (this may hinder collaborations to be established with the MNCs)
- Lack of experience of the local firms in the areas of production, manufacturing, quality control, and engineering
- The MNCs are not so inclined favorably towards knowledge spillover to their local rivals. They may apply specific policies in order to protect their advantages and to prevent spillovers to local firms
- The market share of the local firms diminishes or their operation ceases due to their failure to compete with the MNCs

- The local firms do not have the capability to absorb the potential technology spillovers from the MNCs (especially due to the huge difference between the technological capabilities of the local and foreign firms)

2.7 Theoretical Framework for Case-Study Research

2.7.1 General Information

The relevant literature contains a few studies conducted on the technology transfers. These could be grouped under three main topics in terms of the method employed. The first one is the case studies that are didactic and illuminating in that they provide highly valuable, embedded information even if there is a lack of data and information on the subject being studied. The second one is the econometric studies conducted based on various data at the industry and firm level. The last one is the studies where the first two types above are used together in a complementary way such as our study.

Recently, there has been a growing interest in the econometric analysis of the technology transfers realized to the firms in the developing countries. When the international literature is reviewed, it is seen that many studies conducted on the econometric estimations of the technology transfers realized to the firms in developing countries are in general based on the observations obtained from the manufacturing industry statistics published regularly or from the annual industry reviews gathering input-output data. In general, the following results have been obtained from these studies: If the local firms in developing countries conduct more activities in the international markets, if the year of foundation of the MNC affiliates in developing countries is recent, if the domestic market of the developing country is larger, and if the share of the foreigners in the affiliation established in developing country is higher; the technology transfers used by the local firms are more important (Grosse, 1996). However, the findings obtained from such econometric studies are less illuminating than the findings obtained from the substantial case studies.

Case studies provide important clues about the firms that will contribute to our understanding of the transfer process deeper and better, and especially help us to extract the deep-buried information on the subject being studied, if there is a lack of econometric data concerning the manufacturing industry, and help us to grasp the sector as the subject of the study better (Pack, 2005). Such kinds of studies where the researchers collect the relevant data by themselves ensure both that the complex subjects are understood better and that more substantial data are generated (Bigsten et al., 2002). When we look at the empirical studies conducted on the technology transfer, we could see that most of the studies on certain sectors have been conducted in the form of case studies, since there is no detailed data at micro level regarding the firms. We also see that many of the micro- and significant data have been obtained from the substantial case studies conducted on the firms (Pack, 2005). Case studies not only are informative but also present a profound definition of the factors that determine the international technology transfer. For instance, the question whether the FDI made by *Intel* in Costa Rica in 1990s brought along any technology transfer or not was analyzed by Larrain et al. (2000). This study is informative and illuminative in that the authors made interviews with top managers of *Intel* and authorities of Costa Rican government and tried to understand their ideas, concerns, and motivations. Therefore, the information at the firm level that will contribute to understand better the findings on the details and processes of the technology transfer are obtained by means of case studies. However, it is apparent that supporting the studies conducted in the form of case studies—as is the case with *Intel*—with econometric studies is important (Keller, 2004). In general, it could be said that the econometric and case studies complement each other.

2.7.2 Technology Transfers at Intra- and Inter-Firm Level

In the literature, many channels through which domestic firms may benefit from technology transfers from MNCs are discussed. Some studies point out the two main channels of international technology transfer: (i) joint venture partnership of local firms with foreign firms and (ii) "contractual agreements" signed with foreign firms such as "licensing agreements", "turnkey contracts", "technical, management, or service contracts", and "international subcontracting" (see Sect. 2.1.4) (Pack & Saggi, 2001; Techakanont, 2002, p. 13). Furthermore, some studies specify three other main channels by emphasizing the relationships between source and recipient: (i) "arm's length trade of technology", (ii) intra-firm technology transfer, and (iii) inter-firm technology transfer. Figure 2.4 shows the inter- and intra-firm technology transfer channels that occur between source and recipient.

- The first channel is the "arm's length trade of technology", in which the recipient acquires the technology through agreements signed with the independent technology source, and in return, make royalty payments for such technology (Techakanont, 2002).
- The second one is *the intra-firm technology transfer*, in which there are strategic and interdependent relationships between the source and recipient (see Fig. 2.4). In this case, the source has some equity partnership with the recipient in the form of joint venture, or with foreign affiliate in host country (Techakanont, 2002, p. 19). In other words, some equity of the recipient is owned by the source. In this case, source (MNC, foreign firms, or parent foreign company) provide various technology transfers and training activities to local workers of the foreign affiliate in the host country because the affiliate must be successful in order to get high profits, and this depends on the relationships between such actors and the quality of the transfers provided (Hobday, 1995; Kim, 1997).

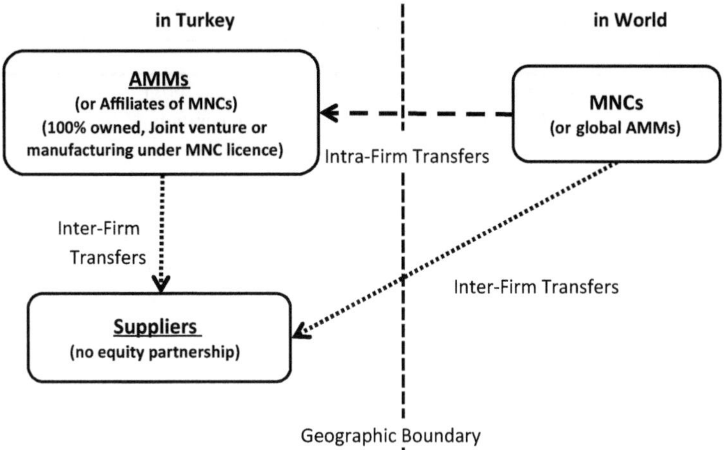

Fig. 2.4 Intra- and inter-firm technology transfers. *Source*: By the author, based on ideas of Techakanont (2002, pp. 19–21). *Note*: The dashed line shows intra-firm linkages between firms, while the dotted lines show inter-firm linkages between such firms

- The third one is *the inter-firm technology transfer*, in which the relationship occurs between the buyer and supplier, and there is not any equity partnership between such independent actors (see Fig. 2.4) (see Lall, 1980). This kind of relationship is also called buyer-supplier relationships. The relationship between such actors starts after the supplier firm is evaluated according to the buyer's own criteria, and if the supplier is selected and approved by the buyer, then an official relationship such as a business contract takes place (Techakanont, 2002, pp. 20–24). To be selected by a buyer, supplier firms should have some technological capabilities and skills in order to produce the demanded products at desired quality and cost level (see Sects. 6.5 and 6.6). In this case, buyers provide various specific technologies to their suppliers without asking for money in order to benefit from products supplied by demanding higher quality standards, improvements, and lower costs. Therefore, local suppliers may gain a chance to access such technologies, and they must upgrade their capabilities according to the specifications of buyers in order to work together continuously.

As discussed above, it is expected that MNCs should provide more technology transfers at *intra-firm level* relative to *inter-firm level*.

2.7.3 Forms of Technology Transfer at Intra- and Inter-Firm Level

Forms of technology transfers can be classified as formal and informal. MNCs actively transfer technology to their suppliers in formal ways to ensure that the

Table 2.3 Forms of technology transfer at intra- and inter-firm level

At intra-firm level		
Technology level	Yamashita (1991)	Kuroda (2001)
Operation	1—Operation	1—Operation
	2—Maintenance	2—Maintenance
	3—Quality control	3—Quality control
	4—Production management	4—Production control
Improvement	5—Improved technology	5—Process improvement (Kaizen)
	6—Molding	6—Development of mold/die/jig
	7—Design	7—Development of equipment
Development	8—Product development	8—New process technology
	9—Equipment development	9—Engineering of new design
		10—R&D of new products
At inter-firm level (Wong, 1991)		
– Advice on plant layout, equipment selection and operations planning		
– Advice/training on quality management system and other "good manufacturing practices" (GMP)		
– On-site audit of plant operation and troubleshooting of specific productivity problems		
– Loaning of equipment and machinery, either temporarily or permanently		
– Training of supplier staff through formal courses/seminars or informal consultations/visitations		
– Product design specification and performance requirements		
– Early supplier involvement in prototype development and value engineering stage		
– Informal sharing of technical information and ideas among the technical staff of both companies		
– Exposure to MNC system of managing and organizing manufacturing activities and observation of MNC		
– Provision of a stable source of income to finance the investment		
– Sourcing of technical experts to solve specific technical problems encountered by the supplier		
– Advanced indications on future quality/performance/features requirements and targets		
– Testing and diagnostic feedback on quality and other dimensions of performance of supplier's products		

Source: Extracted from Techakanont (2002, pp. 24–26)

products and services supplied meet the technical specifications in terms of quality, cost, durability, and reliability. Formal ways of the transfers can be summarized as FDI, the purchase of foreign machinery and turnkey plants, foreign licenses, consultancies from special engineering firms, technical service/assistance/specifications provided by foreign firms, technical licensing agreements with foreign firms, providing of blueprints, and more generally mergers and acquisitions (M&A) of low or high-tech foreign firms and so on. On the other hand, local suppliers can try to increase their capabilities by using informal mechanisms in order to supply competitive products to MNCs. Major informal forms for technology transfers are OEM arrangements, international literature, labor turnover, observation, reverse engineering, imitation, etc. Although new technologies could be obtained only through formal channels, informal technology transfers are seen most important in the literature for developing the technology capabilities of local firms (see Ernst & Kim, 2002; Kim, 2001 for further details).

Table 2.3 displays the forms of the technology transfers identified through case-studies conducted in automotive, textile, and electrical machinery industries in ASEAN countries at both intra- and inter-firm level (Techakanont, 2002, pp. 22–24). The authors categorized the form of technology transfers into three groups at intra-firm level: "operation technology", "improvement technology", and "development technology (means the creation of new knowledge)". As can be seen in the table, the researchers also classify technology transfers into nine and ten types depending on their observations (Techakanont, 2002, p. 23). The table also shows 13 direct forms of technology transfers, occurring from buyers to suppliers, observed and identified in these studies at inter-firm level. Studies find that suppliers should benefit from such transfers because of having formal relationship with buyers. MNCs send their expert staff to give advice to their suppliers in host country about production, quality control methods, etc., and train the local personnel of their suppliers. By these assistances, suppliers improve their manufacturing technologies, quality control practices, design, R&D, distribution, and delivery methods. These kinds of direct transfers (inter-firm transfers) have occurred between firms and are not commonly observed. One of the aims of this study is to reveal the forms and intensity of the KTTs that are created between such firms in terms of production, production process, and training.

2.7.4 Theoretical Framework

It is pointed out that process of the technology transfer is completed when the recipient learns, transforms, and internalizes the technology transferred by source into their own technology (or knowledge). In other words, it can be seen as the process of "the internalization of knowledge" successfully from the source to their own. As a result of these transfers, positive effects should be expected or observed on the recipient side such as capability increases, performance increases or "productivity improvement" that "can take any form of reduction of defect rate", "delivery time", "average costs", and "cycle time" (Techakanont, 2002, pp. 27–28) (see Sect. 4.9). However, it is not very easy to reveal these relationships at inter-firm-level because they are not easily observable, or formal data or information is not available. This situation justifies our research design and methodology in this field by conducting case-study approach for data collection at firm level. In the context of the case study, we aim to analyze the KTTs and their effects from two perspectives:

- One looks only at the suppliers in Turkish automotive industry as the recipient of the technologies. In other words, we analyze the technology transfers in detail from customers (AMMs in Turkey, MNCs in world, and other firms) to their suppliers at *inter-firm level* (see Fig. 2.4). We conducted face-to-face questionnaire survey with the suppliers in Turkey in order to collect firm-level quantitative data. By this survey, we also want to reveal suppliers' characteristics, technology capabilities, competition structure, cooperated partners, R&D and innovation activities, etc. There is one caveat to this method; it can reveal technology transfers that occur from other customers, not from a specific customer. This is rational because suppliers have relationships with multiple customers, so they are exposed to technology transfers from many customers. Therefore, it is not possible to analyze the transfers provided by the selected specific customer (at least by this method).

- The second perspective looks at the AMMs (as buyers or customers) operating in Turkish automotive industry as the source of the technologies provided to suppliers at inter-firm level. In this context, we also analyze the technology transfers from MNCs in home country (global automotive manufacturers or parent companies) to their affiliates (AMMs) in Turkey by taking into account the first two types of channels discussed above at both *intra-firm level* (if AMM in Turkey is 100% owned by MNCs, or it is an affiliate of MNCs) and *arm's length trade of technology* (AMMs in Turkey manufacture under MNC license) (see Fig. 2.4). For this aim, we conducted in-depth semi-instructed interview method with the top-executives of AMMs in Turkey to collect qualitative data (see Chaps. 3 and 6). The interview findings are also used for the quantitative survey findings, obtained from recipients and related to inter-firm level technology transfers, in a complementary way.

2.7.5 Studies Conducted for Foreign Countries

A few studies on technology transfer conducted on various countries, which apply the (similar) data collection methods employed in this study, are reviewed below.

Giroud (2003) reviewed in his study on Malaysia whether the foreign main industry companies operating in the electronics sector provide KTTs to the local firms through backward linkages. In this study, both the questionnaire and personal interview methods were used together to collect data at the firm-level. In this study, 320 foreign companies in the sector were approached via mail in 1996, and 95 of them completed a questionnaire survey (a response rate of 30%). Also, in-depth interviews were made with 11 foreign companies operating in the sector, based on their origin (4 American, 3 European, 2 Japanese, 1 Thai, and 1 Singaporean). Unlike the path we have followed in our study, Giroud studied, by means of implementing questionnaire survey via mail only to the foreign companies operating in the sector, what kind of transfers they made to their suppliers on products, production processes, financial support and training, and how often. Under the same scope, he made personal interviews with 11 foreign companies he selected among the surveyed companies. In other words, he studied the KTTs provided by the foreign companies operating in the main sector to the local suppliers only from the viewpoint of the foreign companies realizing those transfers, without contacting any local supplier firms. He tried to examine the questionnaire and interview results first through descriptive analysis; and then, by using logistic regression analysis method, he reviewed the effects of the independent variables (age of the firm, export rate, size of the firm, purchasing strategy, firm origin) determined based on the data he obtained for 95 firms upon the KTTs. The results of the analysis showed that the foreign companies realized KTTs to the local suppliers via training, but those transfers were inversely proportional to the duration for which the firms had been operating in the sector. In addition, it was discovered that there existed a positive relationship between exports and transfers, and it was determined that the size of the firm, the origin of the foreign companies, and the characteristics of the products manufactured were the most important factors explaining the transfers realized by the foreign companies. These findings are highly consistent with our findings in the case study research for Turkish automotive industry that will be discussed in next chapters.

Other studies—Techakanont (2002); Techakanont and Terdudomtham (2004)—analyzed the evolution of the technology transfer between the main industry companies, and local suppliers in Thailand automotive sector. Also, they attempted to analyze the level of the collaboration between the local firms and the main industry companies and the types of the collaborations which the local firms needed to establish with the main companies so that they developed their technical skills.

Techakanont (2002) used the case study and questionnaire methods in the study. He conducted personal interviews with the top-level managers assigned in the Thai affiliate of a Japanese global automotive manufacturer in 2000. Also, he implemented the questionnaire survey to 26 supplier firms of this Japanese automotive manufacturer that were located in Thailand.

Techakanont and Terdudomtham (2004), on the other hand, attempted to analyze—by using the same research method—what kind of developments occurred in the relationships between the main companies and the suppliers in the Thai automotive sector. To this end, during 2002 and 2003, personal interviews were made with the managers of five companies in automotive main industry that were Japanese in origin and that were located in Thailand. They also applied questionnaire survey to 100 firms that were the suppliers of these five companies interviewed, and answers were obtained from 15 supplier firms. The study found out, as a result of the analyses conducted in the form of case study, that the level of the technology transfers between the firms was quite high both in content and in quality, that the difficulty levels of the transfers were determined by the content of the technologies transferred, and that the success of the transfers was directly proportional to the technological capabilities of the suppliers.

Berger's (2005) study reviewed the effects of the MNCs on the innovation and R&D activities of the firms operating in manufacturing sector of Thailand.[19] The author used the qualitative and quantitative techniques together. The study employed the readily available quantitative data obtained from the researches on R&D and innovation in 2000 and 2002 on the basis of questionnaire surveys conducted—observing the OSLO and Frascati guides of OECD—by the National

[19] The study followed the same method as us in collecting qualitative data and it constituted the basis for us to formulate our questions on innovation and R&D.

Science and Technology Development Agency of Thailand. These surveys were implemented on the firms that were operating in nine different sectors in the manufacturing industry of Thailand and that had a turnover exceeding a certain limit. The survey in 2000 was implemented on 2,166 firms and 1,019 of them responded, while the survey in 2002 was implemented on 3,945 firms and obtained answers from nearly 1,500 firms. The author preferred to focus on the case study method and did not make logistics regression analysis. Semi-structured personal interviews were conducted with the R&D managers, production managers, and general managers of totally 20 companies in the automotive and hard-disk sectors in the year 2003, and the results of the study were analyzed statistically via various variance analyses, and the interviews made with the companies were evaluated by means of case study method.

2.7.6 Studies on Turkish Automotive Industry

The number of the studies concerning Turkey, the main focus being on KTTs, is quite limited in the international literature. However, there exist some studies conducted on the productivity, MNC-based technology spillovers, and automotive industry, using some quantitative and qualitative methods.[20] The number of the studies conducted concerning the relationships between the automotive main and supply industry in Turkey is also few. The available studies are rather based on certain areas of the automotive industry, such as R&D or innovation, and most of the studies review the automotive sector not individually but together with the other sectors. Due to the fact that quality and detailed data cannot be obtained at the firm-level, the analyses are tried to be put forward in the form of case studies. We observe that a large majority of the available studies in Turkey were realized during the last decade. No empirical study examining the technological capabilities of the supply industry firms and reviewing the types, intensity channels, and determinants of the KTTs provided by MNCs to domestic firms at both inter- and intra-firm level qualitatively (case study) and quantitatively (econometrical) in detail together, along with the topics discussed above, has been encountered yet; however, it has been observed that there are studies in the national literature which are similar to our study in terms of method. These could be summarized as follows.

Bedir (1999), in order to collect data, implemented a questionnaire survey on the Turkish automotive main and supply industry firms in his study, where the main and supply industry relationships were reviewed. Under this scope, questionnaire forms were sent to 16 main industry companies and 120 supply industry firms (75 of them being TAYSAD member and 45 being other firms operating in the sector) via SPO in 1996. Twelve of the main industry companies and 79 of the supply industry firms

[20] See Pamukçu (2003), Bertinelli et al. (2006), Lenger and Taymaz (2006), Wasti et al. (2006, 2009), Aksoy (2008), Pamukçu and Taymaz (2009), Dayar and Pamukçu (2011).

responded to the questionnaires. The findings obtained from the study were analyzed only descriptively.

Wasti, Kozan, and Kuman (2006) used the questionnaire survey and personal interview methods in order to collect data. In their study, they attempted to determine the types of the relationships between the buyers and suppliers in the Turkish automotive industry. In the study executed in 2002, 16 main industry companies as the members of OSD and all the supplier firms as the members of TAYSAD were chosen as sampling. Ten main industry companies agreed to participate in the study and a questionnaire survey was implemented by conducting semi-structured personal interviews with 51 purchasing managers of those companies. Seventy-two supplier firms who were TAYSAD members agreed to participate in the study and only the questionnaire survey was applied to those firms. Anova and Scheffe tests were used in analyzing the data.

In another study, Wasti, Kozan, and Çınar (2009) used the same data and attempted to determine the probability of the relation-specific investments made by the main industry to the supply industry in the Turkish automotive sector, by means of the multi-regression analysis method.

Çelikel (2009), on the other hand, employed the case study method in her study, where she reviewed the factors affecting the R&D collaborations between the automotive main companies in Turkey and MNCs as their partners. In this context, case analysis studies were conducted with the affiliates of the three MNCs that were operating in the Turkish automotive main industry (Tofaş, Ford-Otosan, and Hyundai-Assan) and their parent companies abroad, namely Fiat, Ford, and Hyundai, respectively. Çelikel obtained the data she used for the case analyses from the in-depth, face-to-face interviews she conducted with 40 people (R&D managers, engineers, and top-level managers) from those six companies, and put forward her findings descriptively.

Ekmekçi's (2009) study, where the factors determining the knowledge transfer from the FDI flows in the Turkish automotive industry to the local supplier firms are analyzed, is another study which uses a similar case study method with Çelikel (2009). In the study, seven main companies and seven supply industry firms with 100% local capital in Turkey, which sell parts to at least one of these seven main companies, were interviewed face-to face in order to gather data, and a descriptive analysis was performed.

Chapter 3
Research Methodology and Design

This chapter describes the research methodology and design of the study in detail. After the introduction, the second section includes some information about the automotive industry in Turkey, which includes basic economic indicators such as production, foreign trade, sales, capacity, employment, place of industry in world, AMMs, and the supply industry in Turkey. In the third and fourth sections, the scope, purpose, and main research questions of the study are discussed. Then, research methodology is justified. In the sixth section, the basic research methods adopted for data gathering are explained in detail. Lastly, some concepts and differences for the firms constructed and used in the analyses are described.

3.1 Introduction

As stated before, the Turkish automotive industry is analyzed as a case study to understand mainly the complex nature, dynamics, and determinants of KTTs at both inter-firm (occurring through backward linkages from customers[1] to their suppliers) and intra-firm level (occurring to AMMs in Turkey from multinationals as their partners). In addition, we want to reveal the characteristics, technology

This study was supported by a research grant from both TUBITAK under 1002-short term R&D funding programme and Faculty Development Program of METU (OYP). Moreover, the professional supports of OSD and TAYSAD (two main representatives of Turkish automotive industry) have played an instrumental role in the success of the questionnaire survey application (see Appendices C and D for their letters of support).

[1] We will use the four terms "customers", "buyers", "AMMs", and "MNCs" interchangeably throughout the study to refer to the same concept since approximately all of the AMMs operating in Turkey are MNC affiliates and/or they are manufacturing under foreign license/brand of global AMMs; by the way, majority of the surveyed firms are direct suppliers of at least one AMM in Turkey and they carry out most of their sales to these customers, so these four terms are used interchangeably (see section "AMMs in Turkey", Sects. 3.7, 4.5, and Appendix H for details).

A. Sönmez, *Multinational Companies, Knowledge and Technology Transfer*, Contributions to Management Science, DOI 10.1007/978-3-319-02033-4_3,

capabilities, innovation and R&D activities of the suppliers, their place in global value chain, and their cooperation activities with other partners. Moreover, we will try to analyze the effects of such activities and characteristics of the suppliers on such transfers. Finally, the effects of these transfers on the performance level of the suppliers are analyzed. We examine and look at our objectives from two perspectives; one is from suppliers' perspective as a recipient of these transfers (Chaps. 4 and 5), and the second one is from customers' perspective as a source of the transfers (Chap. 6). To collect detailed data at the firm level, the methods that have been used are:

1. A detailed questionnaire survey that has been designed to collect quantitative data from the suppliers. Its application is conducted through face-to-face interviews with top-executives of the 166 automotive suppliers operating in Turkey. The database of supplier firms is established from mainly TAYSAD members and various industry unions (BTSO, DOSAB, UIB, GOSB, and NOSAB).
2. Semi-structured in-depth interviews that are conducted with 19 top executives of 11 AMMs in Turkey to collect qualitative data from customers.

3.2 Automotive Industry in Turkey

Automotive industry in Turkey was established in the mid-1950s as an assembly industry. A number of MNCs set up majority joint ventures with Turkish partners and entered into the market in the late 1960s. Until the 1980s, the share of the industry in Turkey's total exports was nearly zero due to the policy of import substitution development strategy. However, with the change in industry policy to outward-oriented industrialization in 1980, the share of the industry in total exports started to increase slowly and reached 1 % on average for 1980–1990. Moreover, after the full capital account liberalization in 1989 and following the signature and formation of the Customs Union agreement between Turkey and the EU in 1996, three global AMMs from Japan and South Korea invested in Turkey and launched production in the mid-1990s (Toyota in 1994, Honda in 1997, and Hyundai in 1997).[2] Turkish automotive industry has taken significant steps since 1960s; these can basically be summarized in five main headings (MITI, 2011):

- In the 1960s, assembly manufacturing of tractors and commercial vehicles for *import substitution*,
- In the 1970s, "localization" and "manufacturing of passenger car" for producing some components and parts in domestic market,

[2] See Şenses and Taymaz (2003), Taymaz and Yılmaz (2008b), SPO (2007), and MITI (2010, 2011) reports on the industry in Turkey for detailed information.

- In the 1980s, capacity and technology investments in the industry,
- In the 1990s, restructuring and integration with the global industry for global competition,
- In the 2000s, entrance to *sustainable global competitive process* by creating higher value added design and manufacturing for world markets.

3.2.1 Basic Economic Indicators

World Automotive Production

Figure 3.1 displays the total world motor vehicle production by the type of vehicles produced over the period 1999–2010. As can be seen in the figure, total production was nearly 70M units in 1999, and 50M of that were personal cars and 20M were commercial vehicles. In 2010, total production increased nearly 14 % compared to 1999 and reached 80M units. Nevertheless, it seems that the production of commercial vehicles remained constant in 20M units, and this increase occurred only in personal cars by 20 %. This shows that the total demand for personal cars is stronger than commercial vehicles.

Figure 3.2 shows the share of countries in world motor vehicle production over the period 1999–2010 and Fig. 3.3 displays the largest 20 countries' production units in total motor vehicle production in 2010. According to these, 65 % of the total world production in 2010 was carried out by only six countries. It seems that China has the biggest share in total world production by nearly 24 % (two times bigger than the nearest rival country) with more than 18M units in 2010. This country is followed by Japan (12 %), USA (10 %), Germany (8 %), South Korea (6 %) and India (5 %). The most striking finding is that shares of the major automotive

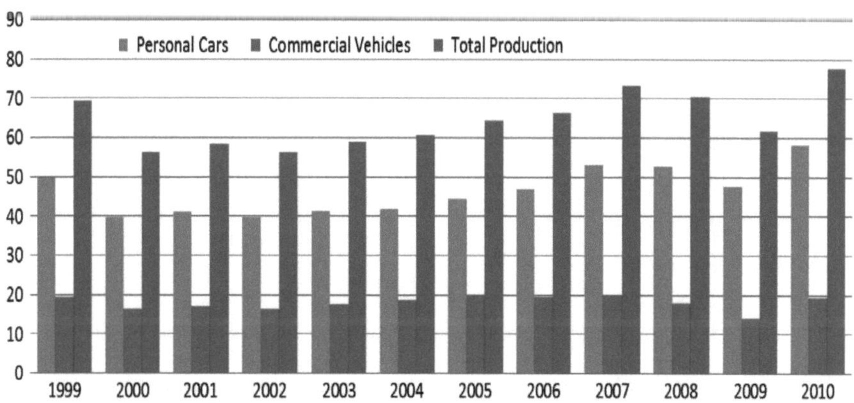

Fig. 3.1 World motor vehicle production by type and year 1999–2010 (million ea.). *Source*: Author calculations based on OICA statistics

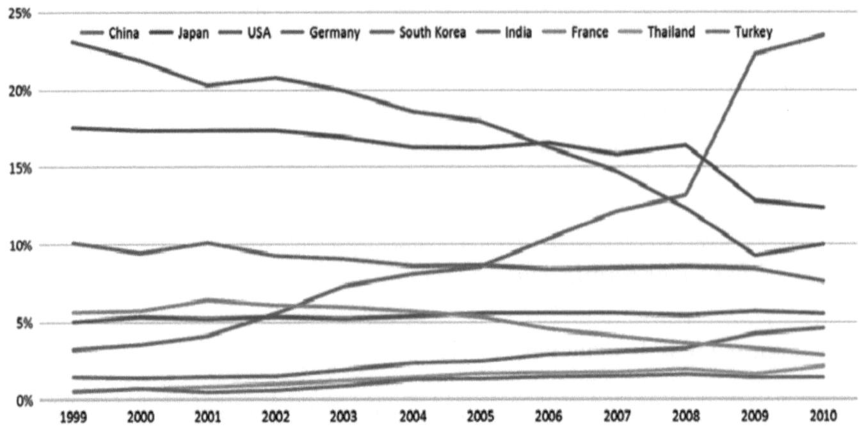

Fig. 3.2 Share of countries in world motor vehicle production by year 1999–2010. *Source*: Author calculations based on OICA statistics

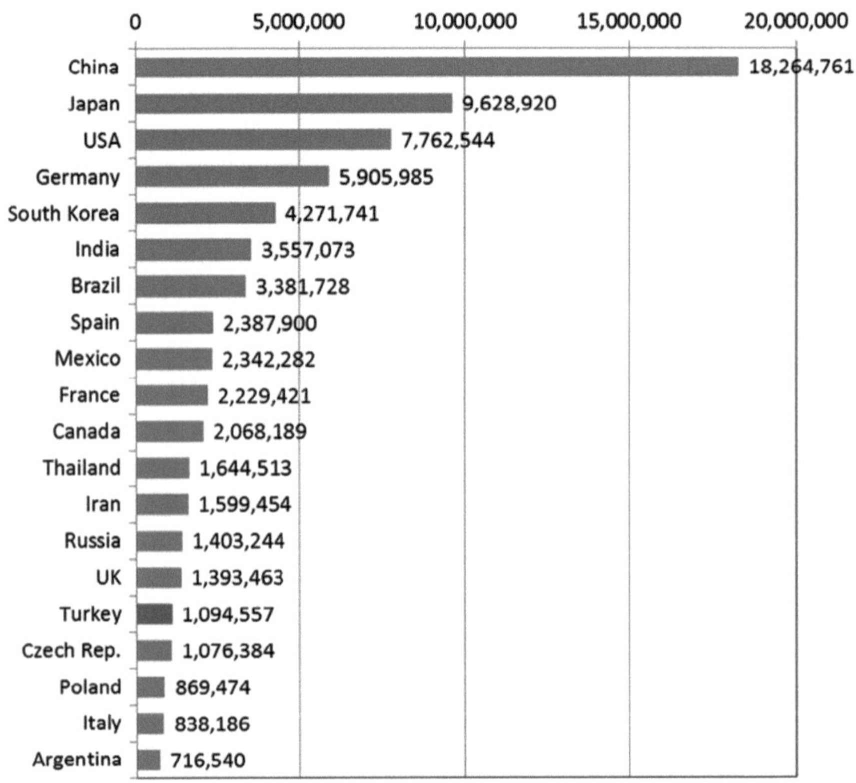

Fig. 3.3 World motor vehicle production by country in 2010. *Source*: Author calculations based on OICA statistics

countries (USA, Japan, Germany, France, and Spain) declined over the period; on the other hand, China increased her share at an increasing rate. This situation is a result of the manufacturers' transferring their production bases from developed countries to China and also reflects that China is the new production base of the world for the global AMMs. In this process, two countries come to the fore: India and South Korea. South Korea was in the seventh place in 1999 by 5 %, and she had succeeded in preserving its place in terms of production share, and today she is the fifth biggest country in terms of production in the world surpassing France, Spain, and Canada. On the other hand, India was in the 12th place in 1999 by 1.5 % share; however, she increased her share nearly three times, and today she is the sixth biggest country by 4.6 % share in total world production. Combined with the previous finding about China, we can say that world motor vehicle production has shifted from the West to the Asian countries.

Production and Place of Turkey in the World

Figures 3.4 and 3.5 display the share of Turkey's motor vehicle production in the world and total production units respectively according to personal cars and commercial vehicles categories over the period 1999–2010. In terms of the number of units produced, Turkey was ranked 24th in the world with total amount of nearly 298,000 vehicle production (222,000 personal cars and 76,000 commercial) corresponding to a total share of 5 ‰ in 1999. Moreover, Turkey was ranked tenth among European countries for the same year. In the last decade, Turkey managed to triple her share (1.41 %) and rose to 16th place in the world with nearly 1.1 million units produced (603,394 personal cars and 491,163 commercial vehicles) in 2010 (see Fig. 3.3). Turkey became the fifth largest producer among European countries after Germany, Spain, France, and UK. Besides, Turkey's share has risen to a larger extent in commercial vehicles rather than in cars. In the

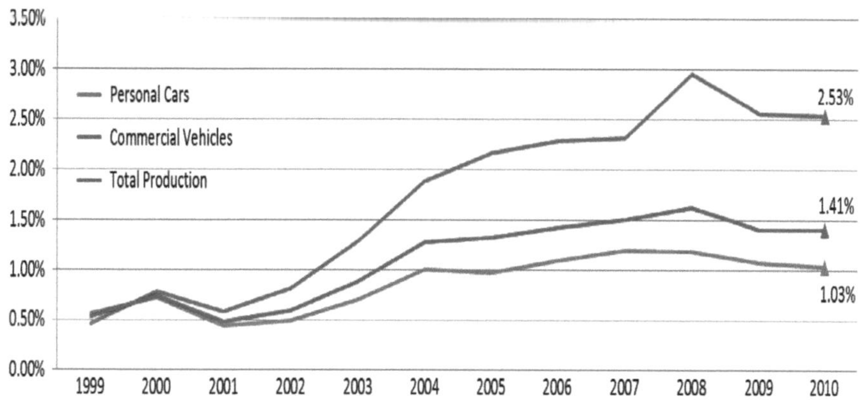

Fig. 3.4 Share of Turkey's motor vehicle production in world production. *Source*: Author calculations based on OICA statistics

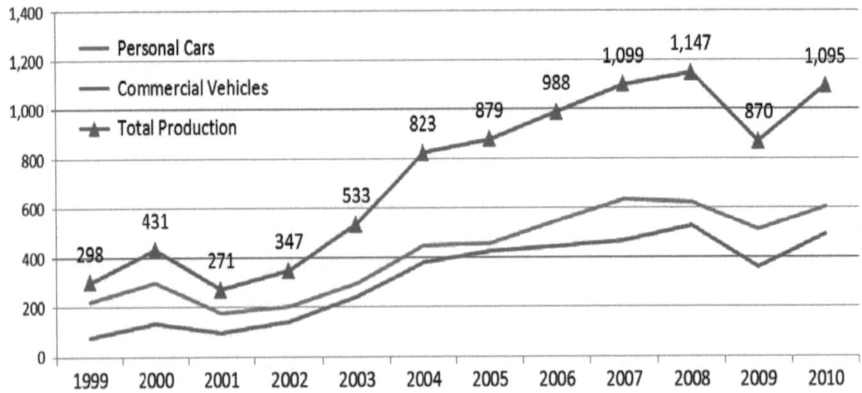

Fig. 3.5 Motor vehicle production of Turkey by year 1999–2010 (×1,000). *Source*: Author calculations based on OICA statistics

world ranking in 2010, Turkey was at the 9th position in the personal car production and 17th in commercial vehicles production with the shares being equal to 1 % and 2.5 %, respectively (see Fig. 3.4).

AMMs in Turkey

Today, there are 15 AMMs operating in Turkey and these are also the members of OSD. Entire list of the AMMs together with their geographic locations, establishment dates, and their statistics in terms of sales, capacity, production, foreign capital structures can be seen in Appendix H.

When the foreign capital structure of the AMMs are analyzed, it is seen that five of them are completely local-capital firms (FS = 0 %), while two of them, namely Toyota and Honda, are with 100 % foreign capital. Four of the remainder firms are foreign capital firms (joint venture with minority foreign ownership) with ratios ranging between 29 and 42 %, and the other four are foreign capital firms (joint venture with majority foreign ownership) with ratios ranging between 51 and 99 % (see Table 3.1 and Appendix H). Moreover, the most striking characteristic of the AMMs is that approximately all of them are MNC affiliates or manufacture under foreign license independent of local or foreign capital structure[3] (see Appendix H); therefore, to use the terms AMMs and MNCs interchangeably throughout this study to refer to the same concept is justified.

[3] It was detected from the interviews that two AMMs which are not seen as manufacturing under a foreign license (B.M.C and Türk Traktör) in Appendix H use foreign licenses to manufacture some parts and/or import their powertrain parts and components from MNCs.

Table 3.1 Share of AMMs in total capital, capacity, production and sales by foreign share (%)

	N	Capital (TL)	Capacity (2011)	Production (2010)	Sales (TL) (2010)
FS > 50 %	6	38.70	43.36	44.74	47.29
29 % < FS < 42 %	4	29.99	47.85	52.08	44.76
FS = 0 %	5	31.31	8.79	3.18	7.94
Total	**15**	**100**	**100**	**100**	**100**

Source: Author calculations based on OSD statistics, see Appendix H

When we look at the types of vehicles manufactured by firms, we see that light, medium, and heavy commercial vehicles (such as bus, minibus, midibus, pick up, truck, and farm tractor) are generally produced by five local ownership firms (FS = 0 %). Manufacture of personal cars is performed by five firms, which have partnership with MNCs as joint venture. The firms manufacturing personal cars are Tofaş (FS = 37.8 %), Oyak-Renault (FS = 51 %), Hyundai (FS = 70 %), Toyota, and Honda (FS = 100 % for both).

When the province in which firms operate is analyzed, four of them operate in Kocaeli, three of them operate in Bursa, three in Sakarya, two in Ankara and remainder three firms operate in Istanbul, Izmir, and Tekirdag. It has been observed that firms manufacturing personal cars are located in Kocaeli, Bursa and Sakarya; firms manufacturing commercial vehicles such as buses and trucks are mainly located in Ankara, Istanbul, Izmir, and Kocaeli.

The establishments of the local-capital firms (except for Hattat and Temsa) date back to early- and mid-1960s. It is observed that personal car manufacturers TOFAŞ and Oyak-Renault were both established in 1971, and the other three foreign-capital AMMs were established in 1994 (Toyota) and 1997 (Honda and Hyundai).

Table 3.1 presents a number of indicators pertaining to the share of AMMs in total capital, capacity, production, and sales according to their foreign share classification.

According to this;

- Total sales of the 15 AMMs for 2010 were approximately TL 28B; 47 % of it was realized by six firms whose foreign capital ratio is above 50 % (majority-owned foreign firms), 45 % of it was realized by four firms with foreign capital ratios ranging between 29 and 42 % (minority-owned foreign firms), and only 8 % was realized by the five local firms (FS = 0 %).
- When the capital structure of these firms is analyzed, their total capital sum is TL 3.1B; the share of the majority-owned foreign firms is 39 % in total capital, the share of the minority-owned foreign firms is 30 %, and the capital share of the local firms is 31 %.
- The Table 3.1 also shows the shares of the firms in total capacity and total production. Total capacity of the firms was 1.6M units in 2011; total production amount was 1.1M units in 2010. As can be observed from the table, four AMMs whose foreign capital ranges between 29 and 42 % had the highest rates regarding the last two indicators (48 % for capacity and 52 % for production), firms whose foreign capital is more than 50 % followed them (43 % for capacity and 45 % for production).

In summary, although capital ratios of the local and foreign capital AMMs seem to be approximately equal, the AMMs with foreign capital have capacity (91 %) ten times bigger than the AMMs with local capital (9 %), they realize 97 % of the total production and 92 % of the total sales. This is significant for indicating the importance of the foreign capital (MNCs) in automotive main industry in Turkey and its superiority in the industry. This situation also explains once more why we use the "MNCs" and "AMMs" terms interchangeably in the study for the main companies in Turkey (see Sects. 3.7, 6.1, and 6.2 for more details on the AMMs).

Automotive Supply Industry in Turkey

Automotive supply industry produces finished goods, semi-finished goods, components, systems, OEM parts directly or indirectly for the AMMs operating in domestic and overseas markets and/or for the replacement market. The suppliers in this industry generally conduct these activities in accordance with the technical documents specified by AMMs. There is not reliable statistical information on the number of firms which operate in the automotive supply industry. According to the various reports regarding the industry, it is estimated that number of suppliers in supply chain is about 1,100, and the number increases to 2,000–3,000 with sub-sectors (see Table 3.7 for details). However, it is predicted that the number of the direct suppliers that produce parts directly for the AMMs is much less and about 300–350[4] (see Sect. 6.5.2 for the exact numbers in terms of AMMs interviewed). These suppliers are largely clustered in Marmara, Aegean, and Central Anatolia regions in parallel with the AMMs (see section "Questionnaire Design" and Sect. 6.5.6 for more details). Main production groups manufactured by domestic automotive supply industry may be categorized as complete engine and engine parts, internal and external tires, powertrain mechanisms, brake systems and parts, hydraulic and pneumatic mechanisms, suspension parts, safety mechanisms, rubber and tire parts, frame mechanism and parts, forged and cast parts, electrical equipment and lighting systems, accumulators, car glasses, seats and plugs.

Total Employment of the Automotive Industry in Turkey

The automotive main industry in Turkey employed about 50 K people in 2010 according to OSD statistics. On the other hand, it is estimated that supply industry employs nearly 200 K and this number reaches to 500 K people together with sub-sectors, distribution, marketing, service and sales networks.

[4] TUBITAK (report for, 2023 vision), SPO (2007), and MITI reports (2010, 2011).

Foreign Trade of the Automotive Industry in Turkey

According to TurkStat statistics,[5] export of automotive industry which was about US$181M in 1991 increased by 13.8 % in 2008 compared to 2007 (US$17B) despite the significant decrease in the last quarter of 2008 due to the global crisis and broke a historical record by rising approximately to US$20B in 2008 (see Figs. 3.6, 3.7, and 3.8). This is very significant since it indicates that the export of automotive industry in 2008 has increased 110 times compared to 1991. According to the same data, personal cars ranked first among automotive export items with US $7.5B in 2008. Figure 3.8 shows the evaluation of the shares of five largest industries in total exports of Turkey's. According to this, the share of the automotive industry in total exports was only 4 % in 1996 and then continued to rise, and from 2000 onwards it rose at an increasing rate and finally by passing three major industries (basic metals, textile, and food), it ranked first among all industries by 13 % with US$15B exports value in 2010. In contrast with export values, the share of the industry in total imports of Turkey changed between 5 % (2001) and 12 % (2004) and became 9 % on average over the period 1999–2010.

These figures point out that the industry acquired a very important place in the world motor vehicle industry and performed very well in terms of exports. Moreover, it has experienced significant performance, output and productivity growth over the last decade and has enhanced its competitiveness in global markets transforming it today into one of the most dynamic, successful, and important industry in Turkey.

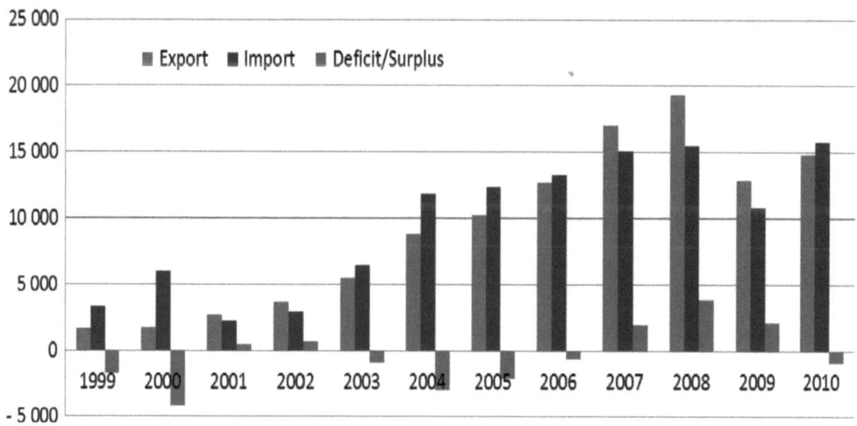

Fig. 3.6 Export and import values of Turkey's motor vehicle industry by the years 1999–2010 (million USA dollars). *Source*: Author calculations based on TurkStat statistics

[5] Author calculations based on TurkStat foreign trade statistics database, NACE 34 ISIC Rev. 3 for motor vehicle industry, BEC 51 for personal cars.

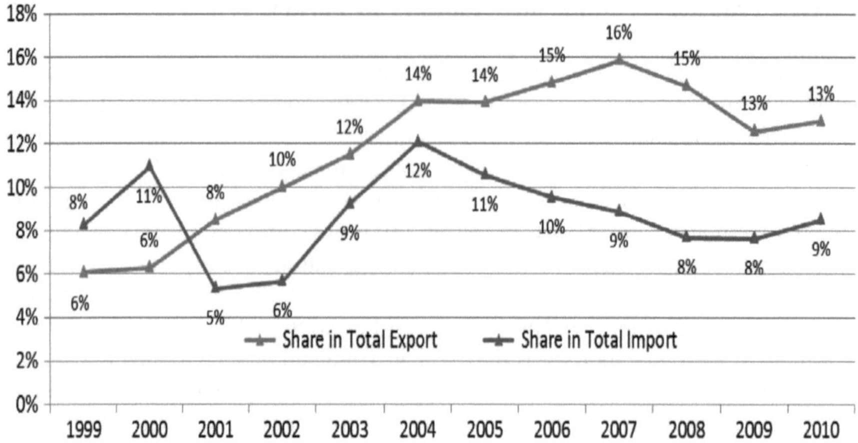

Fig. 3.7 Share of Turkey's motor vehicles industry in total foreign trade by the years 1999–2010. *Source*: Author calculations based on TurkStat statistics

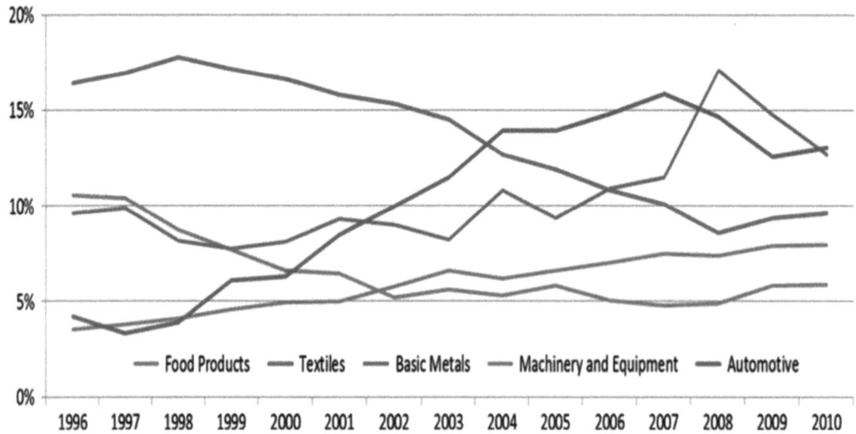

Fig. 3.8 Evolution of industry shares in total exports of Turkey: 1996–2010. *Source*: Author calculations based on TurkStat

To sum up, figures presented above point to the excellent performance of the Turkish automotive industry at least since the 1990s. However, this performance will be very difficult to sustain due to the increase observed in the number of competitors in the world market and the emergence of new low-cost locations to which multinationals will certainly be attracted. Therefore, if Turkey wants its automotive industry to continue to be the major engine of economic growth, and an important source of employment creation, she will have to upgrade it from being solely a production base to an *excellence center* providing high-quality services in terms of R&D, design, test, and innovation. The role of the transfers and collaboration activities with multinationals will undoubtedly play an important role in this transition.

3.3 Scope of the Case Study

In general, the main reason for analyzing the automotive industry is that it is one of the industrial sectors in which globalization process is realized in the fastest way among the current industries in the world. Although there are a few global manufacturers (global AMMs) left due to take-overs and mergers worldwide, these firms highlight their global production networks by making production in different regions and countries around the world, compete with each other at a high level and realize the great deal of R&D and innovation activities when compared with other industries. Besides, the relations between global manufacturers and their suppliers are considerably strategic and intense compared to other industries.

The case-study will target Turkish automotive industry. The main reasons behind the choice of this specific industry can be summarized in this way: (1) it accounts for a significant share of GDP and exports of Turkey, (2) it is among the largest recipients of FDI and is dominated by MNCs, (3) it is highly competitive on a global scale, and (4) as well as being an industry drawing most FDI, it is also highly supported by Turkish governments for strategic purposes.

In sum, the most important reason for the analysis of Turkish automotive industry is the significant place of this industry in Turkish economy which has been achieved so far as discussed in the previous section. It has become one of the primary industries in Turkey by making a fast progress especially in the last 10 years (as of 2000), thus it has a strategic importance today. This situation reveals itself with an important share in total export of manufacturing industry, R&D expenditures, and employment of Turkey. Moreover, the statistical findings revealed that automotive industry (NACE 34) has taken precedence over all industries in Turkish manufacturing industry (see Sect. 2.2.1) in terms of the share of foreign firms in the production, gross output, value added, and in the number of all firms (Sönmez & Pamukçu, 2012: 35). Many foreign firms operating in the industry, presence of strong strategic relations between these and local firms, intense R&D and innovation activities in the industry, and various incentive policies of the government towards the industry may be indicated as other factors playing an important role behind the choice of industry.

The current state of the industry can also be summarized with the information available in the website of TAYSAD as follows[6]:

- Annually 1.5M ea. production capacity,
- Second bus producer, first light commercial vehicle producer, third biggest truck market in the European Union,
- Sixteenth biggest automotive producer of the world in 2010,
- Sixth automotive producer in the EU and the presence of light commercial vehicle market bigger than the total of eight EU member states (Czech Republic, Romania, Slovenia, Poland, Hungary, Slovak Republic, Estonia, Latvia, and Lithuania).

---

[6] Source: http://taysad.support.com.tr/altmenu.asp?AnaId=1411&def_dil_id=149

Due to the above-discussed reasons, it may be assumed that MNCs regard Turkey as the production base for Europe and Middle East in the recent years.

The following issues can also be added to the reasons for analyzing the Turkish automotive industry:

- Turkey's transformation into an automotive production base with 15 automotive factories due to the fact that many AMMs have globally preferred Turkey as production base for their Middle East and Europe markets,
- Turkey's being the third biggest vehicle manufacturer of the Eastern Europe,
- The presence of many MNCs making production in the industry and the production of these firms on a global scale,
- High level of innovativeness and competitiveness of the firms in the industry,
- Realization of cooperation activities in terms of R&D and innovation of strategic purposes between AMMs and direct suppliers.

3.4 The Purpose of the Study and Research Questions

In line with the main purposes of the study discussed before (see Sect. 1.2), the purpose of the case-study is basically to examine the complex nature of KTTs at inter-firm level occurring through backward linkages from customers to their suppliers in Turkey's automotive industry, and to explore the firm-related factors that affect these transfers. However it is not limited with these, it also aims to explore suppliers' major markets, their technology capabilities related to design—R&D—production, their position in the global value chain, their cooperation activities with other firms, benefits of being a direct supplier, their R&D and innovation activities, and performance levels. Furthermore, it aims to obtain a detailed summary of the story and to look at these events in terms of the AMMs, especially to reveal the AMM and supplier relationships. We also want to investigate the transfers at intra-firm level occurring from MNCs as foreign partners to their affiliates in Turkey. Hence, research questions are formulated from two sides: for suppliers and AMMs. In summary, it is specifically aimed to obtain important information and to find answers for the following questions:

For Supplier Firms:

- What are the general characteristics of the suppliers?
- What types of transfers have been provided through backward linkages from customers to their suppliers at inter-firm level? And what is the frequency or intensity of these transfers?
- What is the market and competition structure of the suppliers?
- What is the level of technological capabilities of the suppliers related to production and design?
- What is the position of suppliers in global supply chain?
- What are the reasons for collaborating with customers and other firms?
- What are the importance and benefits of being a direct supplier of AMM?

- What are the R&D and innovation activities of the suppliers? And absorptive capacity of the suppliers?
- What are the effects of the assistance received from customers on the performance level of the suppliers?
- What characteristics of the suppliers (such as age, size, export intensity, foreign ownership, being a direct supplier, possessing R&D department, carrying out innovation activities, and technology capabilities) have played an important role in all the above-discussed activities especially in the transfers? In other words, what are the determinant factors that played an important role in all these activities especially in the transfers?

For AMMs:

- What are the general characteristics of AMMs in terms of production, R&D, innovation and technology?
- What are the critical technologies and sources of these technologies (explicit or tacit) used in the production?
- What kind of technology activities do AMMs carry out with other partners?
- Have the transfers been provided to AMMs by their partners (MNCs or global AMMs as foreign partners, parent companies, other MNCs)? If yes, what are the channels of the transfers realized by MNCs as foreign partners to the AMMs in Turkey at intra-firm level?
- Who are the direct suppliers of AMMs?
- What are the criteria to be a supplier of an AMM?
- What are the strategic relationships of AMMs with direct suppliers?
- Do AMMs provide any kind of the transfers to these suppliers? If yes, what are the channels of these transfers realized by AMMs at inter-firm level?
- What is the role of AMMs in the development of automotive industry in Turkey?
- What is the role of government in the activities of AMMs especially related to the transfers and collaboration with local firms?
- What policies can be implemented to attract more transfers from AMMs and to increase local suppliers' technological capabilities and competitiveness level compared to other rival countries?

This study provides answers to these questions.

3.5 Research Methodology

In this study, we analyze the Turkish automotive industry as a case study to understand mainly the complex nature of the transfers occurring through backward linkages from AMMs to their suppliers. Case study is defined in the encyclopedia of survey research methods in this way: *it is an in-depth analyses and intensive description of a single individual, group, firm and organization based on collected data from a variety of sources (observations, questionnaires, interviews, documents, direct observation, participant-observation, archival records, etc.*) (Lavrakas, 2008). It is also defined by Yin in a more formal way: *An empirical inquiry that investigates a contemporary phenomenon within its real-life context; when the boundaries between phenomenon and context are not clearly evident; and in which multiple sources of evidence are used* (Yin, 1989: 23).

Case studies have been frequently used by researchers especially in social sciences for the purpose of getting comprehensive and detailed picture of the research problem that is under investigation, and it contributes to knowledge of this problem (Hancock & Algozzine, 2006). Actually, it is a method of qualitative research approach. As defined, it aims to explore a number of factors that may affect a situation, or to explore possible sources of the problem. This generates very valuable, rich and useful information about the phenomenon under investigation especially when there is no data and information about the problem or little is known about the issue being studied. It can be a guide for quantitative study later, and it can illuminate new ideas or build a new source of knowledge for further research. However, it requires more time and resources to collect data, to analyze and to present the results compared to other methods.[7]

In order to answer the previously raised research questions, it is necessary to investigate the specified problem and such questions at firm level; however there is not any data and information on that at national or international level. Moreover, the topic is too complex, the problem and variables are not well known and not specified in the literature. They need further in-depth exploratory research to analyze and to explain the full picture and factors that affect the transfers to suppliers. Therefore, the study, as the scope and nature of that, requires the collection of detailed data and information at the firm level about the suppliers and AMMs operating in the industry. Hence, the research methodology adopted for this study is mixed-methods research design based on the nature of the problem, purpose of the study and research questions. This research design uses a combination of both quantitative (questionnaire survey) and qualitative (in-depth interviews) research methodologies for data collection and analysis in order to address the proposed research questions.[8] Moreover, qualitative data based on face-to face interviews with the top-executives of AMMs in Turkey are used in a complementary way.

---

[7] See Bailey (1987), Hamel, Dufour, and Fortin (1993), Gomm, Hammersley, and Foster (2000), Carson, Gilmore, Perry, and Gronhaug (2001), Travers (2001), Yin (2003), George and Bennett (2005), Anfara and Mertz (2006), Hancock and Algozzine (2006), Ellet (2007), Gerring (2007), Lavrakas (2008), Creswell (2009), Merriam (2009), Mills, Durepos, and Wiebe (2009), and Sekaran and Bougie (2009) for detailed information on research methodologies in terms of case-study.

[8] See Bailey (1987), Tashakkori and Teddlie (1998, 2010), Thomas (2003), Neuman (2006), Greene (2007), Bergman (2008), Teddlie and Tashakkori (2008), Creswell (2010), and Biber (2010) for details on research methods in social sciences.

3.6 Data Collection Methods

In this study, only primary firm-level data, which was collected from the suppliers and AMMs, were used. As defined, it was collected by both quantitative and qualitative research methods. The quantitative data were obtained from suppliers through face-to-face questionnaire survey; on the other hand, the qualitative data were obtained from AMMs through face-to-face semi-structured in-depth interviews. This section analyzes these data collection methods, and gives detailed information about the steps of data gathering procedures.

3.6.1 Quantitative Data: Face-to-Face Questionnaire Survey with Suppliers

This section provides detailed information mainly on questionnaire design, sample selection and sampling procedure. Moreover pilot study, application of fieldwork, survey results and response rates are explained. Lastly, it presents information on data input, the representativeness of the sample, and validity & reliability tests of the data.

Questionnaire Design

For the purposes of the study described in Sect. 3.4, detailed information on the design and implementation of the survey that mainly aimed to explore the factors that affect the transfers to suppliers are presented below (see Fig. 3.9 for flowchart).

The first phase of the study commenced with the design of the draft questionnaire form in the second half of the year 2007. Preparation period of the draft form lasted approximately 2 years in total since there were not any similar studies conducted previously on the studied issue and industry. The first version of the questionnaire form was generated by getting inspiration from many resources following an in-depth scientific literature survey, which lasted for almost 1 year. Another objective of the study is to enable the econometric analyses conducted in the light of the data to be acquired from the above-discussed questionnaire survey. Therefore, the content of the questionnaire was planned with a view to conducting econometric analyses. The questionnaire form was generated by keeping to the information acquired from many studies conducted before in the field of automotive industry or different sectors, and theoretical framework.

The main and important sources referred in the construction of some sections and items in the questionnaire form are presented below:

- Giroud (2003) for the transfers related to product, production process, financial assistance and training activities,
- Berger (2005) on external collaboration for R&D and innovation activities,

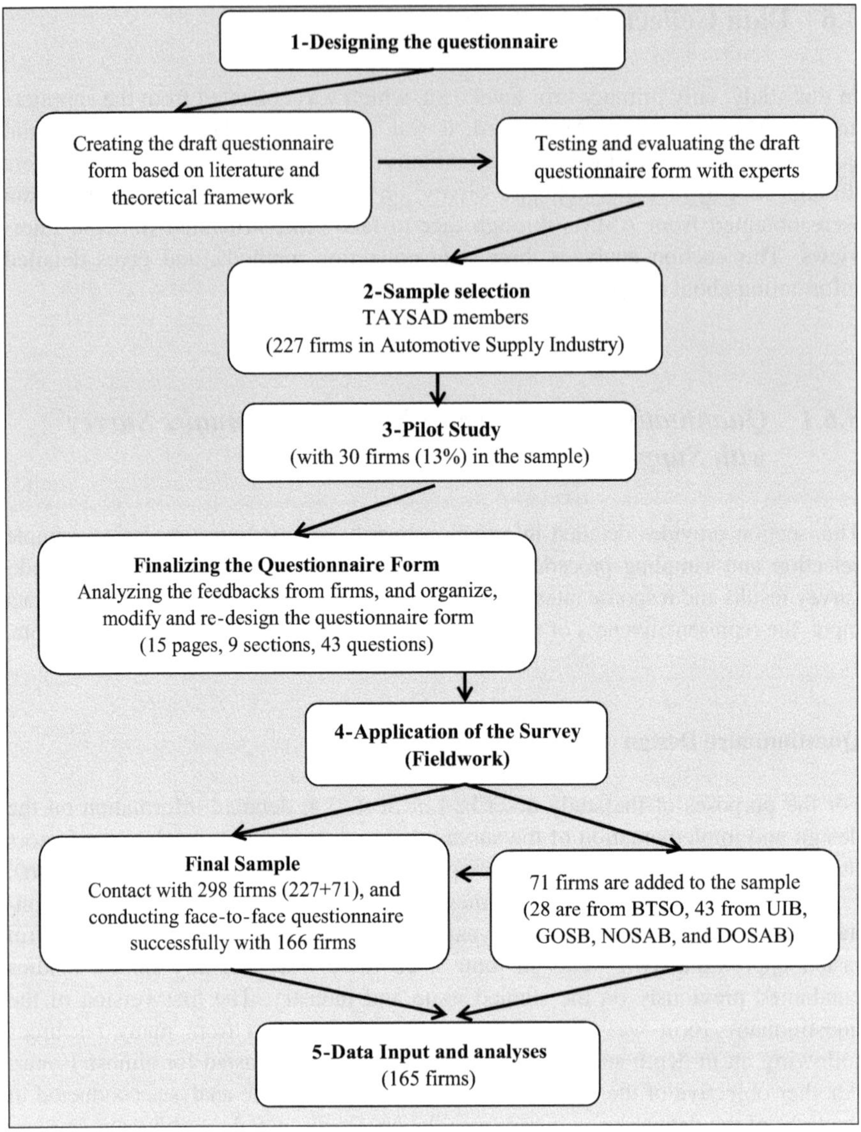

Fig. 3.9 Flowchart for the questionnaire survey

- Techakanont (2002) for design, performance, technology agreements and technology sources,
- Lorentzen, Mollgaard, and Rojec (2003) and Mollgaard and Lorentzen (2004) for supply chain, market and competition structure,
- Akarsoy (2005) for future projections on technology acquisition.

After the draft of the questionnaire form was completed, experts, technology consultants, and academicians, who are knowledgeable in the industry and have conducted studies on the industry previously, were interviewed in order to consider both the characteristics of automotive industry in Turkey and accurate comprehension of questions by the people who will respond to the questions. Furthermore, evaluations regarding the issues such as difficulties to be encountered during the application, what can be done in order to enhance response rate and obtaining preliminary information about the industry were also done with those people.[9] As a consequence of the feedbacks acquired from those evaluations, the form was reviewed, and modified upon recommended alterations and revisions. This period lasted 6-months, and the draft questionnaire form was finalized in order to be used in the pilot study at the end of a 2-year period.[10]

Majority of the questions in the questionnaire are formed according to types such as closed-ended multiple choice, yes/no, Likert scale, and ranking questions.[11] Most important reasons indicated also in the literature for using close-ended multiple-choice questions in the survey are as follows;

- The demand for gathering quantitative data and appropriateness of the data for statistical analyses electronically,
- Non-interpretable quality of the questions for the respondents,
- The possibility of a mutual comparison between firms,
- The possibility of applying it to many firms simultaneously
- Shorter duration required for completion both for respondents and researchers,
- Firms deeming it more proper regarding its format,
- Higher privacy compared to the open-ended questions by the firms and high response rate.

[9] The author participated in the workshop entitled *main industry and supplier relations* held in Bursa by UIB in the last week of October 2008 and had the chance to meet with top-executive representatives of the AMMs and suppliers in Turkey. As a result of the interviews and evaluations done with these people, some very important ideas regarding the study and survey were obtained, and the questionnaire form was also evaluated and reviewed accordingly. Furthermore, the opportunity for observing buyer and supplier relations closely was obtained in these meetings.

[10] The number of the version of draft questionnaire form prepared during the design is approximately ten. Many modifications regarding the general structure, sections, questions, and items of questionnaire form were made during this process. The reasons why the process of questionnaire finalization and design took such a long time are as follows briefly; (1) many firms discontinued production in the industry due to 2008–2009 global financial crisis and/or many top-executives who were requested to be interviewed could not be accessed in the application period due to this crisis, (2) quantitative questions which would be asked to the firms for a specific year were re-organized permanently due to the effects of the crisis, (3) difficulties were encountered while looking for the Turkish equivalents of the English terms (jargons) used in the international literature which would express the same meaning in the most appropriate way, (4) the questionnaire was tested in order to ensure that it is comprehended similarly by the people who will answer them, and (5) the questionnaire form was continuously revised according to the methods to be followed in order to obtain confidential information regarding the firms in a most appropriate way, and reach a high response rate.

[11] *Others* and *no response* options were also added to items and questions to give the respondents a chance to express their opinions on the issues.

Before proceeding to the general application phase of the survey, sampling was made through selecting the firms to which the questionnaire would be applied, and pilot study was carried out with the firms selected from the sample in order to perform the pre-test of the draft questionnaire form and detect the problems which were not anticipated by the researcher regarding the application. Finally, draft questionnaire form was revised according to feedback acquired from the pilot study.

Sample Selection

As stated in the previous sections, our questionnaire survey will target firms in Turkish automotive supply industry as a target population. Within the scope of the study, it was aimed to analyze the supplier firms which play an important role in the supply industry, being the direct suppliers of the AMMs, pioneering the development of the industry, integrated to global economic system, and manufacturing with an international quality. Thereby, the survey concentrates on medium and large-sized firms,[12] institutionalized manufacturing firms, export oriented firms, firms that produce OEM parts, and especially direct suppliers of AMMs. Furthermore, it is anticipated that such firms attach more importance to the study due to their professional, international management, and organization structures.

In the context of the above-discussed sample framework, in the application of the survey, we believe that considering all firms in the supply industry as a target population will not be correct in terms of the purpose and scope of the study.[13] The main reasons of this can be summarized as follows:

In Turkish supply industry;

- The number of firms making production on a small-scale (less than 50 employees) is quite high: According to the 2008 statistics of TurkStat,[14] the ratio of the small-scale enterprises in automotive supply industry is approximately 90 % (see Table 3.7).
- The number of firms producing with institutional quality is low, and/or a lot of firms, being family corporations, have low level of institutionalization.
- Realization of an important part of sales in different sectors (defense, white appliances, electrical, electronic sector, etc.) by some firms despite the fact that they appear as operating in automotive sector.
- Great majority of firms operating for replacement market.
- The majority of the firms producing and operating for raw material and services.
- Low number of firms performing R&D and innovation activities.

[12] We take into account employee number as an indicator of firm size.

[13] One of the important reasons for not applying the survey on all firms operating in the industry is that a study on such a scale requires a longer period of time, a larger project team and budget.

[14] Author calculations based on annual ISSS database, 2008, TurkStat.

- Low number of firms producing according to internationally-accepted quality standards.
- Low number of firms producing for international market (exports).
- Pretty low number of direct supplier firms producing OEM parts for AMMs (see Sect. 6.5.2).
- Low number of firms performing cooperation activities with AMMs.

In brief, although it is expected that more than 2,000 firms operate in the automotive supply industry (see section "Automotive Supply Industry in Turkey", Table 3.7, and section "Representativeness" for details), it is believed that the number of firms which have acceptable production standards, manufacture original parts directly to the AMMs and export by competing in international markets is approximately 300–350 (it was also proved by the results of the interviews with AMMs, see Sect. 6.5.2). Other firms mostly constitute small-scale production facilities as summarized above, and they generally make production for the replacement market (MITI, 2010; SPO, 2007; TUBITAK, report for, 2023 vision). Due to these reasons, member firms of TAYSAD were selected as the main sample among the suppliers operating in the supply industry for the purpose of collecting quantitative data and information. Main reasons for including TAYSAD members in the study scope can be summarized as follows (August, 2009)[15]:

- TAYSAD was established in 1978, and it is the sole and most important representative of the Turkish automotive supply industry.
- With 259 members, TAYSAD represents 65 % of the output for the automotive supply industry and 70 % of the industry's exports.
- In 2008, 29 members were in *Turkey's Top 500 Industrial Enterprises* with approximately total TL 8.5B sales value from production, 41 members were in *Turkey's Top 1,000 Exporting Firms* with approximately US$3.2B export value, 18 members are in *Biggest Companies in Bursa Research* with TL 4.2B sales value from production.[16]
- Members employ nearly 80,000 people and with sub-suppliers this figure reaches approximately 125,000 people.
- 58 members are in the status of foreign firm in various foreign ownership status.
- Majority of the members are direct suppliers of AMMs operating in Turkey.

Moreover, almost all of the important and large firms operating in the supply industry of Turkey, the direct suppliers of the AMMs and the foreign-capital firms are gathered in Istanbul, Bursa, Kocaeli and Izmir provinces where automotive main industry firms are located. When analyzed regionally, 75 % of suppliers are located in Marmara Region, 13 % are located in Aegean Region, 7 % are located in the Central Anatolia Region and 5 % are located in other regions (SPO, 2007;

[15] Source: TAYSAD's web page; http://www.taysad.org.tr/www/en/default.asp?x=hakkinda

[16] These three research reports are conducted and published yearly by The Istanbul Chamber of Industry (ICI), Turkish Exporters Assembly (TIM) and Bursa Chamber of Commerce and Industry (BTSO), respectively.

TUBITAK, report for, 2023 vision). In other words, Marmara region is the main land of the Turkish automotive industry. Thus, 227 firms which are the members of TAYSAD, and operate in one of these four provinces or Marmara region were taken as the basis in making sampling for suppliers, and other 32 member firms of TAYSAD were excluded from the survey.[17] One of the important reasons for excluding these firms is that it is detected through telephone contact with these firms that they are small-scale firms, generally make production for replacement market, produce raw material, and conduct service activities, etc. within the framework of the above-discussed criteria. Another important reason is that since excluded firms operate in 14 totally different provinces; greater budget, team, and time are required for face-to-face interviews with these firms.

Pilot Testing

Pilot study was carried out by the author in a 2-month period between August and September 2009. Thirty (13 %) firms were selected randomly for the pilot study of the questionnaire from the sample (227) composed of TAYSAD members operating in four provinces.

In the pilot study, preliminary information was primarily given to the selected firms on the telephone and the contact information of the top-executives who may participate in the questionnaire survey was obtained. Then, the top-executives of the firms were contacted, and detailed information regarding the objective, scope, and method of the study was conveyed. Following the phone interviews, a cover letter summarizing the objective, scope, and method of the study (see Appendix A), support letter of OSD (see Appendix C), and draft questionnaire form were sent both to the e-mail addresses of the people concerned and firm addresses within special envelopes bearing their names by cargo. It was emphasized both during the phone interviews and in cover letters sent to the firms that all the firm-specific information requested within the scope of the study is confidential.

As is known, quality and success of a study is closely related to its design and application method. Importance attached to the questionnaire and research by the firms and top-executives bears great significance for acquiring reliable information at the end of the study. Therefore, special emphasis was laid on the application methods of the questionnaire in order to enhance feedback rate and prove for the firms that the study is really professional; and great importance was attached to the preparation of questionnaire form, cover letter, envelopes and other documents. For this purpose, envelopes of best quality were used, contact information of the individuals was printed on address labels prepared in the electronic environment and affixed onto the envelopes. Colored print-outs of the cover letter and support

[17] The excluded 32 members of TAYSAD are in 14 different provinces: Aksaray (1), Ankara (4), Düzce (1), Eskişehir (1), Hatay (1), Konya (5), Kütahya (1), Manisa (5), Mersin (1), Sakarya (2), Samsun (1), Sivas (1), Tekirdağ (7), and Trabzon (1).

letters were obtained and questionnaire form was printed on a blue-color, high-quality A4 paper. Furthermore, a pen with METU logo was added into the envelope as a gift to extend our sincere thanks for their participation in the survey, and a stamped self-addressed return envelope, which is ready to be sent, was also added into the envelope. Additionally, a web site including the online version of the questionnaire was prepared in order to carry out the application of the survey in a more effective way and shorter time and test the performance of it through Internet. Online ID address specific to each firm was given and the online access of the firms to the questionnaire form was obtained through these addresses in order to provide privacy for the firms.

Top-executives were called 2 weeks after the envelopes were sent by cargos to the firms in order to remind them about the questionnaire. In the first 4 weeks, five firms gave feedback. One of the firms preferred to fill in the questionnaire form by performing a face-to-face interview with the author. Individuals who did not give feedback were contacted again and were reminded about the survey. Although the individuals had stated that they would participate in the survey during the phone calls, they afterwards told that they could not allocate time for completing the survey due to some excuses such as having a tight schedule, being on vacation, abroad or participating in a meeting, and requested an extension. Complete feedback was received from totally 8 firms (Bursa 3, Istanbul 3, Kocaeli 2) and 27 % response rate was reached at the end of the period allocated for pilot study.[18]

At the end of the pilot study, it was figured out that questionnaire application by means of cargo, e-mail, and online methods would not be very efficient due to low response rate and concerns about time. Therefore, it was decided that application of the questionnaire via face-to-face interviews with the authorized staff from the firms would be more effective and important both for the success and quality of the research. Another important deduction is that quantitative data of the firms regarding 2009 was negatively affected from the global financial crisis that began at the end of 2008. Although the questionnaire would be conducted in 2010, it was decided that quantitative data to be gathered about the firms would be concerning 2008 for this reason. Data regarding the employee figures of the firms was also asked for 2007, in order to test whether the crisis affected the data of 2008 or not. However, any finding indicating that the data of 2008 was affected negatively from the crisis was not obtained during an analysis of the data regarding 2007–2008 employee years at the end of the general questionnaire, and the data concerning 2008 was used in the analyses.

[18] It was revealed during the pilot study that unpaid leave was applied and the production of many firms ceased in the industry in this period due to the effects of 2008–2009 global financial crisis, so these are the most important causes for low response rate.

Designing the Final Questionnaire Form

Following the pilot study, the questionnaire form was reviewed again in the light of the feedback taken from the firms and the data collected. Some questions and items were excluded from the questionnaire, parts which were understood partially or misunderstood were re-organized and replaced, new questions and items were added to some sections, and the structure of the questionnaire form was re-built within this scope, and the revision was finalized.

Final version of the questionnaire form which was re-designed following the pilot study consists of 15 pages, 9 sections, and 43 questions. Sixteen of the questions are closed-ended multiple choice, eleven are numerical, eight are on an integer Likert scale, six are binary (yes/no), and two are ranking (see Appendix G for the survey questionnaire).

The questionnaire form is generally designed to reveal how often and what type of transfers are provided by customers to suppliers, common characteristics of suppliers, production, technology and design capabilities of them, the partners of suppliers in various collaboration activities, benefits received from being a direct supplier, the R&D and innovation activities of the suppliers, suppliers' place in the global supply chain, and their performances (see Appendix F for a detailed-questionnaire plan). The structure of the questionnaire form is summarized under nine main sections, each dealing with diverse aspects of the KTT process:

- **Section 1**: General information on the characteristics of suppliers including number of employees, total sales (turnover), foreign ownership, establishment date, share of export, etc.
- **Section 2**: Types of the transfers (related to product and production process) and also financial transfers (assistance) provided to suppliers by their customers.
- **Section 3**: Types and modes of training activities provided to suppliers by their customers.
- **Section 4**: Market and competition structure of suppliers.
- **Section 5**: Technology capabilities of suppliers related to production and design.
- **Section 6**: Information on supply chain, input sources of suppliers.
- **Section 7**: Cooperation activities of suppliers with their customers.
- **Section 8**: R&D and innovation activities of suppliers.
- **Section 9**: Performance and future projections of the suppliers.

Application of Fieldwork

Fieldwork of the questionnaire survey was performed via face-to-face interviews with firms during a 7-month period between February and August 2010. As is known, such questionnaire surveys are applications which are generally challenging, time-consuming, and which require greater budgets financially. Especially face-to-face interviews with the top-executives of the firms are harder with respect to the difficulty of providing their participation to the survey, making appointments on proper dates, and requesting firm-specific private data required by the research. Other difficulties may be sorted as follows:

- Application of the questionnaire to many firms,
- Operation of the firms to which the questionnaire will be applied in many provinces and locations,
- Time-consuming process during the contact with the top-executives of the firms and making an appointment with them.

When the above-discussed difficulties are considered, taking professional assistance becomes obligatory in order to access many firms within a specific time period, and perform face-to-face interviews. Thus, a professional research company located in Istanbul and having relevant experience was consulted for the application of the questionnaire.[19]

In the first phase of the fieldwork, in February 2010, meetings with the authorities and pollsters of the company which would perform the questionnaire survey were held in the office of the cooperated company located in Istanbul. In these meetings, participants were initially informed in detail by the author about the objective, scope, method of the study and who would be interviewed and how they would be interviewed and explanations thereof were made. Secondly, questionnaire form and application methods were mutually evaluated and discussed with the participants. Furthermore, the duration of the survey, the list of the firms to participate in the questionnaire, and the number and qualities of the pollsters were specified. Since the people who would be interviewed face-to-face are top-executives, great importance was attached to the phases of questionnaire application. For instance, many issues such as the arrangement of appointments, attendance at the appointments on the specified date and time, informing the firm beforehand in case of delays, pollsters to attend the appointments being granted the degree of Master of Arts and at least Bachelor of Arts and being experienced, and appearance of pollsters in the interviews were taken into account.

Application phases of the questionnaire and the way of communication with the firms were realized by adhering to the instructions specified by the author previously. Questionnaire application with the firms consisted of the following steps briefly:

1. The firms were briefly informed on the telephone and the names, positions and e-mail addresses of the authorized individuals were learned (R&D manager, production manager, general manager, etc.). Furthermore, it was verified whether the main operation field of the firm was automotive industry, along with the addresses and contact information of the firms.
2. Authorized individuals whose contact information and names were obtained in the first phase were called and more detailed information regarding the objective and method of the study was provided and an appointment was requested. Cover

[19] The company that provided cooperation for the questionnaire application is a research enterprise (incorporated company, Co. Inc.) which has been engaged in many national and international projects; and a specialist and pioneer in this field. Supplier firms which were surveyed were additionally informed about this company. Nevertheless, confidentiality agreement was signed with this cooperated company due to data privacy committed by us to the firms.

letter, support letters of OSD and TAYSAD were sent to the e-mail addresses of these people, as well (see Appendices A, C, and D).
3. Contacted authorized individuals were called again on the pre-defined date, and an appointment on an appropriate date and time was made.
4. The relevant firm was visited by the pollster on the appointment date, and authorized individuals were interviewed face to face, and the questionnaire was applied[20] (see Appendix E for detailed instructions).

Firstly, 227 firms which are the members of TAYSAD and operate in Istanbul, Kocaeli, Bursa, and Izmir were contacted during the application phase of the questionnaire.[21] The need for increasing the number of sample firms arose due to the fact that some firms rejected the survey, and the main operation field of some firms deviated from the automotive industry during the upcoming phases (in February and March 2010) of the fieldwork. In this context, following sources were utilized respectively in order to enhance the number of sample firms:

- *250 Large Firms Research* report in 2008 published by Bursa Chamber of Commerce and Industry (BTSO)[22] (65 firms),
- *Performance sequential member list of motor vehicles and supply industry exporters association* in Bursa Uludağ Exporter Unions' (UIB) (the period of January-December 2009)[23] (2,489 firms),
- Lists of the member firms located in Nilüfer Organized Industrial Zone (NOSAB) (290 firms) and Demirtaş Organized Industrial Zone (DOSAB) (383 firms) in Bursa,
- List of the member firms located in Gebze Organized Industrial Zone (GOSB) (127 firms) in Kocaeli.

More than 3,000 firms on these lists were separately scanned by the author via various media such as corporate web pages and telephones of the firms in order to search for their compliance with the survey.[24] The automotive suppliers were

[20] As is the case in international studies, the managers contacted for the survey were given the guarantee that all kinds of information and data provided for the survey would certainly be kept confidential, and both the names of the interviewed people and the title of the firm would never be cited in publications such as a report or article taking the concerns of the firms regarding their competitors and competition environment into account. Therefore, information regarding the interviewed firms and managers were not released in the study.

[21] Contact information and address of the firms were obtained from TAYSAD.

[22] In this report there are 14 different industries; automotive industry is in the first row with 65 firms. The turnover of these 65 firms is TL 15.8B and export value is US$ 7.6B. However, these firms include two AMMs, TOFAŞ and Oyak-Renault, the turnover and export value of these TL 10B and US$ 5.8B, respectively.

[23] In this list, there are 2,489 members from 42 different provinces in Turkey. This list also includes TAYSAD members, AMMs and firms from various industrial zones.

[24] Firstly, the web pages of the firms were checked and up-to-date information such as organization structure, sales, productions, references, quality certificates, customers were used. When the required information was not available or when deemed necessary, firms were called and information was obtained.

Table 3.2 Distribution of the sample by union, province and corporation type

Union	Province		Corporation Type		Total	Share in Total (%)
			Co. Inc.*	Ltd. Co.**		
TAYSAD	Kocaeli		69	20	89	29.87
	Istanbul		61	18	79	26.51
		Anatolia	33	12	45	15.10
		Europe	28	6	34	11.41
	Bursa		34	3	37	12.42
	Izmir		19	3	22	7.38
		Total	**183**	**44**	**227**	**76.17**
BTSO	Bursa		24	4	**28**	**9.40**
UIB NOSAB DOSAB GOSB	Bursa		14	5	19	6.38
	Istanbul		9	5	14	4.70
		Anatolia	5	3	8	2.68
		Europe	4	2	6	2.01
	Izmir		3	2	5	1.68
	Kocaeli		5	-	5	1.68
		Total	**31**	**12**	**43**	**14.43**
	General Total		**238**	**60**	**298**	**100**

Source: Author calculations based on survey results
*Incorporated Company, **Limited Liability Company (LLC)

evaluated based on various factors such as not being a member of TAYSAD, being medium or large-scale, having high institutionalization level, operating mainly in the automotive industry, being export-oriented, being a direct supplier, having a foreign partner or being a foreign firm, and sustaining the operation within this scope. At the end of these evaluations, a total of 71 firms (28 from the list of BTSO and 43 firms from the other lists) were added to the sample (see Table 3.2). The total number of firms in the sample is 298 after the addition of these firms ($227 + 71 = 298$).

The number and distribution of the firms in the sample are shown in Table 3.2 and Fig. 3.10 in detail. Accordingly, 76 % of the firms in the sample consist of TAYSAD members, and remaining 24 % consist of the members of other five unions. Furthermore, 80 % of the firms are incorporated companies (Co. Inc.), and 20 % are limited liability companies (LLC). When the distribution of the firms according to the provinces is analyzed, 32 % of them are located in Kocaeli, 31 % are located in Istanbul, 28 % are located in Bursa, and 9 % are located in Izmir (Fig. 3.10).

The author was continuously in touch with the firms and pollsters applying the questionnaire survey and closely monitored the survey during the application phase. Many of the top-executives who were contacted within the scope of the questionnaire survey requested to meet with the author directly and obtain information regarding the research. Notification explaining that "author may always be contacted during each phase of the research regarding all problems and questions

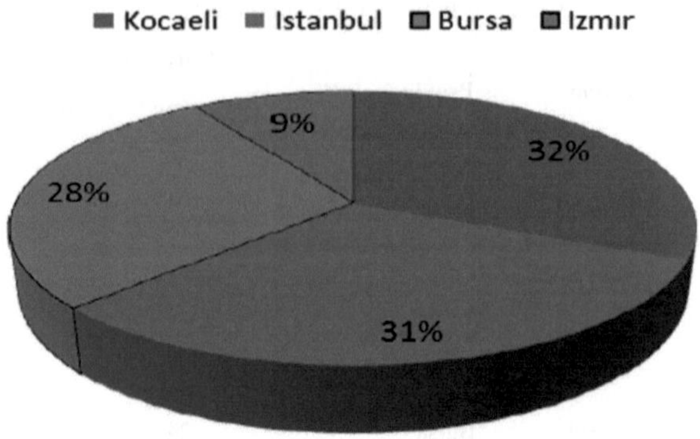

Fig. 3.10 Distribution of sample by provinces. *Source*: Author calculations based on survey results

of the firms" played an important role in this situation. Thereby, the author had the chance to interview with the top-executives who were the respondents of the questionnaire by phone, and acquired detailed information which could not be obtained through the questionnaire.

Since the questionnaire form consisted of more than one section (quantitative data such as turnover, number of employees, export; and qualitative data such as technology transfers, market structure, technology capabilities, cooperation activities, R&D, and innovation activities), some firms stated that it was more convenient to interview also with the other authorities responsible for that particular section in order to have satisfactory answers related to these sections. Therefore, more than one executive manager in charge of different departments of some firms was interviewed face-to-face to give answers to those sections. The titles of the interviewees of the surveyed firms are general director, production manager, factory manager, R&D manager, sales or marketing manager, deputy managers, product development manager and firm owner. Interviews for the questionnaire lasted between 60 and 90 min on average.

It was not an easy process to access the firms and arrange the interviews during the survey. Firms were called many times in order to make the appointments, and the appointment dates set before were postponed to a later date many times due to the unavailability of those to be interviewed. Although it was guaranteed by us that all kinds of information regarding the firms and respondents would be kept confidential, some firms were considerably hesitant about the issue. Although some top-executives stated that they were willing to participate in the survey, they had to obtain permission from the senior management due to the confidential information included in the survey regarding the firms. Some firms also requested additional time in order to decide whether to participate in the survey or not due to the firm policy and confidential information. Despite all these difficulties, attention attached by the top-executives played an important role in the success of the survey. All the top-executives who participated and did not participate in the survey cited that they were closely interested in the study and regarded it as an original and important research about the industry. Some firms explained that they were closely interested in the results of the study and they wanted to participate in the survey in order to obtain these results.

Survey Results

At the end of a 7-month fieldwork, 298 sample firms were contacted, and face-to-face questionnaires were conducted successfully with the top-executives of 166 firms in total, and 55.70 % response rate was obtained (Table 3.3). Response rate of the questionnaire is significantly high when compared to the studies carried out at the national and international level on the industry considering both the existing and above-discussed difficulties regarding the questionnaire surveys. Moreover, when it is considered that the questionnaire form includes questions which required detailed, comprehensive, and confidential information on the firms, it might be claimed that the study was carried out with success. Professional methods followed in the application phase of the questionnaire survey, attention paid to the survey by the top-executives of the firms, individual contacts of the academicians who were closely interested in the industry, face-to-face interview method, supports of TUBITAK, and two important representatives of the industry (OSD and TAYSAD) played an important role in the success of the survey.

Table 3.3 shows the distribution of response rate regarding the survey results according to union and province, while Table 3.4 shows the rate according to province. Accordingly, 82 % of the 166 surveyed firms are members of TAYSAD,

Table 3.3 Distribution of survey results by union and province

Union	Province		Realized Sample			Total Sample	Response Rate (%)
			Corporation Type		Total		
			Co. Inc.*	Ltd. Co.**			
TAYSAD	Izmır		18	2	20	22	90.91
	Bursa		24	3	27	37	72.97
	Istanbul		36	7	43	79	54.43
		Anatolia	22	5	27	45	60
		Europe	14	2	16	34	47.06
	Kocaeli		35	11	46	89	51.69
		Total	**113**	**23**	**136**	**227**	**59.91**
BTSO	Bursa		16	2	**18**	**28**	**64.29**
UIB NOSAB DOSAB GOSB	Bursa		12	0	12	19	63.16
	Istanbul		0	0	0	14	0
		Anatolia	0	0	0	8	0
		Europe	0	0	0	6	0
	Izmir		0	0	0	5	0
	Kocaeli		0	0	0	5	0
		Total	**12**	**0**	**12**	**43**	**27.91**
		General Total	**141**	**25**	**166**	**298**	**55.70**

Source: Author calculations based on survey results
*Incorporated Company, **Limited Liability Company (LLC)

Table 3.4 Distribution of survey results by provinces and unions

Province	Union			Total Sample	Realized Sample*	Response Rate (%)
	TAYSAD	BTSO	UIB, NOSAB DOSAB, GOSB			
Izmir	22	-	5	27	20	**74.07**
Bursa	37	28	19	84	57	**67.86**
Kocaeli	89	-	5	94	46	**48.94**
Istanbul	79	-	14	93	43	**46.24**
Anatolia	45	-	8	53	27	**50.94**
Europe	34	-	6	40	16	**40.00**
Total	**227**	**28**	**43**	**298**	**166**	**55.70**

Source: Author calculations based on survey results
*The realized sample (166) consists of TAYSAD members except 30 firms in BURSA

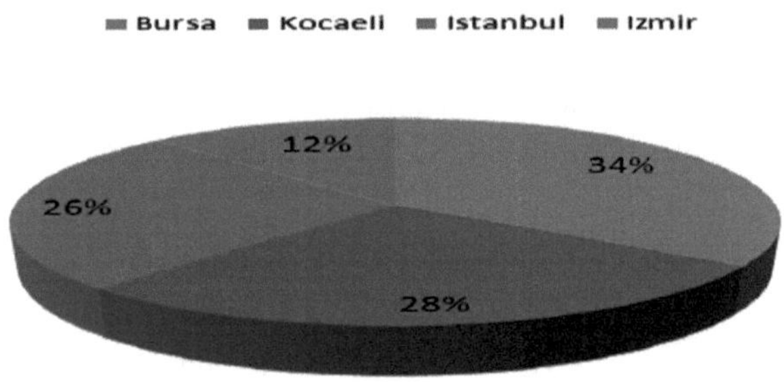

Fig. 3.11 Distribution of survey results by provinces. *Source*: Author calculations based on survey results

18 % are the members of other unions. When the corporation types of the surveyed firms were analyzed, 15 % were Ltd. Co., 85 % were Co. Inc. Participation rate of the firms which fall into the classification of Co. Inc. (59 %) was higher than the firms that were regarded as Ltd. Co. (42 %). When the response rate is analyzed on the basis of provinces, the highest rate belongs to Izmir (74 %), respectively Bursa (68 %), Kocaeli (49 %), and Istanbul (46 %) followed it (Table 3.4). In addition, Fig. 3.11 presents the distribution of surveyed firms according to provinces. Accordingly, 34 % of the firms are located in Bursa, 28 % in Kocaeli, 26 % in Istanbul, and 12 % in Izmir.

The Main Reasons of Unrealized Sample

Table 3.5 shows the distribution of the reasons for unrealized sample of 132 firms (44 %) which were contacted but could not be applied the questionnaire under three categories (non-quota, reject and cancel) according to union and province. Accordingly, 22 (7 %) firms were excluded by us under non-quota category due to the fact that they did not meet the specified criteria (see section "Sample Selection") in the application phase of the survey. Other 25 (8 %) firms which fall into reject category

Table 3.5 Distribution of unrealized sample by union and province

Union	Province		Non-quota	Reject	Cancel	Total	Share in Total (%)
TAYSAD	Kocaeli		7	12	24	43	32.58
	Istanbul		11	6	19	36	27.27
		Anatolia	4	3	11	18	13.64
		Europe	7	3	8	18	13.64
	Bursa		0	5	5	10	7.58
	Izmir		0	0	2	2	1.52
		Total	**18**	**23**	**50**	**91**	**68.94**
BTSO	Bursa		3	1	6	**10**	**7.58**
UIB NOSAB DOSAB GOSB	Istanbul		0	0	14	14	10.61
		Anatolia	0	0	8	8	6.06
		Europe	0	0	6	6	4.55
	Bursa		0	1	6	7	5.30
	Izmır		1	0	4	5	3.79
	Kocaeli		0	0	5	5	3.79
		Total	**1**	**1**	**29**	**31**	**23.48**
	General Total		**22**	**25**	**85**	**132**	**100**

Source: Author calculations based on survey results

Table 3.6 The reasons of unrealized sample

Non quota (7 %)	• Main operation field of the firm is not automotive industry • Firm is closed down and/or ceased its activities; it only has an office • Firm sells only raw material • It provides after-sales service, quality control, etc. but does not make production
Reject (8 %)	• Firms state that they do not want to participate in the survey directly or due to company policy • Firms do not participate in the study indicating that they have a tight schedule • Firms deem it inconvenient to share the information regarding their firms due to foreign linkages and competition • Firms deem it inconvenient to participate since the survey includes confidential information regarding the firm
Cancel (29 %)	• Foreign firms state that they should initially consult the main firm located abroad and this process is significantly time-consuming • Firms with foreign partners want to consult their partners and this process is significantly time-consuming • Firms state that the permission of the administrative board must be taken firstly in order to provide the confidential information regarding the firm, and this process is significantly time-consuming • Although contacted top-executives are interested in the survey, they have to obtain the permission of the board of directors and this process is significantly time-consuming • While some top-executives were interested in the survey previously, they showed lack of interest in them in the later stages • Failing to arrange a proper appointment date since the people to be interviewed are continuously engaged in meetings, they are abroad or they are out of the city during the survey process and/or continuous postponement of the appointments by firms

stated that they could not participate in the survey due to the reasons summarized in Table 3.6. Cancel category indicates 85 firms (29 %) which stated that they would participate in the survey, and they were continuously contacted during the survey process however could not finalize the questionnaire survey due to the reasons summarized in Table 3.6.

Accordingly, the ratios of the firms falling in the categories of cancel, reject, and non-quota were respectively 64 %, 19 % and 17 % within the unrealized sample (132). The rate of the firms who accepted to participate in the survey but could not be applied the questionnaire due to above-discussed reasons under cancel category is 29 % (85 firms) in the total sample (298). The high rate regarding the cancel category may be interpreted as a positive indicator of the quality and success of the survey as well as showing that firms (or top-executives) who wanted to participate in the survey but could not be applied the questionnaire due to above-discussed reasons notably supported and they were interested in the study.

The main reasons for unrealized sample identified during the survey process can be summarized as follows in Table 3.6.

Data Input and Missing Values

In September 2010, data collected at the end of the questionnaire survey were organized and clarified, and then recoded and transformed in order to make them compatible with the use in statistical software (STATA) and analyses, and transferred into the electronic environment. The data were also re-checked and examined in order to detect whether there was an error in the data, and data input. Then, 166 cases whose data input was carried out were analyzed case by case and evaluated in terms of missing values. It was detected that one of the cases includes too much missing data in the transfer questions for the analyses to be carried out in further phases, and it was decided to exclude the case from the scope of the analysis.[25] It was detected that there were a few missing values in the items or questions regarding the other cases analyzed, but they did not have any potential to pose a problem for the further analyses to be performed. Calculating new data to compensate for the missing values was not preferred, and it was deemed proper to leave it in the same way. Consequently, analysis sections of the survey were conducted based on 165 cases (firms).

After data input and analysis procedures, it was detected that 132 (80 %) of the 165 surveyed firms were direct suppliers of one or more than one AMMs operating in the automotive industry of Turkey and furthermore, there was foreign capital in 45 firms (27 %) with a minimum rate of 10 % (see Sect. 4.1.1). When the previously-discussed objectives and difficulties of the study are considered together, these rates may also be deemed successful.

[25] The said firm (case) refused to respond to the questions regarding the transfers.

Representativeness

Automotive industry is divided into three sub-sectors as 341 (manufacture of motor vehicles), 342 (manufacture of bodies (coachwork) for motor vehicles; manufacture of trailers and semitrailers) and 343 (manufacture of parts and accessories for motor vehicles and their engines) according to NACE (Rev. 1.1) classification. NACE 341 covers AMMs and generally represents automotive main industry. On the other hand, NACE 342 and NACE 343 represent automotive supply industry. In compliance with the discussed aim and scope of the study, we specifically focused on the relatively most developed suppliers in the supplier industry (NACE 342 and NACE 343 sectors) by using our sample frame (see section "Sample Selection"). Therefore, as a result of the sample method (judgement) evaluation, it is revealed that surveyed firms consist generally of medium- and large-sized firms, first-tier supplier firms, firms that export by competing in the international market, have high level of technology, possess various quality certificates, perform R&D and innovation activities, have a high level of efficiency, have high level of cooperation with MNCs, manufacture OEM parts directly to the automotive main industry, and firms that are pioneers of the industry (see Table 3.7 and Sect. 4.1).

In Table 3.7 surveyed firms were compared with the NACE 342 and NACE 343 sectors, which are the target population according to some indicators. When the number of surveyed firms is compared with the target population, it represents 13 % of the firms included in NACE 342 sector (column V), 7 % of the firms included in NACE 343 sector (column VI) and 4.4 % of the automotive supply industry when the two sectors (NACE 342&343) are regarded together (column VII). On the one hand, when the number of the main sample is considered (298 firms), the rates approximately double the amount and become equal to 8 % of the automotive supply industry. On the other hand, if we take the total employment and total turnover of the supply industry into account, the representation rate of the sample is much higher, and successfully represents 55 % of the total employment (column VII), and 66 % of the turnover (column VII) of the automotive supply industry respectively.

It must be pointed out here that focusing on such a limited number of firms (298 firms) does not affect negatively the quality of our study. On the contrary, it can be claimed that the survey has a successful rate having a high representation percentage regarding the sample formed in accordance with the previously-discussed objectives of the study due to many reasons summarized as follows:

- Small-sized enterprises (less than 50 employees) were significantly widespread in the automotive supply industry of Turkey (see section "Sample Selection"). According to 2008 statistics of TurkStat, the ratio of the small-sized enterprises in automotive supply industry is approximately 90 %.[26] The statistics in Table 3.7 showing that the firms operating in NACE 342 and NACE 343 sectors

[26] Author calculations based on annual ISSS database, 2008, TurkStat.

Table 3.7 Basic indicators for surveyed firms and automotive supply industry

	I	II	III	IV	V	VI	VII
Indicators	Surveyed firms (2008)	NACE 342[a] (2008)	NACE 343[b] (2008)	(II + III)	I/II	I/III	I/IV
Number of firms	165	1,225	2,525	3,750	0.13	0.07	0.04
Total employment	47,366[a]	10,839	75,556	86,395	4.37	0.63	0.55
Total turnover (TL)	8,871,695,911[b]	1,605,652,768	11,929,854,438	13,535,507,206	5.53	0.74	0.66

Source: Statistics were calculated by the author from survey results and TurkStat's Annual Industry and Service Statistics Survey database (ISSS database) in 2008 for [a]NACE 342 (manufacture of bodies (coachwork) for motor vehicles; manufacture of trailers and semitrailers) and [b]NACE 343 (manufacture of parts and accessories for motor vehicles and their engines) (Rev. 1.1). Data for the years 2009 and 2010 were not yet published by TurkStat

Notes: The number of the observations are 161 for employment (a), and 143 for turnover (b) due to missing values in the surveyed sample (see Appendix H for details)

employ respectively 9 and 30 people on average confirm also this hypothesis. Besides, 91 % of the surveyed firms within the scope of the questionnaire are at least medium (50–250 employees) and large-sized firms (250 and more employees) and employ 294 people on average, 13 times bigger than that of the sector (see Sect. 4.1 for details on surveyed sample).

- More than 90 % of the firms included in automotive supply industry have the characteristics specified in section "Sample Selection", and they are not convenient for the analyses to be carried out within the frame of the study due to the reasons specified also in the same section.
- We can say that all the members of TAYSAD were surveyed. 82 % of the 166 surveyed firms are the members of TAYSAD. As indicated previously (section "Sample Selection"), members of TAYSAD represent 65 % of the automotive supply industry output and 70 % of the export. To that end, 227 firms which are members of TAYSAD, and operate in four provinces were contacted, and the questionnaire survey was carried out successfully with 136 of them. The representation of TAYSAD members with 60 % in the survey was realized. Moreover, communication with the other 32 members of TAYSAD located in other 14 provinces was established; however, it was identified that these firms did not already have the desired characteristics defined in the study.
- According to TAYSAD statistics, foreign capital was available in 58 of the TAYSAD members at varying rates; the rate of the foreign firms in the total number of TAYSAD members (259) is 22 %. On the other hand, there are 45 foreign firms among the surveyed firms included within the scope of our questionnaire; their rate in the total of surveyed firms is 27 %.
- As discussed before, there is not reliable statistical information and resources on firms which operate in Turkish automotive industry regarding the number of the large-sized firms, direct supplier firms, developed firms, firms produced OEM parts to AMMs, and firms that manufacture at international level. However, according to various reports, studies regarding the sector, and information acquired from OSD and TAYSAD, it is predicted that the number of supplier firms which have international manufacturing standards in automotive supply industry of Turkey, manufacture OEM parts directly to the AMMs, and export by competing in the international market is about 300–350 (MITI, 2010; SPO, 2007; TUBITAK, report for, 2023 vision). As will be described in a more expanded way in Chap. 6, it was also detected by the author as a consequence of the interviews carried out with the AMMs that the number of the suppliers (average 338, see Sect. 6.5.2 and Table 6.1) having the above-discussed qualifications was approximately equal to the figure provided above. Majority of the other firms have small-sized production facilities, and they generally produce for replacement market, and have the characteristics specified in section "Sample Selection".
- We also think that non-response bias is minimal because of the above-discussed reasons.
- Moreover, we surveyed the firms located in the Marmara Region which is the main land of the Turkish automotive industry.

In summary, while the surveyed firms represent a relatively small percentage (between min 4.4 % and max 13 %) of the Turkish automotive supply industry (NACE342 and NACE343) when compared to the total number of the firms, the representation rate of the surveyed sample is much higher in terms of total employment and turnover of the sector, and it successfully represents more than half of the target population considering the aim and scope of the study along with the reasons specified above.

Validity and Reliability Tests

This section examines the validity and reliability of the scales used in the survey to measure the various concepts.

In this study, we take two types of validity into account: *content* and *construct validity*. It is considered that our questionnaire has *content validity* because, as discussed in detail before, survey was constructed based on in-depth review of the literature, advices and reviews of the experts and academicians, characteristics of Turkish automotive industry, and pilot testing (see Yin, 1989). The *construct validity*[27] of the scales will also be tested by using Factor Analysis (or principle component analysis) as described in Chap. 5 in detail (see Hair, Black, Babin, Anderson, & Tatham, 2006).

The reliability of the scales in the questionnaire is assessed by using *internal consistency* measured with *Cronbach's alpha* (α). *Internal consistency* estimates whether several items that are constructed to measure the same concept generate similar scores. In other words, it is an indicator of how well the various items measuring the same issue are (Giroud, 2003). It is generally measured by *Cronbach's alpha* which is a coefficient of reliability and ranges from 0 to 1; higher alpha values indicate higher reliability (Cronbach, 1951; George & Mallery, 2003). As a rule of thumb, alpha coefficient of a scale above 0.70 is seen sufficient by researchers, although 0.8 is more preferable (Devellis, 2003; Nunnally, 1978; Nunnally & Bernstein, 1994). Coefficients could be sensitive to the number of items in the scale and it is common to find low values for short scales (Pallant, 2007: 95). In this case Briggs and Cheek (1986) recommend that it would be better to use the *mean inter-item correlation* statistics since it is not affected by the number of items that make up the scale. According to them, optimal *mean inter-item correlations* should be between 0.2 and 0.4.

The *Cronbach alpha* coefficients of the scales tested together with contents of the scales are presented in the questionnaire form in Appendix G. In addition to alpha coefficients, the *mean inter-item correlations* with both min and max statistics for the items were also calculated for the scales with fewer than ten items and presented only in Appendix G.

---

[27] It refers to whether a scale measures the theoretical construct for which the scale is designed (Sekaran, 2003).

Table 3.8 Internal reliability of the scales

Section in the questionnaire form	Analyzed as a dependent variable in section	Scales	Cronbach's alpha (a)	Mean inter-item correlation (if # of items <10)
2	"Determinants of the KTTs Related to the Production Process" in Chap. 5	1. KTTs related to production process	0.853	
	"Determinants of the KTTs Related to the Product" in Chap. 5	2. KTTs related to product	0.785	0.422
	"Determinants of the Financial Transfers Provided by Customers" in Chap. 5	3. Financial transfers	0.712	0.389
3	"Determinants of the Training Activities Provided by Customers" in Chap. 5	4. Training activities provided by customers	0.835	0.627
	[a]	5. Training modes	0.721	0.234
5	5.5.2	6. Technology capabilities related to design, R&D and production	0.751	
	5.4.4	7. Technology agreements	0.726	
	5.4.5	8. The sources of technologies acquired and/or used	0.779	
	"Determinants of the Partners in Cooperation for the Co-design Activities Related to Products" in Chap. 5	9. Partners in cooperation for the co-design activities related to products	0.730	
7	5.5.3	10. Inter-firm cooperation activities	0.925	0.58
	5.5.4	11. Improvements in production capability	0.929	
	5.5.5	12. Benefits of being a direct supplier	0.951	
8	5.5.6	13. Cooperation level with partners in innovation activities	0.718	0.336
9	5.5.7	14. Performance increases related to production capability	0.745	0.379

(continued)

Table 3.8 (continued)

Section in the questionnaire form	Analyzed as a dependent variable in section	Scales	Cronbach's alpha (a)	Mean inter-item correlation (if # of items <10)
	[a]	15. Factors affecting production and/or developing new technologies	0.795	

Source: Author calculations based on survey results
[a]The scales 5 and 15 did not yield significant results as a dependent variable and therefore they were not analyzed in Chap. 5

A summary of the *Cronbach's alpha* and *mean inter-item correlations* of the 15 scales used in both Chap. 4 (for descriptive analyses) and Chap. 5 (as a dependent variable for econometric analyses) are shown in Table 3.8. The results indicate that most of the alpha coefficients are generally high or have a *mean inter-item correlation* between 0.2 and 0.4. Coefficients are ranging from 0.712 to 0.951 indicating satisfactory internal consistency. On the other hand, the *mean inter-item correlations* for the short scales are ranging from 0.234 to 0.627. Although Cronbach's alpha coefficients achieved the acceptable scores above 0.7, mean inter-item correlations for the three scales are beyond the optimal range 0.4. According to Briggs and Cheek (1986) this situation indicates that these three scales may be redundant, however it is decided to retain these scales.

Critique of the Questionnaire Survey Method

It is necessary for the criticism of the present survey to specify and consider the following disadvantages regarding the questionnaire surveys no matter they are included in the literature or not:

- One of the most important points is the reliability of the data collected through questionnaire. Questionnaire surveys are based on the personal evaluations of the respondents. Therefore, there is always the possibility of the information to be manipulated intentionally or unintentionally.
- Since the questionnaires are responded generally by a person, data may be dependent on the professional position in the firm, personal opinions and evaluations of that person.
- Despite the fact that confidentiality of the data is guaranteed, firms may provide misleading information due to competition, and rival firms or respondents may have a tendency to give excessively positive responses regarding their firms.
- In spite of the fact that the questionnaire is tested and designed carefully, it may not be comprehended similarly due to the backgrounds (reasons such as

educational and cultural levels, department they are in charge, professional background, and experience) of the respondents or the forms may be regarded to be very scientific.

- It may be possible that various questions or items are not described clearly or described deficiently in the questionnaire. There may be items which are not added to the categories of the questions but regarded necessary by the industry.
- It is possible that the respondents wanted to fill in the questionnaire as soon as possible, or responded quickly without thinking about the questions due to time limitations.

The concerns that we tried to summarize above constitute the leading criticism for such kind of surveys including the questionnaire surveys made by the public authority. As the questionnaire application process was tried to be explained in detail in the beginning of this section, it is anticipated that such kind of criticisms and problems in the present survey will be negligible. Following issues might be among the most important reasons for this situation:

- Questionnaires were carried out by means of face-to-face in-depth interviews in order to eliminate misunderstandings regarding the questions, ensure that accurate people answer the questions and collect reliable information.
- Questionnaire form was designed by following a time-consuming and detailed literature research, consulting the opinions of the experts of the industry and academicians studying on the industry, and it is tested by carrying out a pilot study.
- Respondents consist of top-executives of the firms, they are highly-trained, and educated since they both work in the firm for a long time and therefore they are experienced, and they are closely acquainted with the industry. The importance of the accuracy and reliability of the data to be provided by them is known by the respondents, but this situation was also reminded to them during the individual interviews. Attention paid by these people for the study was already on a high level, and the attention became the main criterion in the success of the study. Due to these reasons, it is claimed that the information is notably reliable, and they were based on first-hand account.
- It was specified and guaranteed that all kind of data and information provided for the survey would be kept confidential.
- The options *other* were added to the scales in the questionnaire form due to the possibility that there may be items which were not identified by us.

3.6.2 Qualitative Data: Face-to-Face Interviews with AMMs

In addition to the questionnaire based quantitative data, the author conducted semi-structured interviews with 19 top level executives of the 11 AMMs in Turkey between March and August 2010. The qualitative data based on these interviews is also used in a complementary way with quantitative data based on survey. This section describes the semi-structured in-depth interview process conducted, manufacturers to be interviewed, interview process, results, and representativeness of the interviewed sample.

Semi-structured Interview Method

The second phase of the research for collecting data and information was composed of semi-structured face-to-face interviews with the top-executives of AMMs operating in Turkey within the previously-discussed scope of the study. This method is qualitative research method carried out by means of face-to-face interviews with the participants. The method plays an important and complementary role in order to analyze the issues thoroughly, reveal embedded information, and cover the deficiencies to emerge in the questionnaire survey, acquire in-depth data and information which could not be obtained via questionnaires. Furthermore, it is possible to discuss each question thoroughly with the participants by using this method.[28] Considering these qualifications, it was suggested that semi-structured interview method with top-executives of AMMs would contribute significantly to the research and therefore it was preferred.

It was also planned that interviews in addition to the questionnaire survey would be made with the supplier firms within the scope of the research. A box was added at the end of the questionnaire form for this purpose, it was explained to the respondents that if they wished to share their opinions and recommendations regarding the study with the author on a proper date via an additional interview, they could put a tick in this box. At the end of the questionnaire survey, it was detected that 50 firms (30 %) accepted additional interview. However, the application of additional and comprehensive supply industry interview was abandoned since the questionnaire survey was already time-consuming, and due to time and budget limitations. A positive factor playing a role here is that the data collected in the questionnaire survey was comprehensive and obtained successfully.

In the interviews with the top-level executives of the AMMs, pre-designed, open-ended, and semi-structured manual was used (see Appendix I). The main reasons for using semi-structured manual in such kind of surveys may be summarized as follows:

- Giving an opportunity for those interviewed to express their opinions regarding the industry or firms freely,
- Gathering information about the issues that the researcher is unfamiliar with and that cannot be anticipated previously but arise during the interview,
- Possibility to make comparisons between the firms on a common basis by using a common manual,

[28] See Carson et al. (2001), Wengraf (2001), Drever (2003), Rubin and Rubin (2004), Seidman (2006), Kvale and Brinkmann (2008), Marshall and Rossman (2010), and King and Horrocks (2010) for details on qualitative survey methods.

- Ensuring that the persons to be interviewed are pre-informed about the questions.

Furthermore, top-executives requested the delivery of the manual to them in order to be informed about the interview questions previously, and assess the compatibility of the questions with the company policy and confidentiality and to get prepared for the questions before the interview.

Headings and questions of the manual used in the interviews were determined within the framework of the research objectives specified in Sect. 3.4 (see Appendix I for the semi-structured interview guide). The manual consisted of 28 questions categorized under five main sections as follows:

- First section includes the general questions in terms of AMMs,
- The second section questions the technological activities of the AMMs, the critical technologies used in manufacturing and the technology resources,
- The third section is composed of questions intended for understanding the strategic cooperation of the AMMs with the other firms in terms of R&D and innovation, and the transfers provided for AMMs by their parent companies or partners.
- The fourth section is composed of questions investigating the relations of the AMMs with the supplier firms, the number of their suppliers, criteria sought for the direct suppliers, cooperation activities performed with the suppliers, and transfers to the suppliers.
- In the last section, opinions and recommendations regarding the effects of public policies, the technological capabilities and competition levels of the suppliers and enhancing the transfers from MNCs as well as predictions and strategies of the AMMs in terms of industry are included.

The manual used was prepared and tested following the examination and revision of the academicians and experts who have conducted studies on the industry, as is the case with the design process of the questionnaire form (see section "Questionnaire Design"). While the questions used in the manual specify the general framework, the interviews reveal comprehensive opinions and recommendations of the participants about the issues to which they want to contribute.

Administration of Interview Process

It was aimed to make individual interviews with all the AMMs operating in Turkey. As stated before, there are 15 AMMs operating in Turkish automotive main industry, and these are also the members of OSD (SPO, 2007) (see Appendix H for the entire list of contacted AMMs).

The in-depth interviews with the AMMs were arranged and conducted by the author in a time period of approximately 6 months between March and August 2010. Support of two of the most important organizations as OSD and TAYSAD, and contributions of TUBITAK as well as individual contacts of people working at universities and various state institutions (SPO, KOSGEB, TUBITAK, TTGV, etc.) played important roles in accessing top-executives of the AMMs and having a successful interview with them.

Top-executives of the 15 AMMs were contacted separately in order to make face-to-face interview for the survey in March 2010 within the scope of the second phase of the study. The following steps were followed briefly in this phase:

1. Relevant departments (general directorate, human resources and relations, etc.) of the AMMs were called by phone, authorities were kept in touch and preliminary information regarding the study was provided for them.
2. A cover letter summarizing the objective, scope, and method of the study including (1) an appointment request, (2) OSD and TAYSAD support letters, and (3) an interview guide were sent to the e-mail addresses of the individuals contacted in the first phase (see Appendices B, C, D, and I).
3. Some of the AMMs called us back and obtained additional information and/or scheduled an appointment depending on the methods preferred by the firms. Some of the AMMs were called again by us in order to remind them about the study on the specified date, and request an appointment. It was not easy to access top-executives of the AMMs, it was required to contact with and inform many people repeatedly. Responses obtained from the contacted firms may be summarized as follows:
    - They will convey the issue to the authorities and respond as soon as possible,
    - They will primarily assess the issue in the board of directors and respond afterwards,
    - They will respond after making a decision about who would be the interviewee,
    - Due to private and confidential information required by the study, they will respond after an assessment of the study's compatibility with the company policy and confidentiality,
    - It is necessary to consult and acquire the approval of the parent company located abroad and then they will respond,
    - Since it is a public corporation, it is necessary that the issue is primarily analyzed by the authorities.
4. Names, positions and contact information of the top-executives to be interviewed were obtained from the AMMs which accepted to participate in the study.
5. These people were contacted, brief information in this respect was given, and an appointment was requested for a face-to-face interview on a further proper date.
6. Top-executives of the 11 AMMs which accepted to participate in the study were visited by the author on the appointment dates and face-to-face interviews regarding the study were conducted.

Results of the Interview Process

Eleven AMMs were interviewed by the author between March and August 2010 in six different provinces (Ankara, Bursa, Istanbul, Izmir, Kocaeli and Sakarya). These interviews were conducted with a total number of 19 top-level executives of AMMs. The titles of the respondents are R&D directorate, R&D manager, R&D and technical coordination manager, product development manager, purchasing program and systems manager, purchasing manager, production manager, production engineering manager, development manager, supplier development manager, logistics manager and organizational development manager. Six interviews were conducted with only one respondent, while the rest were conducted with at least two respondents. The interviewed AMMs are shown in the following Table 3.9.

Other four AMMs contacted did not accept to participate in the study due to following reasons[29]:

- The company policy,
- Parent foreign company does not permit participating in such researches,
- It is not regarded as convenient to participate since the study includes confidential information,
- Lack of interests in the study.

Interviews were carried out in various meeting halls of administrative departments located in the main plants of the AMMs. It was observed that the possibility of accessing the plants from outside was quite limited. It was possible to access the departments in the company of authorized people following the comprehensive security checks in each plant. It was figured out according to the information obtained from the authorities later that only limited number of previously-assigned authorized individuals might enter and exit these departments via identity checks, and the entry and exit of the communication media such as mobile phones and laptops (even though they belong to those authorized people) were not permitted due to any kind of industrial espionage. Furthermore, it was explained that communication devices used therein were equipped with special hardware and under permanent control in order to prevent information leakage. Only few people from outside, like the author, are allowed to enter these units where interviews were held, and it was stated during the interview conducted in R&D department that the first person to enter this department of the company from outside was the author.

Interviews were carried out over an open-ended and semi-structured guide that was sent beforehand in order to inform the relevant individuals about the research. Interviews were conducted in a mutual conversation environment with questions and answers and took between 60 and 120 min for each person interviewed. While

[29] Two of the AMMs which did not accept to participate in the survey are manufacturing personal car, pick up, and minibus with 100 and 70 % FS respectively, while the other two are manufacturing truck, pick up, minibus, midibus, bus, and farm tractor with no FS.

Table 3.9 The list of the Interviewed AMMs

1. ANADOLU ISUZU	7. OYAK RENAULT
2. B.M.C	8. TEMSA GLOBAL
3. FORD OTOSAN	9. TOFAŞ
4. M.A.N TÜRKİYE	10. TOYOTA
5. MERCEDES-BENZ TÜRK	11. TÜRK TRAKTÖR
6. OTOKAR	

the interviews were carried out over a guide, some other issues were also raised, and the issues important in terms of the industry were also discussed depending on the respondents. Thus, more detailed information regarding specific issues about the industry which could not be identified in the literature surveys and similar written resources was obtained, and buyer-supplier relations were comprehended in a better way.

In addition to taking notes, the use of tape recorder was preferred during the interviews depending on the approval of the respondent. The main reason for using a tape recorder is that it allows the researcher to participate in the interviews actively and provides the possibility of studying more freely and comfortably. Furthermore, it prevents the researcher from missing the important points while focusing on taking notes, and ensures an easy analysis of the issues which are not observed during the interview but attached importance later on. The above-discussed reasons for using the tape recorder were explained to the respondents, and it was added that a tape recorder would not be used if it became inconvenient due to reasons like confidentiality and/or if the people thought they could not express their ideas freely. It was also guaranteed that if the tape recorder was used, the records would be kept confidential and would not be, in any way, shared with the third parties in a way to reveal the identity of respondents and names of the firms. Only three of the respondent firms rejected the use of the tape recorder. In other interviews in which a tape recorder was used, respondents stated that they were accustomed to it and would not be disturbed.

Conversations made after the interviews were examined and analyzed by the author in order to prevent information loss. Conversations were analyzed case by case in detail under titles considered to be important in order to make a common comparison between the AMMs. However, responses were categorized and presented in Chap. 6 as a whole in order to keep the information regarding the firms confidential.

Representativeness and Reliability of the Sample

According to Table 3.10, the AMMs interviewed represent 73 % of the total population. Moreover, as can be observed from the table, the shares of the firms regarding total capital, capacity, production and sales in the industry are respectively 80 %, 83 %, 89 % and 91 %. It indicates that the representation rates of the

Table 3.10 Share of the interviewed AMMs in industry level variables (%)

	N	%	Capital (TL)	Capacity (2011)	Production (2010)	Sales (TL) (2010)
Interviewed AMMs	11	73.3	79.92	82.75	88.96	90.61
Not interviewed AMMs	4	26.7	20.08	17.25	11.04	9.39
Total	*15*	*100*	*100*	*100*	*100*	*100*

Source: Author calculations based on OSD statistics and interview results, see Appendix H

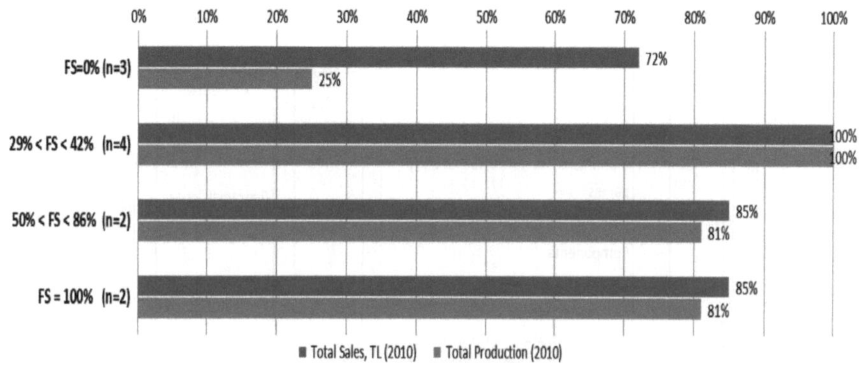

Fig. 3.12 Share of the interviewed AMMs in total sales and production in 2010 by different foreign shares. *Source*: Author calculations based on OSD statistics and interview results, see Appendix H

interviewed AMMs are much higher. When assessed in terms of foreign capital, eight (80 %) of the ten AMMs with foreign capital (FS > 29 %) and three (60 %) of the five AMMs with local capital could be interviewed. Moreover, as can be seen in Fig. 3.12, the shares of the interviewed AMMs in total sales and production in 2010 by different foreign ownership status are also very high. The AMMs with minority foreign ownership are represented fully, and ones with majority foreign ownership and with full ownership are represented by 85 % in total sales and 81 % in total production. The AMMs with no foreign capital are also represented by 72 % in total sales and 25 % in total production as well. When previously-discussed difficulties of the study are considered, these rates may be accepted as successful.

Respondents consisted of highly-educated and experienced top-level executives who were in charge of various duties for years in the AMMs, and have detailed information about both the history of the firm and the relations with the suppliers. These people, who are well-informed about the automotive industry at both national and international level, and who performed cooperation with the MNCs, are also the leaders in the industry. Moreover, attention attached to the study by these people played an important role in the success of the study. Therefore, it may be claimed that the collected data and information are very reliable, first-hand information.

3.7 Descriptions of Some Concepts and Differences for the Firms

This section explains some of the firm-related concepts constructed by the author, and used in the analyses in the next Chapters. With this object in view, a basic structure of the automotive industry and relationships between the concepts used in the analyses are summarized in Figs. 3.13 and 3.14.

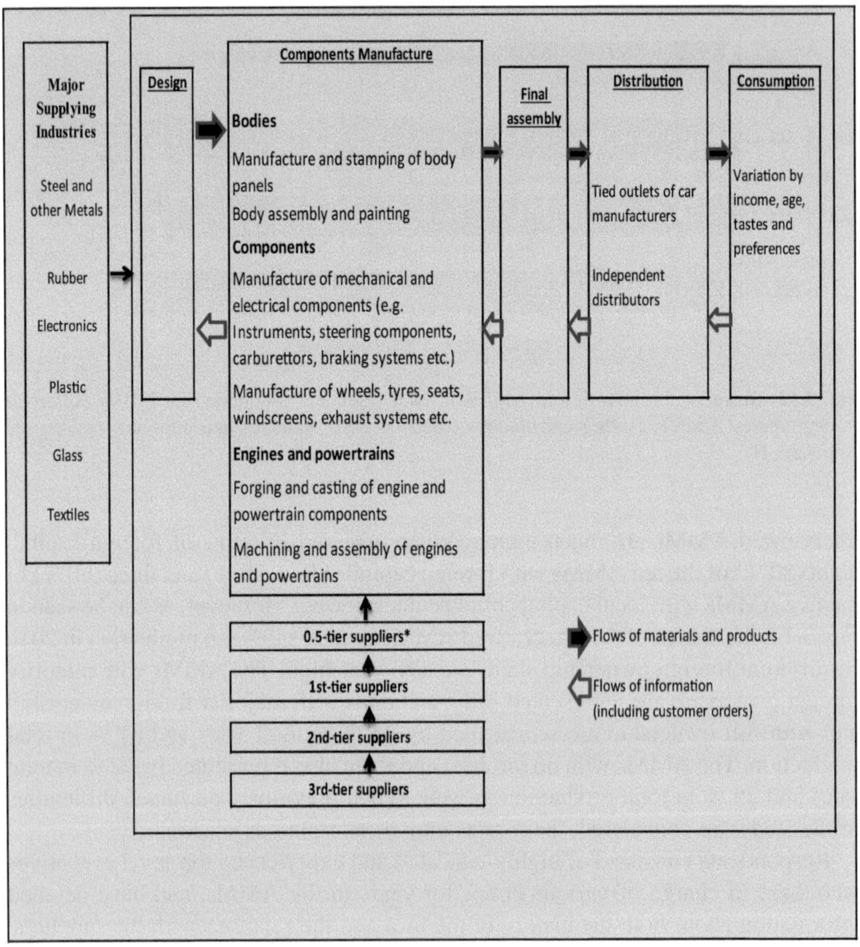

Fig. 3.13 The automobile production circle. *Source*: Dicken (2007: 279). *Added to original table by the author

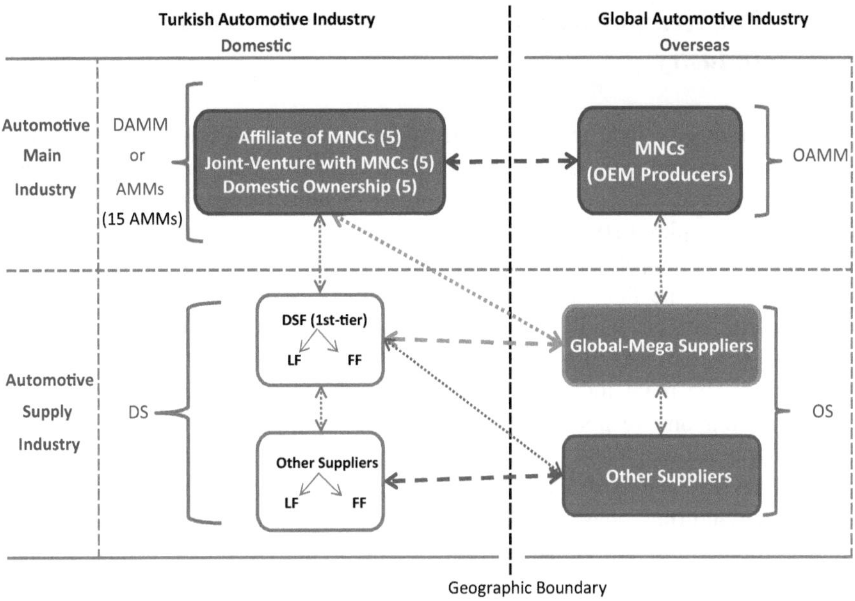

Fig. 3.14 Illustration of the basic structure of the automotive industry and concepts used in the analyses. *Source*: Constructed by the author. *Notes*: *Lines with thicker dashes* show possible intra-firm linkages between firms, while *thinner* ones show possible inter-firm linkages between such firms. "Other suppliers" are those at lower tier-levels. AMMs are classified in three categories according to ownership status: affiliate of MNC, joint-venture with MNCs and domestic ownership. AMMs in domestic ownership category are also referred as "independent local assemblers" although they are manufacturing under MNC license. DSF can be the affiliates of global-mega suppliers, they have an equity partnership with such suppliers, or they can be independent local suppliers with domestic ownership

3.7.1 *Automotive Industry, Automotive Main and Supply Industry*

The terms *automotive industry*, *automotive main industry* and *supply industry* are described as follows in the "Report of Automotive Industry Specialization Commission" prepared within the framework of 2007–2013 nineth development plan of SPO: *Motor road vehicles are vehicles with four or more rubber wheels which are driven with one combustion or explosion motor and manufactured in order to carry load or passenger and run on road traffic according to a specific technical legislation*. The industry manufacturing these vehicles is called *main Industry*. *Automotive supply industry* is *the industry branch producing original and equivalent products, semi-finished products, module and system through domestic or foreign manufacture, directly or indirectly to replacement market in compliance with the technical documents specified by the main industry*. The term *automotive industry* includes all of the two sub-sectors (SPO, 2007). Primary products manufactured by the automotive main industry are personal car, bus, midi-bus, minibus, truck, pick-up and tractors.

3.7.2 The Relationships of Automotive Main and Supply Industry

In fact, automotive industry is an industry branch in which thousands of components manufactured with various characteristics by many manufacturers are assembled and installed. The center of the automotive manufacturing circle (see Fig. 3.13) consists of complex network of relations between main industry (main manufacturers) and component suppliers. There are primarily three processes prior to the phase in which the final installation is performed: manufacturing of car bodies, components, engines and powertrains equipment. The manufacturing of these components is carried out by the main industry and main suppliers. However, the tendency of the main industry in the last decade has also been transferring the manufacturing and responsibility of these components from the main industry to the suppliers completely (Dicken, 2007). In other words, main industry specifies the outlines and standards (design, quality, etc.) of these components and leaves their manufacturing phase to the suppliers. In a global sense, the automotive industry consists of the global AMMs (see the definition of OAMM) and many suppliers in different tiers.[30] According to the Fig. 3.13, there are four tiers of suppliers, however there can be more. The hierarchy between the main industry and suppliers may be summarized as follows.

Global mega suppliers: Global mega suppliers rank first among the suppliers. They directly supply major system components to the main manufacturers. They have strategic and close relationships with the main manufacturers. They follow the main manufacturers who globally take decisions for making investments in various countries, and they also invest both as main suppliers and manufacturers in these countries. They manufacture main system components within the framework of their own R&D, innovation, design and product development methods in line with the advanced technologies; they supply the components to more than one main manufacturer. While their number is very few globally, they have transformed in main MNCs by exceeding many automotive main manufacturers in terms of turnover and size such as Siemens, Bosch, Delphi, Voleo, Magna, and Autoliv. The suppliers are also called as *0.5-tier* suppliers due these characteristics they possess.

(continued)

[30] In the last years, as result of global competition and crises, it has been observed that many manufacturers began to take stakes in other automotive manufacturers. Therefore the number of the automotive manufacturers decreased globally. In addition, many cooperation activities are observed among these manufacturers.

(continued)

First-tier suppliers: They are the firms directly supplying major components to the main manufacturers. Some of them have transformed into global mega suppliers in the course of time. These firms are suppliers with significant R&D, innovation, design, design verification, product development, and test competences.[31] However, they do not have as close relationships with the main manufacturers as mega suppliers, and their global manufacture is more limited compared to the mega suppliers. In other words, the suppliers cannot follow the main manufacturers investing in other countries, and their global facilities are limited.

Second-tier suppliers: They are the suppliers manufacturing components according to the design characteristics specified by the first-tier suppliers, global mega suppliers or main manufacturers. These firms are dependent on process engineering capabilities in order to meet cost and flexibility criteria specified by customers. They must also enhance the capabilities in order to exist in the market, meet required quality conditions and obtain important quality certificates (ISO9000 and QS9000).

Third-tier suppliers: They generally manufacture simple components compared to second and third-tier suppliers. In many cases, basic engineering capabilities are sufficient for these firms. Investment, training, technology and skill levels of these firms are considerably low. It may be claimed that the firms in this tier compete with each other in terms of price in the market.

Source: Based on Humphrey and Memedovic (2003: 22) and Dicken (2007)

3.7.3 Domestic and Overseas Automotive Main Manufacturers (DAMM and OAMM)

The term "Domestic Automotive Main Manufacturers" (DAMM) for firms manufacturing within the borders of Turkey in automotive main industry (see Appendix H), and "Overseas Automotive Main Manufacturers" (OAMM) for firms manufacturing beyond the borders of Turkey are used in the questionnaire form and analyses by keeping to the descriptions provided above (Fig. 3.14). The sole factor used in the classification is the manufacturing performed in Turkey or abroad (geographical factor) independently of the capital structure (local or foreign capital). The AMMs, whose results are analyzed in Chap. 6, were specified in detail previously and interviewed individually within the scope of the research independent of their capital structure fall into "DAMM" category. On the other hand, the

[31] Due to these reasons, we put a special emphasis on the fact that supplier firms used in the survey are 0.5 and 1st-tier suppliers.

global AMMs (Toyota Motor Co., General Motors Co., Volkswagen Group AG, Hyundai Motor Group, Ford Motor Co., Daimler AG, etc.) fall into "OAMM" category, and we also call them MNCs. AMMs in Turkey are also classified in three categories according to ownership status: (1) affiliate of MNC, (2) joint-venture with MNCs, and (3) domestic ownership. AMMs in domestic ownership category can also be referred as "independent local assemblers" although they are manufacturing under MNC license.

3.7.4 Domestic and Overseas Supply Industry

In the study, the term "Domestic Supply Industry" is used for the supply industry firms making production within Turkish borders in the automotive industry, and the term "Overseas Supply Industry" is used for the supply industry firms making production beyond Turkish borders by keeping to the definitions specified by SPO (Fig. 3.14). The sole factor used in the classification is the manufacturing performed in Turkey or abroad (geographical factor) independently of the capital structure (local or foreign capital). The sample, which was specified in detail previously, surveyed within the scope of the study independent of the capital structure and whose results are analyzed in Chaps. 4 and 5 is "Domestic Supply Industry".

3.7.5 Domestic and Overseas Suppliers (DS and OS)

The term "Domestic Suppliers" (DS) is used for the firms operating within the borders of Turkey, and providing raw material and inputs for various sectors. Similarly the term "Overseas Suppliers" (OS) is used for firms located beyond the border of Turkey in the questionnaire form and analyses independent of capital structures (local or foreign) (Fig. 3.14). While these supplier firms may conditionally operate in automotive industry, they may also operate in another industry apart from the automotive industry, providing inputs for automotive industry. For instance, when the surveyed firms are requested to specify the resources of technologies acquired in the next Chapter (see Sect. 4.4.4), by the used items of "DS" and "OS" (see section 5, question 4 in Appendix G) the sub-tier firms providing inputs-raw material (iron-steel, petroleum chemistry, aluminum, plastic, glass, etc.) to the surveyed firm and operating in various sectors in Turkey (DS) or overseas (OS) were intended. On the other hand, in case of "DS" and "OS" items used when the information is requested regarding the main customers of the suppliers surveyed (see section 1, question 4 in Appendix G), it is clear that these main customers are domestic and overseas suppliers in automotive supply industry on a higher tier (as the customer of the supplier being on 0.5 tier or 1st tier according to tier of the surveyed suppliers).[32] In other words, each automotive supplier is a supplier however each supplier is not an automotive supplier.

In terms of DAMM, all domestic and overseas suppliers (including the domestic automotive suppliers) providing inputs for them are "supplier firms".

3.7.6 *Local and Foreign Firms (LF and FF)*

Descriptions of the firms according to their capital structure were also provided in the questionnaire form and analyses. As specified previously, the firms whose foreign capital share (FS) is 10 % or more were accepted as "foreign firm" (FF), and the firms whose foreign capital share is lower than 10 % were accepted as "local firm" (LF). Firms in the surveyed sample of "Domestic Supply Industry" were defined as local (LF) and foreign firms (FF) according to foreign capital structure.

Apart from the above-discussed classification, domestic and overseas classification cannot be made for the other "foreign" term in the questionnaire form. For instance, the item of *KTTs from foreign manufacturers*, which is discussed later, is present among the factors causing increase in the production capabilities of the surveyed firms (see Sect. 4.9.2). What is intended with the term "foreign manufacturers" used within this item is the firms located in Turkey or overseas and whose capital is controlled by foreign people. While these firms may operate in automotive industry, they may also operate in another industry providing service or raw material-inputs to the automotive industry. Especially in the last 10 years, domestic supply industry has been receiving professional assistance from domestic and overseas foreign-capital engineering, consultancy and design firms in order to participate in various projects or enhance capabilities. Likewise, foreign firms which do not operate in the automotive industry but manufacture robots, machinery, and equipment to be used in the production line of the automotive industry are in an intensive interaction with domestic automotive main and supply industry. Thus, firms of domestic supply industry which are in interaction with the foreign firms were asked whether they made the transfers from the foreign manufacturers.

3.7.7 *Parent Company (PC)*

What is meant with the parent company (PC) in the study and questionnaire is the affiliation with the domestic or overseas main company (conglomerate, MNC, etc.)

[32] When it is taken into consideration that the main activity of the surveyed domestic suppliers is manufacturing for the automotive industry, it is anticipated that suppliers which are the main customer of these firms (either domestic or overseas) are supplier firms operating in automotive industry on a higher tier.

in terms of administration or execution independent of the capital structure of the surveyed firm. These are the companies with overseas or domestic location, and they directly or indirectly control all and any commercial activities, management, and/or capital structure of the surveyed firms. A significant part of the surveyed foreign firms (84 %) operating in domestic supply industry is the affiliate (subsidiary) of another foreign main company (MNC) located overseas. Local firms falling into this category are generally one of the group companies affiliated to a domestic conglomerate (see Sect. 4.5.1). By the way, the parent company may operate in automotive industry or another industry. In a similar way, some AMMs operating in Turkey (such as TOYOTA, HONDA, and M.A.N) are the affiliates of global AMMs located overseas; therefore, we also use the term "PC" to refer to these parent companies overseas.

3.7.8 Direct Supplier Firms (DSF)

They are the firms directly supplying parts to the AMMs in Turkey (DAMM), or they are the first-tier suppliers of DAMM. The sample of "Domestic Supply Industry" surveyed was also analyzed in terms of DSF as well as LF and FF difference in Chap. 4. DSF consist of local and foreign firms but a classification according to the capital was not performed due both to space limits and the small number of the firms. Moreover, the DSF can be the affiliates of global-mega suppliers, they have an equity partnership with foreign firms, or they can be independent suppliers (Fig. 3.14).

Chapter 4
Survey Findings: Descriptive Analysis

This chapter is devoted to a broad descriptive analysis of the survey data collected through the questionnaire. Results of the survey are presented and analyzed according to three groups of surveyed firms categorized as local firms (LF), foreign firms (FF), and direct supplier firms (DSF),[1] respectively. Also, the last column in the tables concerns all firms (AF) included in the sample.[2] This distinction is maintained in all the tables, and it does enable us to test the significance of the equality of the mean of different variables between LF *versus* FF as well as between DSF *versus* non-DSF by using Mann–Whitney U[3] test, Pearson's Chi-Square test (from here on chi-square test) or *t*-test depending on both the type of variable (ordinal, categorical and continuous) and the number of observations available. Survey results are analyzed under nine main headings within the framework of the study referred in Chap. 3.

[1] Data on non-DSF will not be presented in the tables due to both space reasons and the small number of the firms in this category (33 firms).

[2] For a detailed description of the firms, see Sect. 3.7.

[3] Mann–Whitney *U* test (also called the Mann–Whitney–Wilcoxon (MWW) test) is a non-parametric test and used for comparing two independent samples. It does not require the assumption that the dependent variable is a normally distributed interval variable (it is assumed at least ordinal). Null hypothesis of the test is that two samples have identical distribution (see Pallant, 2007, p. 210).

A. Sönmez, *Multinational Companies, Knowledge and Technology Transfer*, Contributions to Management Science, DOI 10.1007/978-3-319-02033-4_4,

4.1 General Characteristics of the Surveyed Firms[4,5]

4.1.1 Capital Structure

Table 4.1 indicates foreign capital structure of the suppliers. We define foreign firms as firms with a foreign share (FS) 10 % or more.[6] According to this definition, foreign firms accounted for 27 % (45 firms) of all firms, while local firms accounted for 73 % (120 firms). Generally, about one-quarter of the firms in our sample consists of foreign firms. Therefore, collected data and information is very suitable for analyzing in terms of local and foreign firms.[7]

When we analyze the ownership status of foreign firms within all firms, we see that 13 % of the firms are full ownership (FS = 100 %), 9.7 % are joint venture (FS is between 40 and 69 %), 2.4 % are majority ownership (FS is between 70 and 99 %), and finally 2 % are in the status of minority ownership status (FS is between 10 and 39 %). In other words, nearly 50 % of the foreign firms are in the status of full ownership, 36 % are in the joint venture, and 9 % are in the majority ownership. Therefore, we can say that foreign investors prefer full ownership and joint ventures to minority ownership, because they want to have full control over their firm. In sum, these findings also confirm that foreign firms are under the full control of foreign capital.

In addition, 132 firms out of 165 firms (80 %) in the sample are direct suppliers (first-tier suppliers) of one or more AMMs in Turkey. This confirms that most of the surveyed firms (80 %) are direct suppliers. Share of the local and foreign firms in direct suppliers are 70 % and 30 %, respectively. With respect to foreign ownership status, the same pattern is observed with all firms. Furthermore, the share of the direct suppliers in foreign firms (87 %) is higher than that in local firms (77 %).

---

[4] Turkish economy and especially firms in Turkish automotive industry were negatively affected by the global financial crisis, which started by the end of 2008, through 2009 together with world economies and global automotive industry. As a result of this crisis, it was expected that economic indicators of the firms were not so good through 2009. Our pilot testing for the questionnaire survey also confirmed this situation, as well (see section "Pilot Testing" in Chap. 3). Therefore, although the questionnaire would be applied in 2010, all quantitative statistics (total sales, number of employees, share of export in total sales, R&D expenditures or share of R&D expenditures in total sales) related to year 2008 were asked to firms in the application of the questionnaire survey to prevent the bias in these indicators. In addition, number of employee statistics was also asked related to year 2007 to compare and test whether there was significant difference between the years 2007 and 2008. As a result, any significant difference was not obtained between such years related to these statistics. Hence, we used statistics related only to 2008 in the analyses.

[5] Appendix J provides summary statistics for key indicators in the survey according to firm descriptions.

[6] We define foreign firms as firms if FS is at least 10 % or more in total capital of the firm in accordance with the OECD, UNCTAD, and the IMF's definitions.

[7] There are some countries in which significant portion of direct suppliers are composed of foreign firms: See Bernard and Ravenhill (1995), and Giroud (2003).

Table 4.1 Distribution of surveyed firms by foreign capital and ownership structure*

	All Firms		DSF	
	N	%	*N*	%
Local firms (FS < 10 %)	120	72.73	93	70.45
Foreign firms (FS ≥ 10 %)	45	27.27	39	29.55
Minority ownership (10 % ≤ FS ≤ 39 %)	3	1.82	2	1.52
Joint venture (40 % ≤ FS ≤ 69 %)	16	9.7	15	11.36
Majority ownership (70 % ≤ FS ≤ 99 %)	4	2.42	2	1.52
Full ownership (FS = 100 %)	22	13.33	19	14.39
Total	**165**	**100**	**132**	**100**

Source: Author calculations based on survey results
*FS: Foreign share defined as the share of foreign partner in firm equity
Legend: DSF (Direct Supplier Firms)

Origin of Foreign Capital

When we look at the origin of foreign capital, we see that there are 14 countries with different ownership status (between minority and full ownership) in surveyed firms (Table 4.2). The origin of many firms is Germany (29 %) with 13 firms; this is followed by the USA (13 %) and France (11 %) with 6 and 5 firms, respectively. Especially, German firms have a large weight and they dominate the foreign firms in the sample. Moreover, half of the full ownership firms are German (7) and Japanese (4), and only Japanese firms prefer full ownership status.

Table 4.2 Distribution of foreign firms by origin and ownership

	Origin								
Ownership status	German	USA	France	Japan	Italy	Spain	Other[a]	*N*	%
Minority ownership	1	0	0	0	1	0	1	3	7
Joint venture	4	3	2	0	1	2	4	16	35
Majority ownership	1	1	0	0	1	0	1	4	9
Full ownership	7	2	3	4	1	1	4	22	49
N	13	6	5	4	4	3	10	**45**	
%	29	13	11	9	9	7	22		**100**

Source: Author calculations based on survey results
[a]England (2), Belgium (2), S. Korea, Taiwan, India, Hungary, Finland and Switzerland

4.1.2 Establishment Date (Age)

Distribution of the surveyed firms by establishment year is presented in Table 4.3. According to this, 83 % of all firms were established after the year of 1970, and most of the foreign firms (36 %) were established 2000 onwards. Mean age for

Table 4.3 General characteristics of surveyed firms

in 2008	LF		FF		DSF		AF	
	N	%	*N*	%	*N*	%	*N*	%
Establishment		***		***				
1930s	1	0.83	–	–	1	0.76	1	0.61
1940s	2	1.67	–	–	1	0.76	2	1.21
1950s	9	7.50	1	2.22	8	6.06	10	6.06
1960s	13	10.83	2	4.44	12	9.09	15	9.09
1970s	39	**32.50**	13	28.89	43	**32.58**	52	**31.52**
1980s	27	22.50	4	8.89	25	18.94	31	18.79
1990s	20	16.67	9	20.00	23	17.42	29	17.58
2000 Onwards	9	7.50	16	**35.56**	19	14.39	25	15.15
Total	**120**	**100**	**45**	**100**	**132**	**100**	**165**	**100**
Employment						*		
10–19	2	1.68	1	2.38	2	1.55	3	1.86
20–49	9	7.56	2	4.76	6	4.65	11	6.83
50–99	22	18.49	10	**23.81**	21	16.28	32	19.88
100–249	41	**34.45**	9	21.43	44	**34.11**	50	**31.06**
250–499	34	28.57	10	**23.81**	39	30.23	44	27.33
500–999	8	6.72	6	14.29	10	7.75	14	8.70
1000 +	3	2.52	4	9.52	7	5.43	7	4.35
Total	**119**	**100**	**42**	**100**	**129**	**100**	**161**	**100**
Total Sales		*		*				
less than 5 mTL	12	11.43	2	5.26	10	8.77	14	9.79
5–9.9 mTL	18	17.14	3	7.89	15	13.16	21	14.69
10–19.9 mTL	9	8.57	5	13.16	10	8.77	14	10.49
20–29.9 mTL	14	13.33	2	5.26	14	12.28	16	10.49
30–39.9 mTL	12	11.43	2	5.26	9	7.89	14	9.79
40–49.9 mTL	7	6.67	3	7.89	8	7.02	10	6.99
50–99.9 mTL	23	**21.90**	11	**28.95**	29	**25.44**	34	**23.78**
100–199.9 mT	4	3.81	3	7.89	7	6.14	7	4.90
200 mTL +	6	5.71	7	18.42	12	10.53	13	9.09
Total	**105**	**100**	**38**	**100**	**114**	**100**	**143**	**100**
% of Export in Total Sales		*		*				
0	8	6.67	6	13.33	12	9.09	14	8.48
1–10	28	**23.33**	4	8.89	25	**18.94**	32	**19.39**
11–20	17	14.17	5	11.11	21	15.91	22	13.33
21–40	19	15.83	8	**17.78**	20	15.15	27	16.36
41–50	15	12.50	4	8.89	17	12.88	19	11.52
51–70	19	15.83	7	15.56	19	14.39	26	15.76
71–90	12	10	7	15.56	15	11.36	19	11.52
90–100	2	1.67	4	8.89	3	2.27	6	3.64
Total	**120**	**100**	**45**	**100**	**132**	**100**	**165**	**100**

Source: Author calculations based on survey results
Notes: ***, **, * denote significance level at the 1 %, 5 %, and 10 %, respectively (chi-square, two-sided). Fisher's Exact test has been used if one or more cells have an expected count ≤ 5.
Legend: LF (Local Firms); FF (Foreign Firms); DSF (Direct Supplier Firms); AF (All Firms)

foreign firms is 19 years, while for local firms it is 29 years.[8] This means that foreign firms are significantly younger than local firms in the sample and there is also a statistically significant difference at 1 % level (chi-square test). With respect to direct suppliers, we observe the same pattern with all firms, majority of the firms (33 %) were established in 1970s, and mean age for these firms is 26 years.

4.1.3 Number of Employees

Table 4.3 displays that 51 % of the surveyed firms are medium-sized (50–250 employees) and 40 % are large-sized (250 and more employees), while only 9 % of the firms are small-sized firms (less than 50 employees). Mean number of employees is 294 for all firms (median[9]: 200), minimum and maximum numbers are 15 and 3,011, respectively.[10] Majority of the foreign firms are large-sized firms (48 %) and this is followed by medium-sized firms (45 %). Mean number (404) (median: 236) for foreign firms is significantly higher than local firms (mean: 255, median: 195). When we look at the direct suppliers, we see that 50 % are medium-sized, and 43 % are large-sized firms. Moreover, mean number is much higher (325) (median: 210) than non-direct suppliers (mean: 168, median: 100), and there is a statistically significant difference between these two group of firms at the level of 10 % (chi-square test).

4.1.4 Total Sales

Table 4.3 displays the distribution of the surveyed firms according to total sales (turnover) in 2008. Irrespective of the firm classification (LF, FF, DSF and AF), most of the firms (in varying rates of 22 % and 29 %) have total value of annual sales between TL 50M and TL 100M. Total sales value of the 143 respondent firms is nearly TL 9B with mean value of TL 62M per firm (median TL 32M), and minimum value is TL 800K while maximum is TL 488M.[11] In the meantime, only 10 % of the firms have less than TL 5M total sales value. Foreign firms have nearly two times higher sales value than local firms, the median and mean values for foreign firms are TL 53.5M and TL 97M, whereas for local counterparts they are TL 29M and TL 49M, respectively (statistically significant at the 1 % level) (chi-square test). In addition, direct suppliers have also higher sales values than local firms (but lower than foreign firms): the median and mean values are TL 38M and TL 67M, respectively.

---

[8] See Appendix J for detailed summary statistics.

[9] The median is given as a measure of central tendency, because it is less affected by outliers.

[10] For details see Appendix J.

[11] For details see Appendix J.

4.1.5 Market Orientation

Table 4.3 displays also the distribution of the firms according to their share of export in total sales (export intensity). It can be seen from the table that most of the firms (92 %) are exporters (less than 9 % of the firms do not export their products), 32 % are also engaged in export-oriented production (export at least 50 % of their sales). Exports accounted for 44 % (mean) of sales for foreign firms (median: 40 %) and only 35 % for local firms (median: 26 %),[12] and the difference between these two groups are statistically significant at 10 % level (chi-square test). Moreover, most of the foreign firms are in the class of 21–40 %, while local counterparts are in the 1–10 % class. Therefore, foreign firms (40 %) are mainly more export-oriented than local counterparts (28 %). With respect to direct suppliers, we see that they have the same pattern with local firms. The mean share of export for direct suppliers is 35 % (median: 25 %), and most of them (19 %) are in the class of 1–10 %.

4.1.6 Technology Level of the Firms

Table 4.4 shows the distribution of the surveyed firms according to their technology levels (high-, medium-, and low-tech). In the technological classification, the technological complexity levels of the major products manufactured/assembled by the firms have been considered.[13] According to this, mainly three major products and their shares in total sales have been analyzed case by case as follows[14]:

- Initially, the first product specified by the firms was taken as the basis in the technological classification (since it is assumed as the most important product manufactured/assembled by the firms). If specified by the firms, the shares of the second and third major products in the total sales respectively, and whether they are of the same product type with the first product were evaluated, as well. As a result of the evaluation, it was realized that the products specified in the second and third order had less share in the sales, and/or these products were almost in the same categories as the products specified in the first order (the firms generally manufacture/assemble similar product types).
- Secondly, the specialist engineers in this field (those employed by the AMMs and/or supply industry) were consulted in the classification of the technology levels involved in the products. Accordingly, a technological classification has been made by taking four factors into consideration:

---

[12] For details see Appendix J.

[13] The number of the firms specified only one, only two or only three products are 26, 17, and 122, respectively.

[14] See section 1 question 7 in the questionnaire survey presented in Appendix G.

Table 4.4 Distribution of surveyed firms according to technological complexity of products manufactured (%)

	LF***	FF***	DSF***	AF
N	**119**	**45**	**131**	**164**
High-technology	**45.38**	**68.89**	**55.72**	51.83
Medium-technology	35.29	20	31.30	31.1
Low-technology	19.33	11.11	12.98	17.07
Total	**100**	**100**	**100**	**100**

Source: Author calculations based on survey results

Notes: ***, **, * denote significance level at the 1 %, 5 %, and 10 %, respectively (Mann–Whitney *U* test, 2-sided).

Legend: LF (Local Firms); FF (Foreign Firms); DSF (Direct Supplier Firms); AF (All Firms)

i. The knowledge and technology level involved in the product
ii. Whether it is a standard product
iii. The level of difficulty in its manufacture
iv. Its place in the production chain (primary, secondary, raw material, etc.)[15]

As seen in the table, more than 50 % of the surveyed firms (AF) manufacture at high-technology level. Yet the rate of the foreign firms manufacturing at the high-technology level (69 %) is more than the rate of the local firms manufacturing at the same level (45 %). In addition, while the rate of the local firms manufacturing at medium and low technology levels is 55 %, the rate of the foreign firms manufacturing at the same levels is 31 %. This indicates that foreign firms manufacture at a higher technology level, and highly statistically-significant difference has been detected between the local and foreign firms at 1 % level.

When we examine the table in terms of the direct suppliers, 56 % of these firms manufacture at high technology level, 31 % at medium, and 13 % at low technology level. In sum, direct suppliers manufacture at a higher technology level than local firms but not higher than foreign firms. Moreover, a statistically-significant difference has also been detected between the direct suppliers and non-direct suppliers (at 1 % level) (Mann–Whitney *U* test).

4.2 KTTs at Inter-Firm Level

In this section, various types of the transfers at inter-firm level accruing from customers to their suppliers in the automotive industry in Turkey are examined, especially those related to (i) *production processes*, (ii) *product,* (iii) achieved—albeit in an indirect manner—through *financial assistance,* and (iv) implemented

[15] E.g. the parts/components as motor, gear box, suspension, braking system, safety systems, and so on (in primary product class) were classified in the high-technology category; the parts as various automotive fasteners, headlight, ventilation ducts, damper, seat, internal trim materials and such in the medium-technology category, and the parts as mudguard, seat cover, indicator, signal arms, mirror and exhaust silencer in the low technology category.

through *trainings* given/provided to suppliers by their customers. A thorough analysis of these various channels of the transfers and their relative importance will shed a light on the importance and the nature of linkages occurring in the industry between customers (buyers) and suppliers.

4.2.1 Transfers Related to Production Process

Table 4.5 presents how often (*frequently, rarely, never*)[16] 13 KTT types pertaining to the production process are provided to the firms by the customers.[17] These transfers regarding the production process are primarily categorized as *provide documentations, know-how,* and *assistance*[18] *for R&D activities—logistic management—quality control methods—design.*

When the items are examined according to the *frequently* transfer degree and in terms of all firms, the *provide documentations* item chosen by 33 % of the firms comes in the first rank (36 % for LF, 24 % for FF, and 33 % for DSF). This item is followed by *assistance for logistic management* with 15 %, *assistance for quality control methods* with 14 %, and *know-how* and *assistance for R&D activities* items each being 10 %. Other eight items were chosen with a rate ranging from minimum 2 % (*assistance for business management*) to maximum 9 % (*supply of raw material*) of the firms. When the items are examined in terms of local and foreign firms, it is found out that higher-quality KTT types were being provided to the foreign firms at higher rates (*know-how, assistance for R&D activities, send expert staff, assistance for productivity-related problems, patent and/or license rights granted, assistance for business management*). The transfers to the local firms are provided mostly in fields such as *provide documentations, assistance for logistic management*, and *supply of raw material*. When the items are examined in terms of direct suppliers, it is found out that *supply of raw material, assistance for design* and *send expert staff* items were being provided to the direct suppliers at higher rates. This is an important point for the direct suppliers to show that they implement common activities with the AMMs. Otherwise, it is observed that direct suppliers bear the same characteristics as the local firms.

---

[16] Frequency of provided KTT types (items) related to production process, product, financial transfers and training is measured using a three-point frequency Likert scale constructed from five-point Likert scale (*frequently, occasionally, rarely, very rarely, never*).

[17] Information on *never* category has not been identified in the table for space reasons but can be obtained for each item by summing the proportions of answers given to *frequently* and *rarely* categories and then subtracting from 100 %.

[18] The implied meaning of the **assistance** used in the items related to production process transfers (assistance related to R&D activities - logistic management - business management - quality control methods - design - productivity problems) refers to every help that is provided by customers such as providing technical, support, training, advice, and KTT.

Table 4.5 Types of KTTs related to the production process

Types of knowledge and technology transfer related to production process	Frequently				Rarely						
	LF %	FF %	DSF %	AF %	LF %	FF %	DSF %	AF %	LF N	FF N	DSF N
Provide documentations	**35.83***	**24.44***	**32.58**	32.73	55.83*	57.78*	57.58	56.36	120	45	132
Assistance for logistic management	17.5*	8.89*	15.91*	15.15	58.33*	55.56*	59.85*	57.58	120	45	132
Assistance for quality control methods	14.17	13.33	13.64	13.94	**65**	53.33	**64.39**	61.82	120	45	132
Know-how	9.32*	13.33*	12.31	10.43	50*	**62.22***	50	53.37	118	45	130
Assistance for R&D activities	9.17	13.33	11.36	10.30	53.33	53.33	54.55	53.33	120	45	132
Supply of raw material	10	8.89	10.61	9.70	29.17	37.78	32.58	31.52	120	45	132
Customer sent its expert staff to stay at your plant for a certain period of time for assistance in solving problems in the production process	5.83	8.89	7.58	6.67	50.00	51.11	50.76	50.30	120	45	132
Assistance for design	6.67	6.67	8.33**	6.67	51.67	**62.22**	56.82**	54.55	120	45	132
Supply of machinery, tools and equipment	4.17	8.89	6.82	5.45	40	40	40.15	40	120	45	132
Assistance for productivity-related problems	4.17	8.89	6.82	5.45	50.83	42.22	49.24	48.48	120	45	132
Customer assigned its expert staff in the establishment of production process of the plant	5.83	4.44	6.82**	5.45	29.17	31.11	31.82**	29.70	120	45	132
Patent and/or license rights granted	2.52	4.44	3.82**	3.05	14.29	13.33	16.79**	14.02	119	45	131
Assistance for business management	1.67	4.44	2.27*	2.42	30.83	35.56	35.61*	32.12	120	45	132

Source: Author calculations based on survey results

Notes: ***, **, * denote significance level at the 1%, 5% and 10%, respectively (Mann-Whitney U test, 2-sided). The items do not add up to 100% because the questionnaire included a further category called "Never", but has not been identified in table for space reasons. However, information on "Never" category can be obtained for each item by summing the proportions of answers given to "Frequently" and "Rarely" categories and then subtracting from 100%. Items are arranged by both "All Firms" (AF) and "Frequently" transfer degree.

Legend: LF (Local Firms); FF (Foreign Firms); DSF (Direct Supplier Firms); AF (All Firms)

If the items are examined according to the *rarely* transfer degree and in terms of all firms, the rate of nine items appears over 40 %, and ranges from minimum 14 % (*patent and/or license rights granted*) to maximum 62 % (*assistance for quality control methods*). While the item *assistance for quality control methods* comes in the first rank in terms of local (65 %) and direct supplier firms (64 %), *know-how* and *assistance for design* items come first for the foreign firms (62 %). This once again verifies that the higher-quality transfers are provided to the foreign firms at a higher rate in terms of the quality of the transfers.

When the transfer degrees are examined together as both *frequently* and *rarely*, the rate of six items is obtained as higher than 60 %, ranging from minimum 17 % (*patent and/or license rights granted*) to maximum 89 % (*provide documentations*). The most important six items are respectively *provide documentations* (89 %), assistances for *logistic management* (73 %)—*quality control methods* (76 %)—*R&D activities* (64 %), *know-how* (64 %)—*design* (61 %).

In sum, the degree of the transfers provided according to the results summarized in Table 4.5 has been detected as *rarely* for each item. The rate of the firms specified that transfers provided *rarely* according to *frequently* transfer degree for each item ranges between minimum two and maximum 13 times. Another conclusion is that the rate of *never* providing transfer for *assistance for productivity-related problems*, *assistance for design*, *assistance for R&D activities* and *know-how* items including an important level of knowledge transfer is very high as being 46 %, 39 %, 36 %, and 36 %, respectively.

A significant difference has been detected between local and foreign firms in terms of the items *provide documentations*, *assistance for logistic management*, and *know-how* items each significant at 10 % level. A significant difference has also been found between the direct suppliers and non-direct suppliers in terms of the items *assistance for logistic management* (at 10 % level), *assistance for design* (at 5 % level), *customer assigned its expert staff in the establishment of production process of the plant* (at 5 % level), *patent and/or license rights granted* (at 5 % level), and *assistance for business management* (at 10 % level) (Mann–Whitney *U* test).

4.2.2 Transfers Related to Product[19]

Table 4.6 shows how often (*frequently*, *rarely*, *never*) five KTT types pertaining to the product are provided to the firms by the customers. These five factors with

[19] The transfers provided by customers with respect to the product were asked to the executives of the supplier firms as to the most important product/s manufactured. Although there is a possibility that transfers with respect to the products might change for each firm depending on the manufactured product or the type of the part, when the fact that survey application and study purposes as well as supplier firms are specialized in one product group (see Sect. 4.1.6) is taken into consideration no product or part classification has been made for the transfers with respect to the product.

Table 4.6 Types of KTTs related to the product

Types of knowledge and technology transfer related to product	Frequently				Rarely						
	LF	FF	DSF	AF	LF	FF	DSF	AF	LF	FF	DSF
	%	%	%	%	%	%	%	%	N	N	N
Technical specifications, original design or technical drawings for products	**55.46*****	**31.11*****	**46.97**	48.78	31.93***	40***	36.36	34.15	119	45	132
Joint operation related to product	25.21	28.89	27.27	26.22	53.78	57.78	54.55	54.88	119	45	132
Product specifications	27.73	22.22	25.76	26.22	38.66	55.56	43.18	43.29	119	45	132
Joint design activity related to product	15.13	13.33	16.67**	14.63	**54.62**	**60**	**57.58****	56.10	119	45	132
Assistance related to product designs	10.17	15.56	12.21*	11.66	52.54	51.11	54.96*	52.15	118	45	131

Source: Author calculations based on survey results.

Notes: ***, **, * denote significance level at the 1%, 5% and 10%, respectively (Mann-Whitney U test, 2-sided). The items do not add up to 100% because the questionnaire included a further category called "Never", but has not been identified in table for space reasons. However, information on "Never" category can be obtained for each item by summing the proportions of answers given to "Frequently" and "Rarely" categories and then subtracting from 100%. Items are arranged by both "All Firms" (AF) and "Frequently" transfer degree.

Legend: LF (Local Firms); FF (Foreign Firms); DSF (Direct Supplier Firms) AF (All Firms)

respect to the product are the provision of *assistance related to product designs,*[20] *joint operation related to product, joint design activity related to product, product specifications*, and *technical specifications, original design or technical drawings for the products.*

When the items are examined according to the *frequently* transfer degree and in terms of all firms, the item *technical specifications, original design or technical drawings for the products* (49 %) ranks first (55 % for LF, 31 % for FF, and 47 % for DSF). This item is followed by the items *joint operation related to product* (26 %), *product specifications* (26 %), *joint design activity related to product* (15 %), and *assistance related to product designs* (12 %). The items which are of higher quality in terms of the knowledge and technology included (*assistance related to product designs* and *joint design activity related to product*) are provided less to the firms by the customers. When the items are examined in terms of local and foreign firms, we can observe that the firms have similar rates regarding the items. An important finding is that the transfers including higher quality knowledge and technology are provided more to the foreign firms compared to the local firms (*assistance related to product designs* and *joint operation related to product*). When the items are examined in terms of direct suppliers, they are observed to have almost the same distribution as the local firms; however, the item *joint design activity related to product* (17 %) is higher both than the foreign and the local firms. This is an important point for the direct suppliers showing that they implement common product design activities with the AMMs.

If the items are examined according to the *rarely* transfer degree and in terms of all firms, the rate for four items appears over 40 %, and ranges from minimum 34 % (*technical specifications, original design or technical drawings for the products*) to maximum 56 % (*joint design activity related to product*). The item *joint design activity related to product* ranks first in terms of three groups of firms (LF, FF, and DSF). This reveals that the level of cooperation between the customers and suppliers in terms of the *joint design activities for the product* is high. This is also the result of the fact that the customers wish to guarantee the quality, defect rate and on-time delivery of the products used in the production, and provided from the suppliers and thus not to have any problems in the production lines. Another finding is that various types of the transfers with respect to the product are intensively provided to the foreign firms (compared to both LF and DSF).

When the transfer degrees are evaluated together as both *frequently* and *rarely*, the rate of the items exceeds 64 %, the highest rate being *technical specifications, original design or technical drawings for the products* item with 83 %. This item is followed by *joint operation related to product* with 81 %, *joint design activity related to product* with 71 %, *product specifications* with 69 % and *assistance*

[20] The implied meaning of the ***assistance*** used in the item related to product transfers (assistance related to product designs) refers to every help that is provided by customers such as providing technical support, training, advice, and KTT.

related to product designs item with 64 % in the last place.[21] According to the results summarized in Table 4.6, the transfer degree of the four items other than *technical specifications, original design or technical drawings for the products* item (detected as *frequently* with 49 %) with the highest percentage has been detected as *rarely*.

A significant difference has been found between local and foreign firms concerning the item *technical specifications, original design or technical drawings for products* (at 1 % level). A significant difference has also been found between the direct suppliers and non-direct suppliers in terms of the items *assistance related to product designs* (at 10 % level), and *joint design activity related to product* (at 5 % level) (Mann–Whitney *U* test).

4.2.3 *Financial Transfers*[22]

Table 4.7 below shows how often five types of financial transfers (assistance) are provided to the firms by the customers. These transfer types are: *unilateral financial aid, loans with low interest rates, contribution to risk capital, pre-financing of machinery, equipment and tools*, and *prepayment for orders before delivery*.

When the items are examined according to the *frequently* transfer degree and in terms of all firms, the items *pre-financing of machinery, equipment and tools* chosen by 11 % of the firms ranks first (9 % for LF, 18 % for FF, and 13 % for DFS). This item is followed by *prepayment for orders before delivery* with 7 %, *loans with low interest rates* with 1 %, and the item *contribution to risk capital* with 0.6 %. The item with the lowest rate is *unilateral financial aid* with 0 %. When the items are examined with regard to local and foreign firms, it is observed that the items *loans with low interest rates* and *contribution to risk capital* are not provided with *frequently* transfer degree to the foreign firms. The fact that foreign firms are generally the affiliates of the multinationals abroad (see Sect. 4.5.1) and therefore have stronger capital structure can be marked as the possible reason behind this. This is also verified by the data regarding the size of the surveyed foreign firms. On the other hand, *pre-financing of machinery, equipment and tools* item is provided two times more to the foreign firms than the local firms. This indicates the fact that foreign firms set up more extensive relations with their customers. When we take a glance upon the items from the direct supplier's point of view, the distribution of the items show the same properties as the local firms: the most important difference

---

[21] In other words, *never* transfer degree of the *assistance related to product designs* item requiring an important level of knowledge and skills has the highest rate with 36 % {100 % − 12 % (frequently) − 52 % (rarely) = 36 %}.

[22] Foreign literature and findings of the earlier case studies conducted on various sectors showed that financial transfers to the suppliers were quite minimal and not preferred so much by the cooperated customers (Giroud, 2003).

Table 4.7 Types of financial transfers (assistances)

	Frequently				Rarely						
	LF	FF	DSF	AF	LF	FF	DSF	AF	LF	FF	DSF
Types of financial transfer	%	%	%	%	%	%	%	%	N	N	N
Pre-financing of machinery, equipment and tools	**8.55***	**17.78***	**13.18*****	11.11	**39.32***	**42.22***	**44.19*****	40.12	117	45	129
Prepayment for orders before delivery	7.56	6.67	7.63	7.32	36.13	31.11	32.06	34.76	119	45	131
Loans with low interest rates	1.69	0	1.54	1.23	6.78	2.22	4.62	5.52	118	45	130
Contribution to risk capital	0.85	0	0.77	0.61	4.24	0	3.85	3.07	118	45	130
Unilateral financial aid	0	0	0	0	7.69	4.44	7.75	6.79	117	45	129

Source: Author calculations based on survey results.
Notes: ***, **, * denote significance level at the 1%, 5% and 10%, respectively (Mann-Whitney U test, 2-sided). The items do not add up to 100% because the questionnaire included a further category called "Never", but has not been identified in table for space reasons. However, information on "Never" category can be obtained for each item by summing the proportions of answers given to "Frequently" and "Rarely" categories and then subtracting from 100%. Items are arranged by both "All Firms" (AF) and "Frequently" transfer degree.
Legend: LF (Local Firms); FF (Foreign Firms); DSF (Direct Supplier Firms); AF (All Firms)

is that *pre-financing of machinery, equipment and tools* item is nearly two times higher compared to local firms.

If the items are examined according to *rarely* transfer degree and in terms of all firms, the rates pertaining to the items range between 3 % (*contribution to risk capital*) and 40 % (*pre-financing of machinery, equipment and tools*). A significant conclusion according to the *frequently* transfer degree is that the item *unilateral financial aid* in the last place with 0 % rises to 7 % at *rarely* transfer degree, and thus ranked third by exceeding both *contribution to risk capital* and *loans with low interest rates* items.

When the transfer degrees are examined together as both *frequently* and *rarely*, the transfer with the highest rate is again the *pre-financing of machinery, equipment and tools* item with 51 %. This item is followed by *prepayment for orders before delivery* with 42 %, *loans with low interest rates* with 7 %, *unilateral financial aid* with 7 % and *contribution to risk capital* item with 4 %. Furthermore, when the items are examined individually, the degree of the transfers made is detected as *rarely* for each item.

A significant difference is found for both test groups in terms of only *pre-financing of machinery, equipment and tools* item between foreign and local firms at 10 % level and direct suppliers and non-direct suppliers at 1 % level (Mann–Whitney *U* test).

Consequently, it has been detected that customers do not generally prefer giving financial assistances to supplier firms; but if they do, these assistances are generally given in the form of *pre-financing of machinery, equipment and tools* and/or *prepayment for orders before delivery*. This finding is consistent with the findings of earlier studies conducted in other countries.

4.2.4 Transfers Through Training

Table 4.8 presents three different types of trainings provided by the cooperating customers and the modes of these as visual (visits to customers' plants), off-the-job (via seminars and courses) and on-the-job training (learning by doing/theoretical and/or applied training) according to the degrees.

Types of Training

Three different types of trainings provided by the customers are: *training on technologies used in production*, *training of production/operation staff* (engineers, technicians, etc.), and *training of management staff* (Table 4.8).

When the degree of training provision is examined according to *frequently* degree, it can be seen that training types are provided to foreign firms at higher rates (compared to both LF and DSF). When we examine the rates of the trainings provided by the customers in terms of *production* and *management personnel*

Table 4.8 Types and modes of training activities provided by customers

	Frequently				Rarely						
	LF	FF	DSF	AF	LF	FF	DSF	AF	LF	FF	DSF
	%	%	%	%	%	%	%	%	*N*	*N*	*N*
Types of training											
Training on technologies used in production	**9.24**	15.56	**12.98***	10.98	36.97	37.78	38.17*	37.20	119	45	131
Training of production/operation staff (engineers, technicians etc.)	7.56	**17.78**	**12.98****	10.37	**53.78**	46.67	**52.67****	51.83	119	45	131
Training of management staff	5.83	11.11	9.09*	7.27	47.50	**53.33**	50*	49.09	120	45	132
Modes of training											
A. Visits to customers' plants (learning by visual)	20.83	13.33	18.94	18.79	60.83	71.11	63.64	63.64	120	45	132
B. Off-the-job training (via seminars and courses)											
at supplier's plant	**47.06**	**42.22**	**47.33**	45.73	44.54	51.11	46.56	46.34	119	45	131
at other private specialized institutes	20.17**	26.67**	22.14	21.95	**64.71****	73.33**	**69.47**	67.07	119	45	131
at customers' plants	6.67*	6.67*	6.82*	6.67	61.67*	**77.78***	68.94*	66.06	120	45	132
On-the-job training (learning by doing/theoretical and/or applied training)											
at supplier's plant	**5.88**	**11.36**	**8.46**	7.36	**42.02**	38.64	**40.77**	41.10	119	44	130
at the costumers' plant											
In Turkey	4.17	4.44	4.55	4.24	35	**44.44**	39.39	37.58	120	45	132
Abroad	1.69**	0**	1.54	1.23	22.03**	40**	29.23	26.99	118	45	130

Source: Author calculations based on survey results

Notes: ***, **, * denote significance level at the 1 %, 5 % and 10 %, respectively (Mann–Whitney *U* test, 2-sided). The items do not add up to 100 % because the questionnaire included a further category called "Never", but has not been identified in table for space reasons. However, information on "Never" category can be obtained for each item by summing the proportions of answers given to "Frequently" and "Rarely" categories and then subtracting from 100 %. Items are arranged by both "All Firms" (AF) and "Frequently" transfer degree.

Legend: LF (Local Firms); FF (Foreign Firms); DSF (Direct Supplier Firms); AF (All Firms)

employed by the firms, *training of production/operation staff* has the highest rates among the three groups of firms (LF, FF, and DSF). In addition to that, direct suppliers have higher rates than local firms but lower rates than foreign firms for three training types.

When the table is examined according to *rarely* transfer degree, *training on technologies used in production* is given more to the direct suppliers with 38 %, *training of production/operation staff* more to the local firms with 54 % and *training of management staff* more to the foreign firms with 53 %. Another finding is that while the *training of management staff* is provided to the foreign firms at higher rates than *training of production/operation staff* item, it is vice versa for the local firms and direct suppliers.

When the training provision degrees are examined together as both *frequently* and *rarely*, *training on technologies used in production* rate becomes 48 %, *training of production/operation staff* 62 % and *training of management staff* 56 %. The highest transfer degree for these three items has been identified as *rarely*.

No significant difference has been found between local and foreign firms with respect to the training types. However, a significant difference has been detected between the direct suppliers and non-direct suppliers in terms of *training on technologies used in production* (at 10 % level) and *training of production/operation staff* items (at 5 % level) (Mann–Whitney *U* test).

Modes of Training

When the training modes are examined according to *frequently* transfer degree, off-the job training (learning by seminars and courses) gets in the first rank (Table 4.8). This kind of training is provided *at supplier's plant* (in-house) (46 %), *at other private specialized institutes* (22 %) and *at customers' plants* (7 %) respectively. A similar distribution is observed among three groups of firms in terms of this training mode (LF, FF and DSF). The second training mode is *visits to customers' plants* (learning by visual). In this training mode, the personnel of the supplier firms visits the factories of the customers for a certain period of time within a pre-planned program and take visual training via expert personnel regarding the product and production process. We have realized that this mode of training is given more to local firms (21 %), direct suppliers (19 %) and foreign firms (13 %), respectively. The fact that this visual mode of training is given less to the foreign firms indicates that their technological capability level is very high compared to other firms. The third training mode is on-the-job training (learning by doing/theoretical and/or applied). When we take a look at where this mode of training is given, *the plants of the supplier firms* (in-house) come first with 7 %. This is followed by *customers' plants in Turkey* (4 %) and *factories abroad* (1 %). On the other hand, the foreign supplier firms take higher amount of training at their own factories (in-house).

When the modes of training are examined according to *rarely* degree or together with both *frequently* and *rarely* degree, the above-mentioned results do not change much, and the most important difference results from the fact that foreign firms have higher rates at both degrees (compared to LF and DSF). When we examine the degrees of the items concerning training modes, *rarely* has been detected as the degree with the highest percentage for each item (except *at supplier's plant* item for LF).

A significant difference has been found between local and foreign firms in terms of giving training *at other private specialized institutions* (at 5 % level) and *at customers' plants* (at 10 % level) related to off-the-job training, and *at customers' plant abroad* (at 5 % level) in terms of the on-the-job training. A significant difference between the direct suppliers and non-direct suppliers has also been detected in terms of giving training *at customers' plants* (at 10 % level) for the off-the-job training (Mann–Whitney *U* test).

In summary, various activities regarding the training are being provided intensively for the firms by the cooperating customers, and much more importance is attached particularly to the training of the *production personnel*. As it has been discussed above, we can claim that it results from the fact that the customers wish to increase the quality level of the products supplied and not to have any trouble in the production line by means of increasing the capabilities and technology levels of their suppliers (quality, on-time delivery, low defect rate and cost, etc.). It is also the result of new product development, improvements related to production process and cooperative activities for joint design between the customers and suppliers. Besides, the fact that training activities are intensively provided to the foreign firms shows that the cooperation between these firms and the customers are stronger compared to the local firms.

4.2.5 The Reasons for Not Receiving Sufficient Amount of KTT

A significant number of the firms almost 75 % (124 firms) think that customers do not provide sufficient amount of KTT (82 % for FF, 73 % for LF, and 75 % for DSF) (Table 4.9). The firms thinking in such a way have identified *realizing our own R&D activities* (48 %) and *realizing co-designer and/or our own design activities* (44 %) items as two of the most important reasons. This indicates that the firms regard their own technological capabilities superior in terms of R&D and design and therefore do not need KTTs much. On the other hand, other three items following these items were also specified as the reason with a high rate (*reluctance of foreign firms in KTT* with 39 %, *reluctance of AMMs in KTT* with 31 % and *technologies are strictly confidential* with 31 %). These three items point out that another important reason for not providing transfers is because customers are not keen on transfers to their suppliers and want to prevent spillover by means of keeping their technology confidential. The last two items which are regarded as important by 19 % of the firms were specified as *not working with AMMs* and

Table 4.9 The reasons for not receiving sufficient amount of KTT

		LF %	FF %	DSF %	AF %
Enough knowledge and technology are not transferred (165 firms)		72.50	82.22	75	75.15
Reasons	Realizing our own R&D activities	**48.28**	**48.65**	**47.47**	48.39
	Realizing Co-designer and/or our own design activities	42.53	**48.65**	46.46	44.35
	Reluctance of foreign firms in knowledge and technology transfer	37.93	40.54	42.42*	38.71
	Reluctance of AMM in knowledge and technology transfer	33.33	27.03	31.31	31.45
	Technologies are strictly confidential	33.33	24.32	34.34*	30.65
	Not working with AMM	22.99*	8.11*	-	18.55
	Needed technologies can be obtained by using reverse engineering methods	21.84	10.81	19.19	18.55
	N	**87**	**37**	**99**	**124**

Source: Author calculations based on survey results
Notes: ***, **, * denote significance level at the 1%, 5%, and 10%, respectively (chi-square, 2-sided). Fisher's Exact test has been used if one or more cells have an expected count ≤ 5.
Legend: LF (Local Firms); FF (Foreign Firms); DSF (Direct Supplier Firms); AF (All Firms); AMM (Automotive Main Manufacturers)

needed technologies can be obtained by using reverse engineering methods. This is important regarding working with the AMMs as a significant way of transferring knowledge and technology (23 % for LF, 8 % for FF).

Distributions regarding the items show almost the same patterns for three groups of firms (LF, FF, and DSF). An important finding is that foreign firms have higher rates for the first two items yet lower rates for the fourth and fifth items (compared to both LF and DSF). While the situation regarding the first two items reflects the superior technological capabilities of the foreign firms, the situation concerning the fourth and fifth items reveals that foreign firms have stronger and more strategic relations with the AMMs.

A significant difference has been found between local and foreign firms in terms of the item *not working with AMMs* (at 10 % level). A significant difference has also been detected between the direct suppliers and non-direct suppliers in terms of the items *reluctance of foreign firms in KTT* (at 10 % level) and *technologies are strictly confidential* (at 10 % level) (chi-square test).

4.3 Market and Competition Structure

4.3.1 Major Markets

The firms were asked to prioritize their major markets from 1 to 5 (1 > 2 > 3 > 4 > 5). Accordingly, Table 4.10 shows the major markets of the firms according to the first three importance degrees. As seen in the table, there are similar

Table 4.10 Distribution of surveyed firms by market orientation (%)

	Degree of Importance											
	1st				2nd				3rd			
Major Markets	**LF**	**FF**	**DSF**	**AF**	**LF**	**FF**	**DSF**	**AF**	**LF**	**FF**	**DSF**	**AF**
N	**120**	**45**	**132**	**165**	**115**	**40**	**127**	**155**	**85**	**30**	**94**	**115**
Domestic Market	**60.83**	**51.11**	**60.61**	**58.18**	24.35	35	27.56	27.10	8.24	6.67	6.38	7.83
European Union†	35	42.22	34.85**	36.97	**53.91**	**47.50**	**55.12****	**52.26**	2.35	6.67	4.26**	3.48
Middle East	1.67*	4.44*	3.03	2.42	7.83*	0*	1.57	5.81	**32.94***	**26.67***	**32.98**	**31.30**
Asia	0.83	2.22	0.76	1.21	4.35	10	6.30	5.81	23.53	26.67	23.40	24.35
USA	1.67	0	0.76*	1.21	6.09	7.50	7.87*	6.45	24.71	20	23.40*	23.48
Africa	0	0	0	0	1.74	0	0.79	1.29	2.35	6.67	3.19	3.48
Other Countries*	0	0	0	0	1.74	0	0.79	1.29	5.88	6.67	6.38	6.09
Total	**100**	**100**	**100**	**100**	**100**	**100**	**100**	**100**	**100**	**100**	**100**	**100**

Source: Author calculations based on survey results

Notes: ***, **, * denote significance level at the 1%, 5%,and 10%, respectively(Mann-Whitney U test, 2-sided).

Legend: LF (Local Firms); FF (Foreign Firms); DSF (Direct Supplier Firms); AF (All Firms).

† Germany (23%), France (17%), England (14%), Italy (9%), Spain (7%) and remaining 21 EU members (30%).

* Brazil (40%), Russia (17%), China (12%), Japan (6%) and 6 other countries (25%).

distributions in terms of all three groups of the firms. *Domestic market* ranks first with 58 % and *European Union* second with 37 % among the markets which are attached primary importance by 165 surveyed firms. The total amount of the both markets is 95 %, and it is concluded that the major markets, in which production and export is made, are domestic and countries of the European Union (EU), respectively. *European Union* is the first with 52 % and *domestic market* second with 27 % among the markets which are attached secondary importance by the 155 firms. The total amount of both markets is 79 %, and they preserve the degree of secondary importance, as well. Therefore, the major market of the firms is the domestic, and it is followed by the EU countries.[23] Major EU countries stated by the firms are Germany (23 %), France (17 %), England (14 %), Italy (9 %), and Spain (7 %), respectively. As to the markets attached third degree of importance by the firms, *Middle East* ranks first with 31 %. It is followed by *Asia* with 24 % and *USA* with 23 %. The total amount of three markets is 78 %, and these markets are regarded as the third important markets by the firms.

A significant difference has been detected between local and foreign firms in terms of *Middle East* market (at 10 % level). A significant difference has also been detected between the direct suppliers and non-direct suppliers in terms of *European Union* (at 10 % level) and *USA* markets (at 10 % level) (Mann–Whitney *U* test).

4.3.2 Main Competitors

Table 4.11 shows the distribution of the firms according to location of their main competitors. Accordingly, similar distribution in terms of three groups of the firms is observed. Most of the firms (67 %) stated that their main competitors were both

Table 4.11 Distribution of surveyed firms by location of their main competitors (%)

	LF	FF	DSF	AF
N	**120**	**45**	**132**	**165**
Both	**68.33**	**64.44**	**66.67**	67.27
Abroad	16.67*	26.67*	21.97*	19.39
In Turkey	15	8.89	11.36*	13.33
Total	**100**	**100**	**100**	**100**

Source: Author calculations based on survey results

Notes: ***, **, * denote significance level at the 1 %, 5 %, and 10 %, respectively (chi-square, two-sided). Fisher's Exact test has been used if one or more cells have an expected count ≤5. All columns add up to 100 %.

Legend: LF (Local Firms); FF (Foreign Firms); DSF (Direct Supplier Firms); AF (All Firms)

[23] The rate of the firms whose main customers are at domestic (DAMM: 45.96 % plus DS: 12.42 %) given in Table 4.20 (58.38 %) are very close to the rate of the supplier firms making production for the domestic markets given in Table 4.10 (58.18 %).

abroad and *in Turkey*. This is followed by *abroad* (19 %) and *in Turkey* (13 %), respectively. Furthermore, foreign firms have higher rates (27 %) than other firms among the firms which have specified that their main competitors are *abroad*, while the rate of the local firms is higher (15 %) than the other firms among the firms which have specified that their competitors are *in Turkey*.

A significant difference has been found between local and foreign firms in terms of *abroad* (at 10 % level). A significant difference has also been detected between the direct suppliers and non-direct suppliers in terms of *abroad* (at 10 % level) and *in Turkey* (at 10 % level) (chi-square test).

4.3.3 Business Environment Related to Competition and Relationship

The firms were asked to evaluate to what extent they agree with three items concerning the business environment on a five-point Likert scale ranging from 1 (strongly disagree) to 5 (strongly agree). These three items are *domestic competition is intense*, *global competition is intense* and *relationships between customers and suppliers are strong*. When the items are examined individually, the median of each is obtained as 4, the lowest mean as 3.57 and the highest mean 4.33. Therefore, those evaluated at the level of 4 or 5 were handled in *agree* category (4≥), those evaluated at the level of 3 in *neither agree nor disagree* category (=3), and those evaluated at the level of 1 or 2 were handled in *disagree* category. In Table 4.12, the distribution of the firms whose answers for the items were in *agree* category (4≥) was examined.

Each firm group's point of view for the intensity of the global and domestic competition corresponds with each other. Most of the firms, which specified that the *global competition is intense,* rank first (90 %). The rate of the firms who agree that *domestic competition is intense* is 71 % with the second rank. These findings match with the section in which we have examined where the main competitors of the

Table 4.12 Perception of competition and relationship

		LF	FF	DSF	AF	LF	FF	DSF
Agree		%	%	%	%	*N*	*N*	*N*
Global competition is intense	4≥	**85.71**	**100**	**92.42**	89.63	119	45	132
Domestic competition is intense	4≥	72.27*	65.91*	68.94	70.55	119	44	132
Relationships between customers and suppliers are strong	4≥	57.98	57.78	60.31	57.93	119	45	131

Source: Author calculations based on survey results

Notes: ***, **, * denote significance level at the 1 %, 5 %, and 10 %, respectively (Mann–Whitney *U* test, two-sided). The items do not add up to 100 % because the questionnaire included further categories called *1—strongly disagree, 2—disagree* and *3—neither agree nor disagree*, which have not been identified in the table. 4≥: Responses indicating the degree of agreeing as *Agree* (4) or *Strongly Agree* (5) are presented here.

Legend: LF (Local Firms); FF (Foreign Firms); DSF (Direct Supplier Firms); AF (All Firms)

firms are located (Sect. 4.3.2). At the same time, mostly the local firms (72 %)/(the least foreign firms, 66 %) agree that domestic competition is intense whereas mostly the foreign firms (100 %)/(the least local firms, 86 %) agree that global competition is intense. The most important reason for this is that most of the foreign firms are export-oriented[24] (see Sect. 4.1.5) and make their production for global markets which are generally more competitive. The fact that foreign firms regard the domestic market less competitive means they consider less the domestic firms as their competitors. This may be explained by the fact that foreign firms are larger scale firms (number of employees, total sales, etc.) (see Sects. 4.1.3 and 4.1.4) and have superior capabilities in terms of technology (see Sects. 4.1.6 and 4.4) compared to local firms.

The rate of local and foreign firms which agree that *the relations between the customers and suppliers are strong* is 58 %; the rate of the direct suppliers for the same is 60 %. That the direct suppliers have a higher rate for this item indicates that they have more extensive cooperation with the AMMs as to the product, production process, training and similar activities.

A significant difference has only been found between local and foreign firms in terms of the item *domestic competition is intense* (at 10 % level). However, between the direct suppliers and non-direct suppliers, no significant difference has been found with regard to the items between two groups (Mann–Whitney-*U* test).

4.4 Technology Capabilities Related to Production and Design

4.4.1 Production Capabilities

The firms were asked to evaluate their own capabilities for production activities when compared with other rival or leader firms in the industry on a five-point Likert scale ranging from 1 (low) to 5 (high). When the items are examined individually, the median of nine out of ten items is obtained as 4 and the lowest mean as 3.29 (*co-designer capability*). Therefore, the items evaluated at the level of 4 or 5 were handled in *high* category (4≥), the items evaluated at the level of 3 in *medium* category (=3), and the items evaluated at the level of 1 or 2 were handled in *low* category. Table 4.13 shows the distribution of the firms that specified the level of their capabilities in *high* category (4≥).

On-time delivery by 93 % is the field in which the firms regard themselves as the most capable, and it is followed by *quality control capability* (87 %) and *testing and analytical capability* (87 %). The field in which the firms consider themselves as the

[24] The foreign firms carry out almost half of their production abroad whereas, local firms export two fifth of the production to the markets abroad (see Tables 4.3, 4.10, and 4.20).

Table 4.13 Distribution of surveyed firms by production capabilities

High		LF %	FF %	DSF %	AF %	LF *N*	FF *N*	DSF *N*
On-time delivery	4≥	**92.50**	**93.33**	**95.45***	92.73	120	45	132
Quality control capability	4≥	86.67	88.89	89.39	87.27	120	45	132
Testing and analytical capability	4≥	85.83	88.89	87.88	86.67	120	45	132
Design capability	4≥	70	68.89	71.97	69.70	120	45	132
Product improvement capability	4≥	62.50	68.89	65.91	64.24	120	45	132
Expertness on CAD-CAM-CAE[a]	4≥	64.17	57.78	65.15*	62.42	120	45	132
R&D capability	4≥	58.33*	64.44*	63.64	60	120	45	132
Automation level in production process	4≥	52.50	60	53.79	54.55	120	45	132
Co-designer capability	4≥	51.67	57.78	56.82*	53.33	120	45	132
Reach to lower prices	4≥	52.50	46.67	51.52	50.91	120	45	132

Source: Author calculations based on survey results
Notes: ***, **, * denote significance level at the 1 %, 5 %, and 10 %, respectively (Mann–Whitney *U* test, two-sided). The items do not add up to 100 % because the questionnaire included further categories from 1 to 3 (1 and 2 = *Low Capability*, 3 = *Middle Capability*), which have not been identified in the table. 4≥: Responses indicating the technological degree of capabilities as *High* are presented here.
Legend: LF (Local Firms); FF (Foreign Firms); DSF (Direct Supplier Firms); AF (All Firms)
[a]*CAD* computer aided design, *CAM* computer aided manufacturing, *CAE* computer aided engineering

least capable is *reach to lower prices* with 51 %. When these items are examined in terms of three groups of firms, the ranking of the firms as to the items are almost similar. Yet the capability level of the firms for each item is different. It can be observed that local firms are more capable in *reach to lower prices*, foreign firms in *testing and analytical capability*, *product improvement capability*, *R&D capability*, *automation level in production process*, *co-designer capability* items and direct suppliers in *on-time delivery*, *quality control capability*, *design capability*, *expertness on CAD-CAM-CAE* items. While the foreign firms consider themselves more capable in fields requiring higher technology, more specialization and knowledge, the direct suppliers consider themselves more capable as to the features required by the AMMs for the suppliers (quality control, on-time delivery, design and expertness on various engineering software).

A significant difference has been found between local and foreign firms in terms of *R&D capability* item (at 10 % level). A significant difference between the direct suppliers and non-direct suppliers has also been detected in terms of *on-time delivery*, *expertness on CAD-CAM-CAE*, and *co-designer capability* items (at 10 % level for each item) (Mann–Whitney *U* test).

4.4.2 International Quality Certificates

Table 4.14 presents the distribution of the firms according to international quality certifications. 82 % of the firms have *ISO/TS16949*[25] quality certificate (93 % for FF, 84 % for DSF and 77 % for LF). This quality certificate which was developed by the major manufacturers (BMW, DaimlerChrysler, Fiat, Ford, GM, PSA) and national trade unions (USA, Germany, England, Italy, France, Japan) in global automotive industry specifies the conditions required for the quality systems to be owned by the global suppliers. The fact that a very high number of the firms (82 %) have this certificate shows that the surveyed supplier firms carry out production for AMMs at global level and possess high-quality capabilities. This quality certificate is followed by *ISO9001* (quality management system) with 60 % and *ISO14001* (environmental management systems) with 53 %, respectively. The certificate which is least owned by the firms is *ISO9002* (quality assurance in production, installation and servicing) with 9 % (10 % for LF, 5 % for FF and 8 % for DSF).

Table 4.14 Distribution of surveyed firms by quality certifications

	LF	FF	DSF	AF
	%	%	%	%
ISO/TS 16949	**77.31****	**93.18****	**84.85*****	81.60
ISO 9001	63.87*	50*	58.33	60.12
ISO 14001	47.06***	70.45***	56.82**	53.37
Specific Certificates of AMM	39.50	40.91	49.24***	39.88
Q1-FORD	14.29	18.18	18.94	15.34
ISO 18000	12.61**	27.27**	16.67	16.56
ISO 14000	13.45	15.91	14.39	14.11
QS 9000	10.08**	22.73**	15.91**	13.50
ISO 9002	10.08	4.55	8.33	8.59
Others[a]	19.33	6.82	6.82	15.95
N	**119**	**44**	**132**	**163**

Source: Author calculations based on survey results

Notes: ***, **, * denote significance level at the 1 %, 5 %, and 10 %, respectively (chi-square, two-sided). Fisher's Exact test has been used if one or more cells have an expected count ≤5.

Legend: LF (Local Firms); FF (Foreign Firms); DSF (Direct Supplier Firms); AF (All Firms); AMM (Automotive Main Manufacturers)

[a]TS8646, ISO18001, QONE, TAI, TÜV CERT, GIO, ABS, RINA, DNV, Valeobin, SQBP, IRIS, Ekotoks, CE, ISO10002, Lloyds register, Germanischer Lloyd, OHSAS

[25] "ISO/TS16949 (quality management systems for automotive production and relevant service part organizations) has been used by the major automotive manufacturers to approve more than 35,000 organizations worldwide that produce and supply parts for the industry" (www.iso.org). "It has been developed by the industry, the International Automotive Task Force (IATF), to encourage improvement in both the supply chain and the certification process. In fact, for the majority of leading vehicle manufacturers, certification to this specification is a mandatory requirement for doing business" (www.bsigroup.com).

The firms have also certificates specific to the AMMs (49 % for DSF, 41 % for FF, and 40 % for LF). That the direct suppliers have these certificates a lot more than (49 %) the other firms is the result of being the direct suppliers of AMMs, and at the same time it indicates that they not only meet the requirements of these manufacturers but also have high quality levels. One of the most important certificates in this respect is the *Q1 certificate of Ford* (Q1) which is one of the MNCs. 19 % of the direct suppliers stated to have this certificate.

A significant difference has been found between local and foreign firms in terms of *ISO/TS16949* (at 5 % level), *ISO 9001* (at 10 % level), *ISO14001* (at 1 % level), *ISO18000* (at 5 % level), and *QS9000* (at 5 % level). A significant difference has also been detected between the direct suppliers and non-direct suppliers in terms of *ISO/TS16949* (at 1 % level), *ISO14001* (at 5 % level), *specific certificates of AMM* (at 1 % level), and *QS9000* (at 5 % level) (chi-square test).

4.4.3 Technology Agreements

Table 4.15 presents the distribution of the firms that perform technology agreements with other firms so as to develop KTTs. Accordingly, it can be observed that more than one technology agreement is performed by minimum 2 % and maximum 45 % of the firms. When we take a look at the distribution of 146 firms claiming to

Table 4.15 Distribution of surveyed firms by technology agreements performed

	LF	FF	DSF	AF
Technology agreements	%	%	%	%
Turn-key projects	**45.10**	43.18	**42.75***	44.52
Purchasing of engineering services	39.22**	**50****	41.99**	42.47
Agreements on the determination of product design and/or technical specifications	36.27*	45.45*	35.11	39.04
Technical assistance agreement	29.41*	38.64*	29.01	32.19
Licensing agreement	24.51***	45.45***	32.06***	30.82
Purchasing of license and/or patent	27.45	31.82	25.95	28.77
Joint venture agreement	20.59***	43.18***	29.77***	27.40
Know-how agreement	22.55	29.55	23.66	24.66
Management contracts	21.57	15.91	16.03	19.86
International subcontracting agreements	9.80	11.36	8.40	10.27
Agreements on personnel exchange	0.98*	4.55*	2.29	2.05
Others[a]	4.90	0	2.29	3.42
N	**102**	**44**	**131**	**146**

Source: Author calculations based on survey results
Notes: ***, **, * denote significance level at the 1 %, 5 %, and 10 %, respectively (chi-square, two-sided). Fisher's Exact test has been used if one or more cells have an expected count ≤5.
Legend: LF (Local Firms); FF (Foreign Firms); DSF (Direct Supplier Firms); AF (All Firms)
[a]Confidentiality and various commercial agreements

perform at least one agreement[26]; *turn-key projects* ranks first with 45 %, and it is followed by the agreement *purchasing of engineering services* with 42 %, and the agreement *on the determination of product design and/or technical specifications* with %39. The agreements which are least performed by the firms are *international subcontracting agreements* (10 %) and *agreements on personnel exchange* (2 %).

While the foreign firms mostly perform *purchasing of engineering services* agreements (50 %), local firms (45 %) and direct suppliers (43 %) perform *turn-key projects* agreements. In addition, the foreign firms have higher rates than the other firms for the 9 out of 11 agreement types (excluding *turn-key projects* and *management contracts*). In particular, *agreements on personnel exchange, joint venture, know-how,* and *technical assistance* concerning the knowledge and technology acquisition and development are performed more by the foreign firms.

A significant difference between local and foreign firms has been detected in terms of *purchasing of engineering services* (at 5 % level), *agreements on the determination of product design and/or technical specifications* (at 10 % level), *technical assistance agreement* (at 10 % level), *licensing agreement* (at 1 % level), *joint-venture agreement* (at 1 % level), and *agreements on personnel exchange* (at 10 % level). A significant difference between the direct suppliers and non-direct suppliers has also been observed in terms of *turn-key projects* (at 10 % level), *purchasing of engineering services* (at 5 % level), *licensing agreement* (at 1 % level), and *joint-venture agreement* (at 1 % level) (chi-square test).

4.4.4 Sources of the Technologies Acquired by Firms

Table 4.16 shows the sources of the technologies acquired and/or used by the firms. Accordingly, the firms acquire technology from more than one source and the most important three sources are *domestic suppliers* (DS) with 50 %, *overseas suppliers* (OS) with 49 % and *domestic automotive main manufacturers* (DAMM) with 47 %, respectively. *Overseas automotive main manufacturers* (OAMM), however, is in the fourth rank with 36 %. These are followed by *private engineering and consultancy firms* (33 %) and *parent company* (PC)[27] (27 %), respectively. The firms specified *universities* (25 %) in the last rank as the source of the technologies acquired.

[26] Nineteen out of 165 surveyed firms stated that they did not make any agreements; therefore the number of the examined firms is 146.

[27] These are the firms located at domestic or overseas (abroad), and they directly or indirectly control all and any commercial activities, management and/or capital structure of the supplier firms. A significant number of the surveyed foreign suppliers consist of the subsidiaries (affiliates) of another foreign parent company located overseas. Local suppliers in this category are generally one of the group companies affiliated to a holding (see Sect. 3.7.7).

Table 4.16 Distribution of surveyed firms by technology sources

Technology sources	LF	FF	DSF	AF
	%	%	%	%
DS	**57.5*****	28.89***	49.24	49.70
OS	51.67	40	43.18***	48.48
DAMM	50	37.78	**52.27*****	46.67
OAMM	35.83	37.78	36.36	36.36
Private engineering and consultancy firms	34.17	31.11	37.12**	33.33
PC	10.83***	**68.89*****	29.55*	26.67
Universities	24.17	28.89	25	25.45
N	**120**	**45**	**132**	**165**

Source: Author calculations based on survey results
Notes: ***, **, * denote significance level at the 1 %, 5 %, and 10 %, respectively (chi-square, two-sided). Fisher's Exact test has been used if one or more cells have an expected count ≤ 5.
Legend: LF (Local Firms); FF (Foreign Firms); DSF (Direct Supplier Firms); AF (All Firms); DS (Domestic Suppliers); OS (Overseas Suppliers); DAMM (Domestic Automotive Main Manufacturers); OAMM (Overseas Automotive Main Manufacturers); PC (Parent Company)

The major technology sources of the local, foreign and direct supplier firms are different.

- The foreign firms specified *PC* in the first rank with 69 % as the source of the technologies acquired. Other important technology sources of the foreign firms are *OS* (40 %), *DAMM* (38 %), and *OAMM* (38 %), respectively. The fact that the foreign supplier firms in our country are the affiliates (subsidiaries) of the MNCs results in the dependence on the parent companies for technology sources. That other most important technology sources of these foreign firms are *DAMM* and *OAMM* might be resulting from the fact that they are the global suppliers of the global AMMs.
- The local firms specified the sources of the technologies acquired as *DS* with 58 %, *OS* with 52 %, and *DAMM* with 50 %, respectively. It can be deducted accordingly that local firms build more cooperation with their suppliers (second or low level-tier suppliers) for technology acquisition and try to acquire these technologies particularly through the purchase of new machinery-equipment, etc.
- The direct suppliers specified *DAMM* with 52 % in the first rank as the source of the technologies they acquired (the firms which specified *DAMM* the most are direct suppliers). It can be interpreted that the cooperation between the direct suppliers and *DAMM* is high, and technology transfer is conducted. On the other hand, that the 52 % of the direct suppliers specified *DAMM* as the source of the technology they acquired does not mean the dependence of the supply industry on the external sources since the technology acquisition is low. Because an important number of the DAMM is multinational affiliates operating in Turkey (see Appendix H). The transfer of technology from these firms to the supplier firms does not mean that they have lost the monopoly of the technology they have. Furthermore, it is very significant whether direct suppliers will be able to eliminate the current dependency by using the transferred technologies yet it is out of this study's scope.

A significant difference has been found between local and foreign firms in terms of *DS* and *PC* (each highly significant at 1 % level). A significant difference has also been observed between the direct suppliers and non-direct suppliers in terms of *OS* (at 1 % level), *DAMM* (at 1 % level), *private engineering and consultancy firms* (at 5 % level), and *PC* (at 10 % level) (chi-square test).

4.4.5 Design Capabilities

Table 4.17 indicates the distribution of the firms by four statements concerning design capabilities. Accordingly, *our firm is entirely responsible for all stages of product design* condition of the firms reflects the end condition (*1-High*) in which the design capabilities of the firms are quite high or developed whereas *all technical specifications, design and quality standards of products produced are determined by customers* condition reflects the other end condition in which the design capabilities of the firms are very low or not developed (*4-Low*). We examine the design capabilities of the firms by trying to identify their degrees according to these two end points.

When the obtained results are evaluated in terms of all firms, *High (1)* condition pointing out that the design capabilities of the firms are quite developed ranks last with 30 %, and *low (4)* the condition showing that they are not developed ranks second with 52 %. Most of the firms (61 %) are in the third degree in terms of their design capabilities. As a consequence, the design capabilities of the most of the firms are low as being in the third (61 %) and fourth (52 %) degrees.

When the design capabilities are evaluated in terms of three groups of firms, similar distributions are observed. The design capabilities of the foreign and direct supplier firms are generally intense in the third and second degrees, and the design

Table 4.17 Distribution of surveyed firms by design capability

		LF	FF	DSF	AF
Design capability		%	%	%	%
Low	4. All technical specifications, design and quality standards of products produced are determined by customers	52.94	50	51.15	52.15
↓	3. Although basic designs are determined by customers, we can add details and/or make joint designing with customer (co-designer cap ability)	**61.34**	**61.36**	**62.60**	**61.35**
↓	2. Although our firm makes **all** or **most** of the designing, customer approval is necessary for final designs	50.42	54.55	53.44*	51.53
High	1. Our firm is **entirely responsible** for all stages of product design	30.25	29.55	30.53	30.06
	N	**119**	**44**	**131**	**163**

Source: Author calculations based on survey results

Notes: ***, **, * denote significance level at the 1 %, 5 %, and 10 %, respectively (chi-square, two-sided). Fisher's Exact test has been used if one or more cells have an expected count ≤5. Items are sorted according to design capability level from 4 to 1.

Legend: LF (Local Firms); FF (Foreign Firms); DSF (Direct Supplier Firms); AF (All Firms)

capabilities of the local firms are intense in the third and fourth degrees. Therefore, the design capabilities of the foreign and direct supplier firms are higher compared to the local firms.

No significant difference has been found between local and foreign firms with respect to the items. A significant difference has been observed between the direct suppliers and non-direct suppliers in terms of design capability degree two (at 10 % level) (chi-square test).

4.4.6 Co-design Activities Related to Products and Partners

105 surveyed firms (64 %) have stated to perform co-design activities with their customers as to the products. In addition to that, 86 of the firms (82 %) have stated that they have been participating in this design process since the very beginning of the work (see Table 4.18). When we evaluate in terms of three groups of firms, local firms participate less in both activities compared to other firms. It has been detected that foreign firms cooperate more with their customers with respect to the co-design activities.

No significant difference between local and foreign firms has been found with respect to the items. However, a significant difference between the direct suppliers and non-direct suppliers has been detected in terms of the items *carry out co-product design activities* (at 5 % level) and *included to this co-design process from its beginning* (at 1 % level) (chi-square test).

Partners

In Table 4.18, when we examine the partners (customers) of the firms with whom they perform co-design activities, *DAMM* ranks first with 77 %. It is followed by *OAMM* with 60 %, *DS* with 31 %, *OS* with 30 %, *private engineering and consultancy firms* with 21 % and *PC* with 18 %, respectively. The most striking finding is that the least-cooperated partner is *universities* (10 %). The data acquired hereunder match with the data obtained in the previous section (see Sect. 4.4.4) under the title of technology sources.

When the partners performing co-design activities are examined in terms of three groups of firms, similar distributions are observed. An important discovery is that foreign firms (52 %) perform higher amount of co-design activities with their parent companies compared to other firms. The foreign firms also perform high amount of co-design activities with *OAMM* (60 %) and *OS* (30 %). 87 % of the direct suppliers designated *DAMM* in the first rank as the cooperated partner. This shows that the level of cooperation between the AMMs and their suppliers pertaining to joint product development activities is high. As it might be seen in the table, another important finding is that the rate of the foreign firms who specified cooperation with the *universities* (3 %) and *private engineering and consultancy firms* (10 %) in Turkey is very low compared to both local and direct suppliers.

In summary, the most important partner of the firms with whom co-design activities are performed is generally AMMs in domestic or overseas (since *DAMM* ranks first, and it is followed by *OAMM*).

The difference between the rate of the positive answers to the questions including a statement concerning co-designer capability of the firms in Tables 4.17 and 4.18 (in order of 61.35 % as third degree in Table 4.17 and 63.64 % for carrying out co-design activities in Table 4.18) and the rate of the positive answers to the *frequently* option of the question *joint design activity related to product* in Table 4.6 (14.63 %) is very striking. There might be two significant reasons for the firms to give different answers to two similar questions.

- Firstly, the questions in Table 4.6 are for understanding to what extent the assistance provided by the customers to the supplier firms regarding the product-related KTTs concerns *joint product design.* Yet the questions in Tables 4.17 and 4.18 are for detecting the design capabilities of the supplier firms and intended for understanding the contribution of the customers (here, execution of the design entirely by the supplier firms and the determination of anything for the design by the customer are two different end points.) Participation in joint product design activity is only an intermediate stage. Therefore, the issues highlighted by two questions and the answers given might be different.

Table 4.18 Co-design activities related to products and partners of surveyed firms

	LF	FF	DSF	AF	LF	FF	DSF
	%	%	%	%	*N*	*N*	*N*
Carry out co-design activities related to products—YES (105 firms)	61.67	68.89	68.94**	63.64	120	45	132
Firm is included to this co-design process from its beginning—YES (86 firms)	78.38	90.32	86.67***	81.90	74	31	90
The Partners							
DAMM	**78.38**	**74.19**	**86.67*****	77.14			
OAMM	58.11	64.52	63.33**	60			
DS	32.43	29.03	27.78	31.43			
OS	29.73	32.26	30	30.48			
Private engineering and consultancy firms	25.68	9.68	22.22	20.95			
PC	4.05***	51.61***	21.11*	18.10			
Universities	13.51	3.23	10	10.48			
N	**74**	**31**	**90**	**105**			

Source: Author calculations based on survey results

Notes: ***, **, * denote significance level at the 1 %, 5 %, and 10 %, respectively (chi-square, two-sided). Fisher's Exact test has been used if one or more cells have an expected count ≤5. The first two items do not add up to 100 % because the questionnaire included further categories called *No*, but have not been identified in the table. However, the information on these categories can be obtained by subtracting the stated proportion from 100 %. The partners are arranged by all firms. Legend: LF (Local Firms); FF (Foreign Firms); DSF (Direct Supplier Firms); AF (All Firms); DS (Domestic Suppliers); OS (Overseas Suppliers); DAMM (Domestic Automotive Main Manufacturers); OAMM (Overseas Automotive Main Manufacturers); PC (Parent Company)

- Secondly, when the options *frequently* and *rarely* in the answers of the question *joint design activity related to product* in Table 4.6 are examined together, the rate of positive answer is about 70 %. This rate is closer to the rates in Tables 4.17 and 4.18. At the same time, if we evaluate these two questions in Table 4.6 and Tables 4.17 and 4.18 together, we can interpret the high rate of positive answers in the last two tables as many firms' failure to conduct the co-design activities with the customers regularly. Due to these two reasons, we do not think that this means the firms gave different answers to two similar questions.

A highly significant difference has been detected between local and foreign firms in terms of the item *PC* (at 1 % level). A significant difference has also been detected between the direct suppliers and non-direct suppliers in terms of *DAMM* (at 1 % level), *OAMM* (at 5 % level), and *PC* (at 10 % level) (chi-square test).

4.5 Supply Chain

4.5.1 Management Structure and MNCs Among Clients

The answers given by the firms to three different questions concerning management structures are presented in Table 4.19. Accordingly, 77 % of the foreign firms are a part of the MNCs, and they are the subsidiaries (affiliates) of any overseas (84 %) and domestic (39 %) firm/group/holding, respectively. That the rate of being a

Table 4.19 Distribution of surveyed firms by management structure and MNCs among clients

	LF	FF	DSF	AF	LF	FF	DSF
YES	%	%	%	%	*N*	*N*	*N*
Are there MNCs among your customers?	**83.19***	**93.18***	**87.02**	85.89	119	44	131
Is your firm a subsidiary (affiliate) firm dependent on any domestic firm/group/holding?	35.83	38.64	41.99***	36.59	120	44	131
Is your firm a subsidiary (affiliate) firm dependent on any overseas firm/group/holding?	0***	84.09***	25.19*	22.56	120	44	131
Is your firm part of a MNC?	0***	76.74***	23.08	20.86	120	43	130

Source: Author calculations based on survey results

Notes: ***, **, * denote significance level at the 1 %, 5 %, and 10 %, respectively (chi-square, two-sided). Fisher's Exact test has been used if one or more cells have an expected count ≤5. The items do not add up to 100 % because the questionnaire included a further category called *No*, but have not been identified in the table. However, the information on these categories can be obtained by subtracting the stated proportion from 100 %.

Legend: LF (Local Firms); FF (Foreign Firms); DSF (Direct Supplier Firms); AF (All Firms)

subsidiary of an overseas and a domestic firm/group/holding exceed 100 % for the foreign firms is because some foreign firms are "joint-ventures", and therefore dependent both on domestic and overseas parent companies. This finding confirms our expectations that the majority of the foreign firms in the Turkish automotive industry are the subsidiaries of MNCs. On the other hand, only 36 % of the local firms are the subsidiaries of a domestic firm/group/holding. When we evaluate the table in terms of direct suppliers, it is observed that 42 % and 25 % of the firms are the subsidiaries of the domestic and overseas firms/groups/holdings, respectively; and 23 % are a part of the MNCs. Moreover, according to Table 4.19, 140 of the firms (86 %) stated that there are MNCs among their clients (93 % for FF, 83 % for LF and 87 % for DSF) (significant at 10 % level between local and foreign firms). This situation is very important because it indicates that survey firms are a part of global production networks and also very capable to produce for MNCs.

A highly significant difference has been found between local and foreign firms in terms of being a part of MNC and being a subsidiary firm dependent on an overseas firm/group/holding (at 1 % level). A significant difference has also been found between the direct suppliers and non-direct suppliers in terms of being a subsidiary firm dependent on any domestic (at 1 % level) and overseas (at 10 % level) firm/group/holding (chi-square test).

4.5.2 Main Customers of Surveyed Firms

Table 4.20 displays the distribution of surveyed firms according to their main customers. In the calculation of main customers, distribution of each firm's total sales according to their customers was analyzed one by one, and we defined the customer as a main customer if firms carry out the highest share of their total sales to this customer. According to this, 133 firms (83 %) carry out more than 50 % of their total sales to only one customer, while the rest 28 firms (17 %) carry out most of their total sales to only one customer in various rates between minimum 30 % and maximum 45 %.

Table 4.20 presents that main customers of surveyed firms are generally AMMs, 65 % of the firms sell their products mainly to *DAMM* (46 %) and *OAMM* (19 %), respectively. The next main customers of the firms are higher level-tier suppliers[28] in automotive industry, 23 % of the firms sell their products mainly to *DS* (12 %) and *OS* (11 %). Therefore, the main customers of the firms are *DAMM*, *OAMM*, *DS*, *OS*, and *PC*, respectively.

[28] Higher level-tier suppliers can be DS or OS. As explained in Sect. 3.7, these firms are the suppliers of AMMs and operate in the automotive industry. If a surveyed firm's main customer is DS or OS, this means that the surveyed firm is a low-tier supplier depending on the tier of this DS or OS. For instance, if DS is a 1st-tier supplier of any automotive manufacturer, then the supplier of this DS is 2nd-tier supplier.

Table 4.20 Distribution of surveyed firms by main customers (%)

		LF	FF	DSF	AF
Main Customers	N	**117**	**44**	**129**	**161**
DAMM		**47.86**	**40.91**	**54.26**	45.96
OAMM		11.97	36.36	18.60	18.63
DS		13.68	9.09	8.53	12.42
OS		12.82	4.55	9.30	10.56
PC		2.56	4.55	3.10	3.11
OC †		11.11	4.55	6.20	9.32
	Total	**100**	**100**	**100**	**100**

Source: Author calculations based on survey results

Legend: LF (Local Firms); FF (Foreign Firms); DSF (Direct Supplier Firms); AF (All Firms); DAMM (Domestic Automotive Main Manufacturers); OAMM (Overseas Automotive Main Manufacturers); DS (Domestic Suppliers); OS (Overseas Suppliers); PC (Parent Company); OC (Other Customers)

†Other affiliates of parent company and retailers.

Main customers of the 60 % of the local firms (117 firms) are AMMs (48 % for *DAMM*, 12 % for *OAMM*), and 27 % are other higher level-tier suppliers (14 % for *DS* and 13 % for *OS*). On the other hand, 77 % of the foreign firms (44 firms) carry out most of their sales to AMMs (41 % for *DAMM*, 36 % for *OAMM*), and 14 % to other suppliers (9 % for *DS* and 5 % for *OS*). Therefore, foreign firms carry out their production mainly to AMMs more than both direct suppliers (73 %) and local counterparts (60 %) do. This situation can indicate that foreign firms are the global suppliers of AMMs, and they are more capable than both the direct suppliers and local counterparts to produce for these firms. In addition, main customers of 36 % of foreign firms are *OAMM*, so that this confirms our comments and also corresponds to our findings related to foreign firms' being more export-oriented because it is assumed that technological capability is very important to produce for AMMs, and especially for *OAMM* since global markets are more competitive (see Sects. 4.1.5, 4.1.6, 4.3, 4.4, and 4.8.3). In terms of direct suppliers (129 firms), main customers of 73 % are AMM as expected (54 % for *DAMM* and 19 % for *OAMM*), and main customers of 18 % are other higher level-tier suppliers (9 % for OS and 9 % for DS). On the other hand, parent companies are the main customers of 3 % of the firms (5 % for FF, and 3 % for both DSF and LF).

When we evaluate the first four groups of main customers (*DAMM, OAMM, DS,* and *OS*) together as a main customer and called this group *automotive industry*, we find out that 91 % of foreign firms and 87 % of local firms sell mainly to this industry. In summary, these statistics show that major customers of the firms are in the automotive industry. Furthermore, this confirms our expectations related to the analysis that major customers of the firms are AMMs, in other words, these findings indicate that most of the firms in the sample are first-tier suppliers (or direct suppliers) of AMMs since the majority of the firms carry out their total sales only to one customer. The share of the firms whose main customers are *OAMM* and *OS* is 29 % in all firms, and this rate is not so much different from the firm rate exporting more than 50 % of their sales (30 %) in Table 4.3.

In addition, the statistics related to the main customers of the firms according to their origin (Table 4.21) and ownership status (Table 4.22) can also be seen in the above Tables. The most striking finding in terms of foreign ownership status is that major customers of the full foreign ownership firms are *OAMM*, whereas those of joint venture ownership firms are *DAMM*.

Table 4.21 Main customers of surveyed firms by origin of foreign capital

	Origin									
Main Customers	Local	Germany	USA	France	Japan	Italy	Spain	Other[a]	*N*	%
DAMM	56	4	1	2	2	2	2	5	74	46
OAMM	14	5	4	2	1	1	1	2	30	19
DS	16	1	0	0	1	1	0	1	20	12
OS	15	0	0	1	0	0	0	1	17	11
PC	3	2	0	0	0	0	0	0	5	3
OC[b]	13	0	1	0	0	0	0	1	15	9
N	117	12	6	5	4	4	3	10	**161**	
%	72.67	7.45	3.73	3.11	2.48	2.48	1.86	6.21		**100**

Source: Author calculations based on survey results

Legend: DAMM (Domestic Automotive Main Manufacturers); OAMM (Overseas Automotive Main Manufacturers); DS (Domestic Suppliers); OS (Overseas Suppliers); PC (Parent Company); OC (Other Customers)

[a]England (2), Belgium (2), S. Korea, Taiwan, India, Hungary, Finland and Switzerland

[b]Other affiliates of parent company and retailers

Table 4.22 Main customers of surveyed firms by foreign ownership

	Ownership status						
Main customers	Local	Minority ownership	Joint venture	Majority ownership	Full ownership	*N*	%
DAMM	56	1	12	1	4	74	46
OAMM	14	0	3	2	11	30	19
DS	16	2	0	0	2	20	12
OS	15	0	0	1	1	17	11
PC	3	0	0	0	2	5	3
OC[a]	13	0	1	0	1	15	9
N	117	3	16	4	21	**161**	
%	72.67	1.86	9.94	2.48	13.04		**100**

Source: Author calculations based on survey results

Legend: DAMM (Domestic Automotive Main Manufacturers); OAMM (Overseas Automotive Main Manufacturers); DS (Domestic Suppliers); OS (Overseas Suppliers); PC (Parent Company); OC (Other Customers)

[a]Other affiliates of parent company and retailers

4.5.3 Primary Sources of the Inputs

The firms were asked to prioritize the primary sources of the inputs (raw material and/or intermediate good)[29] used in the production of the major products (1 > 2 > 3 > 4 > 5). Accordingly, the primary input sources of the surveyed firms according to the first three degrees of importance are shown in Table 4.23.

At each degree, almost the same distributions are observed in terms of three groups of firms. When the first degree sources are examined, *DS* ranks first with 57 % and *OS* with 31 % ranks second. This indicates that the most important input sources of 88 % of the surveyed firms are other low-tier supplier firms operating in domestic and overseas. In addition to that, while 10 % of the local firms specified *DAMM* in the first rank as the primary source, the 9 % of the foreign firms designated *PC* for the same. There was no firm to specify *OAMM* firms as the primary input source. When the second degree sources are examined, *OS* ranks first with 51 % and *DS* with 33 % ranks second. The total amount of both input sources is 84 %, and the most important input sources in the second degree are domestic and overseas supplier firms.[30] An important detection here is that 7 % of the firms (each being nearly 7 % for three group of firms) specified *OAMM* as the most important input source.

In sum, the most important two input sources of the firms are *DS* and *OS*, and as for the third source, these two are followed by *DAMM* for the local firms and *PC* for the foreign firms. That the *DAMM* provided input for the supplier firms might be pointing out a cooperation covering the KTT between these firms. However, the main purpose of this input transfer might rather be non-interruption of routine production activities and it may not include any significant knowledge and technology.

A significant difference has been observed between local and foreign firms in terms of *PC* (at 1 % level), *OAMM* (at 10 % level), and *other* (at 1 % level). There is also a significant difference between the direct suppliers and non-direct suppliers in terms of *DS* (at 10 % level), *DAMM* (at 5 % level), *PC* (at 5 % level), *OAMM* (at 5 % level), and *other* (at 1 % level) (Mann–Whitney *U* test).

Figure 4.1 below displays the distribution of the primary input sources of the surveyed firms based on their main customers (see Sects. 4.5.2 and 4.5.3). According to this, major input source of the firms is *domestic suppliers* (DS) regardless of their main customers. Nearly 60 % of the firms receive their inputs from *DS*, and it is followed by *overseas suppliers* (OS). The firms whose main customers are *OS* use only two input sources: *DS* (59 %) and *OS* (41 %), respectively. Moreover, it is seen that firms also receive their inputs from *DAMM*. The 11 % of the firms whose main customers are *DAMM* receive their inputs from

[29] Raw material: Rubber, plastic, sponge, steel, aluminium, etc.; Intermediate goods: Various parts and components which are previously produced and used in the production of another good.

[30] This situation shows us that the firms that chose DS (OS) in the first degree chose OS (DS) in the second degree.

Table 4.23 Distribution of the surveyed firms according to primary sources of inputs by importance degree (%)

	Degree of Importance											
	1st				2nd				3rd			
Input	**LF**	**FF**	**DSF**	**AF**	**LF**	**FF**	**DSF**	**AF**	**LF**	**FF**	**DSF**	**AF**
Sources N	**116**	**44**	**126**	**160**	**98**	**44**	**117**	**142**	**39**	**29**	**61**	**68**
DS	**56.03**	**59.09**	**56.35***	**56.88**	34.69	29.55	34.19*	33.10	7.69	10.34	8.20*	8.82
OS	31.03	29.55	30.95	30.63	**51.02**	**50**	**49.57**	**50.70**	12.82	10.34	11.48	11.76
DAMM	10.34	2.27	7.94**	8.13	5.10	0	4.27**	3.52	**33.33**	13.79	26.23**	25
PC	0***	9.09***	3.17**	2.50	2.04***	11.36***	4.27**	4.93	15.38***	**51.72*****	**34.43****	**30.88**
OAMM	0*	0*	0**	0	7.14*	6.82*	6.84**	7.04	5.13*	0*	3.28**	2.94
Other †	1.72***	0***	1.59***	1.25	0***	2.27***	0.85***	0.70	20.51***	13.79***	16.39***	17.65
Total	**100**	**100**	**100**	**100**	**100**	**100**	**100**	**100**	**100**	**100**	**100**	**100**

Source: Author calculations based on survey results

Notes: ***, **, * denote significance level at the 1%, 5%,and 10%, respectively (Mann-Whitney U test, 2-sided). Legend: LF (Local Firms); FF (Foreign Firms); DSF (Direct Supplier Firms); AF (All Firms); DS (Domestic Suppliers); OS (Overseas Suppliers); DAMM (Domestic Automotive Main Manufacturers); PC (Parent Company); OAMM (Overseas Automotive Main Manufacturers)

†Other affiliates of parent company, licensor firm and various machine manufacturers.

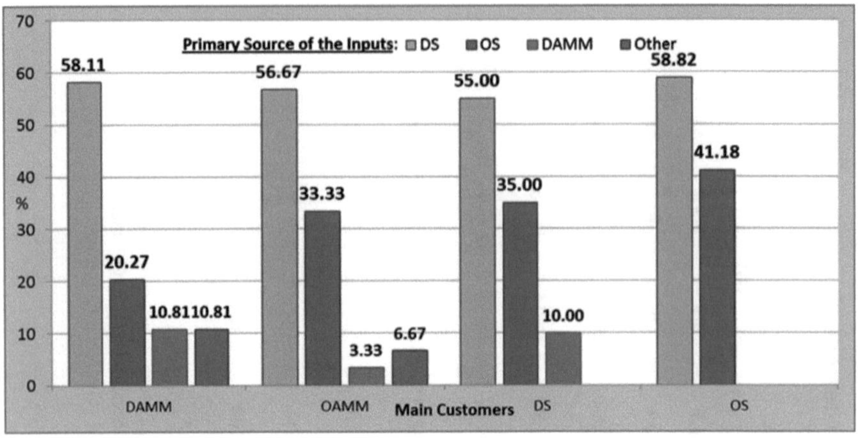

Fig. 4.1 Primary input sources of the suppliers based on main customers. *Source*: Author calculations based on survey results. *Legend*: *DAMM* domestic automotive main manufacturers, *OAMM* overseas automotive main manufacturers, *DS* domestic suppliers, *OS* overseas suppliers, *Other* parent company and other firms

DAMM. This rate is also 10 % for the firms whose main customers are *DS*. In summary, these show that the primary input sources of the firms are *DS*, *OS*, and *DAMM*, respectively. Moreover, it also reveals that the firms whose major customers are DAMM (11 %), DS (10 %), and OAMM (3 %) respectively are being supplied by DAMM.

4.5.4 Purchasing Strategies of the Customers (Being the Sole Supplier of the Customer)

The firms were asked to specify the purchasing strategies of their most important customers (being mainly AMMs, see Sect. 4.5.2). They were asked if the customers preferred to purchase each item from only one supplier or from more than one supplier. According to Table 4.24, 49 % of the foreign firms and 38 % of the local firms stated that their customers preferred to work with only one supplier (not significant, chi-square test). These firms thus become the sole suppliers of their most important customers, as well.

The customers attach so much importance to the production and technological capabilities of their suppliers (manufacturing, quality, engineering, design, just-in-time, defect rate, cost, etc.). Therefore it is very important to be the sole supplier of the customers in one product group when it is considered that a slightest fault in the supply process will cause the production line to come to a halt. In terms of single supplier firms, being the sole supplier of the customer in one group of product also shows that:

Table 4.24 Some indicators of surveyed firms in terms of their customers

	LF	FF	DSF	AF	LF N	FF N	DSF N
Customer prefers to study with only one supplier per item purchased (%)	38.46	48.89	43.85	41.36	117	45	130
	Mean						
Duration of working together with the most-intensive cooperated customer (year)	18.12	15.80	18.59	17.48	118	45	131
Share of Subcontracting Contracts in Total Contracts (%)	13.17	9.45	12.16	12.17	113	42	125

Source: Author calculations based on survey results
Notes: ***, **, * denote significance level at the 1 %, 5 %, and 10 %, respectively (chi-square, two-sided for 1st item, and two independent samples *t*-test, two-sided for 2nd and 3rd items). Fisher's Exact test has been used if one or more cells have an expected count ≤ 5. The first item does not add up to 100 % because the questionnaire included a further category called *customer prefers to study with many suppliers per item purchased*, but has not been identified in the table. However, information on this category can be obtained by subtracting the stated proportion from 100 %. The min. and max. values for the 2nd item are 1 and 55, and they are 0 % and 100 % for 3rd item.
Legend: LF (Local Firms); FF (Foreign Firms); DSF (Direct Supplier Firms); AF (All Firms)

- The level of cooperation with the customers for the products is very high,
- The supplier firm makes high technology production in the field and has high technology capabilities,
- When the number of supplier firms operating in domestic and overseas markets and high level of competition among these is taken into account, the supplier firm has superiority as to the product or there is no any alternative for this product in the market.

Working with only one supplier might have the following reasons in terms of the customers:

- The desire to procure the product from the cooperated supplier firm which was also assisted in the product development,
- Purchasing the product with lower prices by means of procuring it from only one supplier instead of purchasing it from more than one supplier firms (reduction in the supplier's costs due to the economies of scale),
- The product to be procured requires a high cooperation with the supplier firm due to being a strategic or new product and customers' desire to keep the related technologies confidential.

Because of these reasons, it is seen very important to be the sole supplier of the customers in one particular product group.

4.6 Cooperation

4.6.1 The Motivations for Cooperation Activities with Other Firms (Why Do Suppliers Cooperate with Other Firms?)

The firms were asked to evaluate the motivations for cooperating with other firms on a five-point Likert scale ranging from 1 (very unimportant) to 5 (very important). When the items are examined individually, the median of eight out of nine items is obtained as 4 and the median of only one is 3. Furthermore, the lowest mean is 2.95 (*know-how transfer*) and the highest mean is 3.95 (*improving product quality*). Therefore, the items evaluated at the level of 4 or 5 were handled in *important* category (4≥), the items evaluated at the level of 3 in *neither important nor unimportant* category (=3), and the items evaluated at the level of 1 or 2 were handled in *unimportant* category. Table 4.25 shows the distribution of the firms whose answers for the items were in *important* category (4≥).

As can be seen in Table 4.24 in the previous section, the mean duration of working together with the most intense cooperated customer is higher for the direct supplier firms (nearly 19 years) than foreign (16 years) and local firms (18 years). This confirms that direct suppliers conducted more collaborative relationships with AMMs, and also mutual trust is very important in buyer–supplier relationships.

Table 4.25 Why do surveyed firms cooperate with other firms?

		LF	FF	DSF	AF	LF	FF	DSF
Important		%	%	%	%	*N*	*N*	*N*
Improving product quality	4≥	**79.17**	**77.78**	**79.55**	78.79	120	45	132
Learning about new technologies	4≥	75**	60**	71.97	70.91	120	45	132
Opening up global markets	4≥	73.33*	64.44*	71.97	70.91	120	45	132
Entering new technology fields	4≥	74.17*	53.33*	67.42	68.48	120	45	132
Reducing/sharing production costs/risks	4≥	65.83	53.33	63.64	62.42	120	45	132
Carrying out R&D activities	4≥	60.83	46.67	56.82	56.97	120	45	132
Establishing long-term strategic partnership	4≥	52.50	62.22	54.55	55.15	120	45	132
Replacing technologically phased out products with the new ones	4≥	54.17	44.44	53.03	51.52	120	45	132
Know-how transfer	4≥	50.83	42.22	50.76	48.48	120	45	132

Source: Author calculations based on survey results

Notes: ***, **, * denote significance level at the 1 %, 5 %, and 10 %, respectively (Mann–Whitney *U* test, two-sided). The items do not add up to 100 % because the questionnaire included further categories called *1—Very Unimportant, 2—Unimportant, 3—Neither Important Nor Unimportant*, but have not been identified in the table. 4≥: Responses indicating the degree of importance as *Important* (4) or *Very Important* (5) are presented here.

Legend: LF (Local Firms); FF (Foreign Firms); DSF (Direct Supplier Firms); AF (All Firms)

When we examine the items in terms of all firms, the first three items are considered important by at least 70 % of the firms. These are respectively *improving product quality* (79 %), *learning about new technologies* (71 %) and *opening up global markets* (71 %). Other items were evaluated as important by at least 48 % (*know-how transfer*) and maximum 68 % (*entering new technology fields*) of the firms. In addition, a similar distribution in terms of three groups of firms is observed.

One of the most important detections in terms of KTT is that although 49 % of the firms specified *know-how transfer* item as important as the motivation for cooperation, this item is the last in ranking when other items are considered.[31] This relatively low rate can be explained in two different ways:

i. The *know-how transfer* made by these firms to the supplier firms to improve the product quality was covered in the first option in Table 4.25, but not covered in the option of *know-how transfer*,
ii. If the option of *know-how transfer* in the table is associated more with the KTT for the production process, we can explain why other options in the table are dominant by taking into consideration a generally-accepted detection suggesting that production capability in Turkish automotive supply industry is more advanced compared to the design and innovation capability.

The items affecting the cooperation of the firms with the other firms could be ordered as follows:

- The design capabilities of the firms must be developed for *improving product quality* item; the fact that firms specify this in the first rank as the item affecting cooperation might be an indicator of their design capabilities.
- *Learning about new technologies* and *entering new technology fields* might be an indicator of an attempt to access the explicit/tacit technologies that cannot be sold in the domestic market or the world market or, even if it was sold, could not be used efficiently by the supplier firms due to the complexity by means of establishing relations with the AMMs.
- *Opening up global markets* and *reducing/sharing production costs/risks* might indicate that supplier firms avoid the risks of specific markets and try to benefit from the scale economies.
- *Carrying out R&D activities* is a very costly process, besides it requires knowledge. In addition to that, it may also require taking technical support from one firm or many firms since commercial and technical risks are being involved. Furthermore, when a new/improved product was released if it has a buyer, such risks will be reduced and supplier firms will be prompted to decide for implementing R&D activities with other firms. *Establishing long-term strategic partnership* target may be evaluated in the same context.

[31] In other words, the rate of the firms considering *know-how transfer* as "unimportant" (1 and 2 points) ranks first with 32 %.

- *Replacing technologically phased out products with the new ones*, on the other hand, might indicate that the firms can not anymore compete in the market with these products and look for a new/improved product.

A significant difference has been found between local and foreign firms in terms of the items *learning about new technologies* (at 5 % level), *opening up global markets* (at 10 % level), and *entering new technology fields* (at 10 % level). However, no significant difference has been found between the direct suppliers and non-direct suppliers with regard to the items (Mann–Whitney *U* test).

4.6.2 Improvements in the Production Capability as a Result of Assistance[32] Received from the Customers

In this section, improvements in the production capability of the surveyed firms as a result of the KTTs from customers are used as a proxy for the performance increases in the production capability of the supplier firms.

The firms were asked to evaluate on a five-point Likert scale ranging from 1 (very decreased) to 5 (very increased) what kind of *improvement (increase)* and/or *deterioration (decrease)* had been caused in certain fields by the assistance provided by their customers for the last three years. When the items are examined individually, the median of six out of ten items is 4, the median of the remaining is 3, the lowest mean is 2.57 (*profitability*), and the highest mean is 3.67 (*improvement in quality control methods*). Therefore, the items evaluated at the level of 4 or 5 were handled in the category of *increased* (4≥), the items evaluated at the level of 3 in the category of *neither increased nor decreased* (=3), and the items evaluated at the level of 1 or 2 were handled in the category of *decreased*. Table 4.26 shows the distribution of the firms whose answers for the items were in the category of *increased* (4≥).

The ranking concerning items in terms of three groups of firms shows similarity. The rate of the firms which specified an increase in six out of ten items is minimum 51 %. The most important item which was specified to provide increase as a result of customers' assistance is the item *improvement in quality control methods* with 73 %. It is followed by *improve production process and/or capability to find solutions for production problems* with 68 %. Other items were specified as *increased* by minimum 23 % and maximum 58 % of the firms. The two items with minimum increase are *reach to lower output prices* (38 %) and *profitability* (23 %) items. That the rate of the firms which specified that the assistance provided by the customers increased *profitability* item is in the last rank[33] may indicate that

[32] The implied meaning of the ***assistance*** refers to every help such as providing technical support, training, advice, and KTT.

[33] In other words, the rate of the firms which specified the *profitability* item as "decreased" category (1 and 2 points) ranks first with 25 %.

Table 4.26 Improvements in various fields as a result of assistance received from the customers (for the last 3-years)

Increased		LF %	FF %	DSF %	AF %	LF N	FF N	DSF N
Improvement in quality control methods	4≥	**77.5****	**60****	**71.21**	72.73	120	45	132
Improve production process and/or capability to find solutions for production problems	4≥	71.67	**60**	70.45	68.48	120	45	132
Engineering capability related to product	4≥	63.33*	44.44*	60.61**	58.18	120	45	132
Design capability (design new products and processes)	4≥	61.67*	42.22*	59.85**	56.36	120	45	132
Learning about new technologies and production processes	4≥	60**	42.22**	56.06	55.15	120	45	132
Productivity	4≥	57.5*	35.56*	53.79	51.52	120	45	132
Improvement in business management	4≥	47.50	33.33	40.15	43.64	120	45	132
Performance on export	4≥	45.83	33.33	41.67	42.42	120	45	132
Reach to lower output prices	4≥	35.83	42.22	39.39	37.58	120	45	132
Profitability	4≥	22.5*	24.44*	23.48*	23.03	120	45	132

Source: Author calculations based on survey results

Notes: ***, **, * denote significance level at the 1 %, 5 %, and 10 %, respectively (Mann–Whitney *U* test, two-sided). The items do not add up to 100 % because the questionnaire included further categories called *1—Very Decreased, 2—Decreased, 3—Neither increased Nor Decreased*, but have not been identified in the table. 4≥: Responses indicating the degree of improvements as *Increased* (4) or *Very Increased* (5) are presented here.

Legend: LF (Local Firms); FF (Foreign Firms); DSF (Direct Supplier Firms); AF (All Firms)

some firms were not able to use these assistances efficiently in the short term or had to make additional investments.

Local firms stated that greater increase was achieved in all the items except for the last two items (*reach to lower output prices* and *profitability*) compared to other firms. The reason behind this is the fact that local firms are affected more positively by these assistances rather than the other firms which are not capable of absorbing these assistances (the local firms are assumed to have less-developed technology and capability levels compared to other firms, see Sect. 4.1.6). This is also confirmed by the fact that the other two groups of firms (FF and DSF) have higher rates than the local firms in the last two items.

The rate of the firms indicating *decreased* (1 or 2 points) for the other nine items except *profitability* (25 %) item is ranging from 5 to 7 %, this is not evaluated in the table though. The negative effect of the assistance (regarding KTT) on the firms may have been a result of the subjective evaluations not reflecting the truth or imply that some supplier firms lack the capability of absorbing the assistance conveyed by the customers or the cooperated customer is quite selective when providing the assistance.

A significant difference has been detected between local and foreign firms in terms of the items *improvement in quality control methods* (at 5 % level), *engineering capability related to product* (at 10 % level), *design capability* (at 10 % level), *learning about new technologies and production processes* (at 5 % level), *productivity* (at 10 % level), and *profitability* (at 10 % level). A significant difference has also been detected between the direct suppliers and non-direct suppliers in terms of the items *engineering capability related to product* (at 5 % level), *design capability* (at 5 % level), and *profitability* (at 10 % level) (Mann–Whitney *U* test).

4.7 Direct Suppliers of the AMMs in Turkey

In this section, the relations between the AMMs in Turkey and their direct suppliers are examined. Direct suppliers of AMMs use high technology in manufacturing, and meet various quality and manufacturing standards regarding the products, and establish a closer technological cooperation with the AMMs. As stated in Sect. 4.1.1, of totally 132 (80 %) direct suppliers of one or more than one AMMs in Turkey in the sample, 93 are local (FS $<$ 10 %) and 39 are foreign firms (FS $\geq$ 10 %). This section analyses the items in terms of local and foreign direct suppliers.

4.7.1 How the Firms Became a Direct Supplier

This section presents the survey results on how certain firms became the direct suppliers of AMMs in Turkey. As shown in Table 4.27, most commonly the supplier firms attempted and asked for orders (56 %). This is followed by those that applied a long-term plan to work with AMMs (45 %). In the third place come those AMMs approached and offered a business plan (37 %). Least frequently, some firms became direct suppliers as they are referred by other customers (31 %).

The items that are ranked number one and two (Table 4.27) show the firms' own efforts towards being direct suppliers, and are complementary. On the other hand, the last two items show the effect of factors other than their own efforts. Since being the direct supplier of AMMs requires technology capabilities for the products and production process (such as manufacturing, engineering, design, creativity, and technological uniqueness), the first two items reflect that the firms lacking these properties need to develop these aspects and apply to the AMMs by their own efforts. On the other hand, the last two items point out that the firms possessing relatively more of these characteristics can more easily become direct suppliers with the help of AMMs and their business environment. This somewhat explains why the first two strategies are selected by more firms, whereas the last two strategies are selected by fewer firms (it is assumed that not all of the firms have the characteristics required for being a direct supplier). The fact that local direct suppliers have higher rates indicates that they spend greater effort than foreign ones to become direct suppliers.

A statistically significant difference has been found items between local and foreign direct suppliers as regards these two items: *our firm attempted and asked for order* (at 10 % level) and *through advices and introduces of our other customers* (at 10 % level) (chi-square test).

Table 4.27 How the firms became a direct supplier

	Local DSF %	Foreign DSF %	ALL DSF %
Our firm attempted and asked for order	**64.13***	**38.46***	56.49
Our firm applied exhaustive and long-term program to work with AMMs	47.83	**38.46**	45.04
AMM attempted and offered business plan	39.13	30.77	36.64
Through advices and introduces of our other customers	36.96*	17.95*	31.30
N	**92**	**39**	**131**

Source: Author calculations based on survey results
Notes: ***, **, * denote significance level at the 1 %, 5 %, and 10 %, respectively (chi-square, two-sided). Fisher's Exact test has been used if one or more cells have an expected count ≤5.
Legend: DSF (Direct Supplier Firms)

4.7.2 Cooperation Activities with AMMs

Table 4.28 shows the distribution of the direct suppliers who selected the response category *Yes* to seven questions pertaining to the cooperation activities with the AMMs. According to this, 86 % of the firms stated that they had to provide *detailed information to the AMMs about production and/or quality control processes* (88 % for Local DSF and 79 % for Foreign DSF). At the same time, 82 % of the firms specified that *AMM arrange regular visits to evaluate the firm about quality, cost, on-time delivery, etc.* (83 % for local DSF and 79 % for foreign DSF). The criteria AMMs set for working with suppliers are very high; furthermore, they monitor and evaluate the suppliers continuously by applying certain written procedures to prevent any interruptions in the production lines. This is also reflected on the high rates of these two items. Fewer foreign firms responded by *Yes* to these questions, for some of them are exempted from these supervision items probably because they (i) have greater technological capability, (ii) are the global suppliers of the AMMs,[34] and (iii) have been working together with the AMMs for a long time so these supervisions are not required. For instance, direct suppliers who are the global system supplier of more than one AMM under the scope of MNC (such as Bosch, Delphi, Siemens, Voleo, Magna, and Autoliv) are exempted from these supervisions.

64 % of the firms stated that *AMM sends its expert staff to the supplier firm for giving support or assistance about various topics* (67 % for Local DSF and 59 % for Foreign DSF), and 63 % *AMM transfers any knowledge and technology related to product/production process/design* (61 % for Local DSF and 67 % for Foreign DSF). A high number of *Yes* responses to both items is a sign of extensive cooperation between the firms. AMMs' sending their personnel to local direct suppliers at higher rates than foreign direct suppliers shows that, when these firms

[34] The global supplier firms and AMMs conduct extensive cooperation on strategic subjects (such as R&D, innovation, design, and quality) generally at their main headquarters instead of the host countries in which they made investment or their factories are located (see Sect. 6.2).

Table 4.28 Cooperation activities of the direct suppliers with the AMMs

YES	Local DSF %	Foreign DSF %	ALL DSF %	Local DSF *N*	Foreign DSF *N*
Does your firm have to provide detailed information to AMM(s) about production and/or quality control processes?	**88.17**	**79.49**	85.61	93	39
Does AMM arrange regular visits to evaluate your firm about quality, cost and on time delivery etc.?	82.80	**79.49**	81.82	93	39
Does AMM sent its expert staff to your firm for giving support or assistance about various topics?	66.67	58.97	64.39	93	39
Does AMM transfer any knowledge or technology related to product/production process/design?	61.29	66.67	62.88	93	39
Does AMM make regular management and/or governance meetings?	56.99	56.41	56.82	93	39
Does AMM provide regular training activities for your staff?	26.88	20.51	25.00	93	39
Did AMM give any support in the establishment process of your firm? (cooperation in capacity and production planning, in purchasing machines, in technical organization etc.)	6.67	12.82	8.53	90	39

Source: Author calculations based on survey results
Notes: ***, **, * denote significance level at the 1 %, 5 %, and 10 %, respectively (chi-square, two-sided). Fisher's Exact test has been used if one or more cells have an expected count ≤ 5. The items do not add up to 100 % because the questionnaire included a further category called *No*, but has not been identified in the table. However, the information on these categories can be obtained by subtracting the stated proportions from 100 %.
Legend: DSF (Direct Supplier Firms); AMM (Automotive Main Manufacturer)

encounter problems, they need more assistance for their solution. Greater amount of KTTs to foreign direct suppliers confirms once again that the foreign firms have higher technological capabilities. This can also be indicative of the fact that foreign firms produce high-tech products and thus, can absorb these transfers more easily.

57 % of the firms stated that *AMM make regular management and/or governance meetings*. The major aim of these meetings is to increase the coordination and cooperation between the AMMs and suppliers. The main topics in such meetings are production related problems, plans for the new models to be produced, recent developments, and training issues.

25 % of the firms stated that *AMM provide regular training activities for their staff* (27 % for Local DSF and 21 % for Foreign DSF). This rate is close to the total training given to the *production* and *management* staff, and the average of the *off-the-job training* types in the section of *frequent* transfers via training (see Sect. 4.2.4 and Table 4.8). Clearly, local direct suppliers responded with a higher number of *Yes*. This also indicates that the personnel of these firms need more training than the foreign direct suppliers.

Few firms (9 %) said *Yes* to the question *whether the AMM assisted the firm in the establishment stage*. This is because the AMMs prefer working with currently-operating firms in the market who have already proven themselves at home and overseas. When a new product not available in the market and/or within the framework of the new models to be developed is to be manufactured, AMMs prefer to have the previously-cooperated supplier firms manufacture it. In this case a new production line is set up for this product; supplier firms are assisted by AMMs in a variety of situations like production and capacity planning, and machinery procurement and technical organization since this product may involve risks and the supplier firm may not have sufficient sources and equipment. The foreign direct suppliers gave twice as many *Yes* responses (13 %) to this question as local direct suppliers (7 %). This confirms our above-mentioned claim about the characteristics of the foreign direct suppliers that they are global suppliers, produce more technologically, and have greater absorptive capacity. In order to examine any significant difference between the local and foreign direct suppliers as to these seven questions, *chi-square test* was applied, yet no significant difference was found between these two groups.

4.7.3 Benefits of Being a Direct Supplier

The firms were asked to evaluate the benefits of being a direct supplier based on their experience in this field on a five point Likert scale ranging from 1 (very unimportant) to 5 (very important). When the items are examined individually, the median of 13 out of 14 items is 4, and the median of one is 5. The lowest mean is 3.71 (*reducing risks involved with making decision to invest in new technology and/or machines*), and the highest mean is 4.47 (*good reputation and familiarness in sector*). Therefore, the items evaluated at the level of 4 or 5 were handled in *important* category (4≥), the items evaluated at the level of 3 in *neither important nor unimportant* category (=3), and the items evaluated at the level of 1 or 2 were handled in *unimportant* category. In Table 4.29, the distributions of the firms whose answers for the items were in *important* category (4≥) are examined.

As can be seen here, the lowest rating these items received is 67 % (*reducing risks involved with making decision to invest in new technology and/or machines*) and the highest is 96 % (*good reputation and familiarness in sector*). The first four items which are regarded as important by at least 90 % of the firms point out the indirect/external benefits rather than the direct benefits (positive effects) of the AMMs on the suppliers. It can thus be concluded that being a successful direct supplier functions also as a kind of *social capital*. Probably, it also has positive effects on not only sustaining the activities in automotive supply industry (market share) but also taking technical/financial support from AMMs. On the other hand, ten items which are regarded as important with ratings between 67 and 87 % reflect the direct benefits gained by the firms with respect to the developments in test, analysis, engineering, design, quality control, and manufacturing. In other words,

Table 4.29 Benefits of being a direct supplier of the AMM in Turkey

Important		Local DSF %	Foreign DSF %	ALL DSF %	Local DSF *N*	Foreign DSF *N*
Good reputation and familiarness in sector	4≥	**97.85****	**92.31****	96.21	93	39
Developing new business relationships with other firms	4≥	96.77***	82.05***	92.42	93	39
Stable trade	4≥	91.40	92.31	91.67	93	39
Better understand the customers' blueprint and technical specifications	4≥	90.32*	92.31*	90.91	93	39
Improvement in testing and analyzing capabilities/techniques	4≥	87.10**	87.18**	87.12	93	39
Learning/improving new quality control methods	4≥	87.10***	76.92***	84.09	93	39
Learning/improving new production processes	4≥	80.65***	71.79***	78.03	93	39
Attendance to early stage in the design activities and developing prototype	4≥	77.42*	71.79*	75.76	93	39
Learning about new technologies	4≥	79.57**	61.54**	74.24	93	39
Increase in design capability	4≥	76.34**	69.23**	74.24	93	39
Access to international developed markets	4≥	76.34*	69.23*	74.24	93	39
Increase in productivity	4≥	77.42**	66.67**	74.24	93	39
Improving in ability and knowledge about identifying and finding solutions to problems encountered in the production	4≥	70.97	74.36	71.97	93	39
Reducing risks involved with making decision to invest in new technology and/or machines	4≥	63.44	74.36	66.67	93	39

Source: Author calculations based on survey results
Notes: ***, **, * denote significance level at the 1 %, 5 %, and 10 %, respectively (Mann–Whitney *U* test, two-sided). The items do not add up to 100 % because the questionnaire included further categories called *1—Very Unimportant, 2—Unimportant, 3—Neither Important Nor Unimportant*, but have not been identified in the table. 4≥: Responses indicating the degree of importance as *Important* (4) or *Very Important* (5) are presented here.
Legend: DSF (Direct Supplier Firms)

these items reflect the advantages of being a direct supplier for the product, production process, and design.

In sum, that even the item in the last rank was regarded important by 67 % of the firms shows that being a direct supplier has a lot of positive effects on the firms. However, it is important to underline that the AMMs set remarkably high criteria of being a direct supplier, which requires achievement of new targets in the fields specified by the AMMs even after being a direct supplier (e.g. quality improvement, defect rate reduction, delivery time reduction, and cost reduction). In addition to that, the AMMs ask their suppliers to have their own quality certificates and/or some internationally-recognized quality certificates. The supplier firms must fulfill these requirements within a certain period, and those that fail to do so are warned. The firms failing to meet these requirements within a certain period are dismissed from being a direct supplier. Evidently, both the assistance of the AMMs and the efforts spent by supplier firms to remain as direct suppliers play a crucial role in this process.

A comparison of local and foreign supplier firms reveal that local direct suppliers attach greater importance to 9 out of 14 items than foreign direct suppliers do, and the difference as to the items between the local and foreign direct suppliers ranges from 6 to 18 %. Four items in which the difference between the local and foreign direct suppliers is more than 10 % as follows:

i. *Learning about new technologies* (18 % difference),
ii. *Developing new business relationships with other firms* (15 % difference),
iii. *Increase in productivity* (11 % difference),
iv. *Learning/improving new quality control methods* (10 % difference), respectively.

Foreign direct suppliers attach more importance to the remaining five items, and the difference between foreign and local direct suppliers for these items ranges between 0.1 and 11 %. These items are as follows:

i. *Reducing risks involved with making decision to invest in new technology and/or machines* (11 % difference)
ii. *Improving in ability and knowledge about identifying and finding solutions to problems encountered in the production* (3 % difference)
iii. *Better understand the customers' blueprint and technical specifications* (2 % difference)
iv. *Stable trade* (1 % difference)
v. *Improvement in testing and analyzing capabilities/techniques* (0.1 % difference).

All these confirm that strategic relationships/collaborations between foreign direct suppliers and AMMs are the strongest, and foreign direct suppliers benefited from the technological activities much more than the local direct suppliers did. Moreover, foreign direct suppliers are more capable in terms of technology use, and their absorptive capacity is higher than local counterparts (see Sects. 4.1.6, 4.4, 4.7.2, and 4.8.3).

There is a significant difference between local and foreign direct suppliers at significance levels changing between 1 and 10 % in all items except for three (i) *stable trade, (ii) improving in ability and knowledge about identifying and finding solutions to problems encountered in the production, and (iii) reducing risks involved with making decision to invest in new technology and/or machines* items (see Table 4.29) (Mann–Whitney *U* test).

4.8 R&D, Innovation Activities, and Absorptive Capacity

4.8.1 R&D Activities (Technological Input)

Table 4.30 shows the distribution of the firms that answered *Yes* to the questions concerning R&D activities. It was found out that 133 of the 165 firms (81 %) in the sample conducted R&D activities. A total of 99 of these firms are local (83 %) and 34 are foreign firms (76 %). The number of firms that are direct suppliers and conduct R&D activities is 109 (83 %).

It was found out that there was a separate R&D department in 99 (74 %) of the 133 firms that are actually conducting R&D activities. It can be concluded that many of the firms conducting R&D activities prefer to have a separate incorporated R&D department.

The total R&D expenditure of the firms in the sample conducting R&D activities is TL 180M in 2008, and mean value is TL 1.32M. Furthermore, the mean rate of the R&D expenditures in the total sales (R&D intensity) is 2.64 %, and foreign firms (mean; TL 2.56M) expend in R&D nearly three times as much as the local firms (mean; TL 913 K).[35]

Table 4.30 Distribution of surveyed firms by R&D activities

	LF	FF	DSF	AF	LF	FF	DSF
YES	%	%	%	%	*N*	*N*	*N*
Does your firm carry out R&D activities?	**82.50**	**75.56**	**82.58**	80.61	120	45	132
Is there a separated R&D department in your firm?	60	60	62.12	60	120	45	132
Does your firm get any support from government or various institutions related to R&D (KOSGEB—TÜBİTAK—TEYDEB—TTGV—EU etc.)?	53.33	44.44	56.06**	50.91	120	45	132
Are joint R&D activities being carried out with other customers?	44.17	43.18	45.04	43.90	120	44	131
Do your customers help or support your firm related to R&D and design?	45.83	36.36	42.75	43.29	120	44	131
If your firm is a direct supplier of automotive main manufacturer, are joint R&D activities being carried out with these manufacturer(s)?	41.89	45.46	52.27***	42.86	117	44	132

Source: Author calculations based on survey results

Notes: ***, **, * denote significance level at the 1 %, 5 %, and 10 %, respectively (chi-square, two-sided). Fisher's Exact test has been used if one or more cells have an expected count ≤5. The items do not add up to 100 % because the questionnaire included a further category called *No*, but have not been identified in the table. However, the information on these categories can be obtained by subtracting the stated proportions from 100 %.

Legend: LF (Local Firms); FF (Foreign Firms); DSF (Direct Supplier Firms); AF (All Firms)

[35] For details see Appendix J.

About half of the firms (51 %) reported that they got financial support for R&D from the government or various institutions (e.g. KOSGEB, TUBITAK, TEYDEB, TTGV, and EU). When we take a look at the firm groups that get the highest support, direct suppliers are ranked the first (56 %), and this is followed by local (53 %) and foreign firms (44 %). Such projects promote especially cooperation and common activities among firms. In this context, it is expected that direct suppliers and AMMs undertake more common projects and get more support. It is important mainly to increase the cooperation among the firms, reinforce coordination between the personnel of the firms, and reduce R&D risks and costs. A finding we obtained from the interviews with the AMMs is that they participate in quite a lot of R&D projects (e.g. product development, innovation, design) with their suppliers. The fact that government and various institutions provide these supports primarily to the local firms explains why the foreign firms are the last in ranking.

Out of 132 firms, the number of firms who are direct suppliers and make joint R&D with the AMMs is 69 (52 %). It is found out that foreign firms (45 %) conduct more R&D activities with their AMMs than do local firms (42 %). As regards the joint R&D activities performed by the firms with other customers but not the AMMs, direct suppliers once again are placed the first in ranking with 45 %, and they are followed by local (44 %) and foreign (43 %) firms. When the AMMs and other customers are compared, it is figured out that local firms tend to perform joint R&D activities with other customers whereas foreign and direct suppliers do so with AMMs.

The percentage of the firms who specified that, apart from the joint R&D activities, the customers also provide assistance for R&D and design is 43 % (46 % for LF, 43 % for DSF, and 36 % for FF). This is closer to the percentage of the firms that perform joint R&D activities with both AMMs and other customers.

No significant difference has been found between local and foreign firms with respect to items. A significant difference, however, has been found between the direct suppliers and non-direct suppliers in terms of *get any support from government or various institutions related to R&D* (at 5 % level) and *R&D activities being carried out with AMMs* items (at 1 % level) (chi-square test).

Pioneers of R&D Activities

Table 4.31 displays the individuals or organizations that led the firms to start R&D activities. With 84 % of the local firms, starting R&D by *own choice* was ranked the highest. *DAMM* (32 %) and *OAMM* (20 %) were ranked the second and third, respectively. 62 % of the foreign firms specified *PC* in the first rank, which were followed by *own choice* (50 %) and *DAMM* (21 %) in the second and third ranks, respectively. If the item *own choice* and the effects of the affiliated parent companies is excluded, it can be concluded that AMMs significantly affect R&D activities of the firms.

Table 4.31 Pioneers of R&D activities

	LF	FF	DSF	AF
	%	%	%	%
Own choice	**83.84*****	50***	**76.15**	75.19
DAMM that we are the supplier of	36.36*	20.59*	36.70***	32.33
OAMM that we are the supplier of	22.22	14.71	22.02	20.30
PC	3.03***	**61.76*****	20.18	18.05
Foreign partner	0***	17.65***	5.50	4.51
N	**99**	**34**	**109**	**133**

Source: Author calculations based on survey results
Notes: ***, **, * denote significance level at the 1 %, 5 %, and 10 %, respectively (chi-square, two-sided). Fisher's Exact test has been used if one or more cells have an expected count ≤ 5.
Legend: LF (Local Firms); FF (Foreign Firms); DSF (Direct Supplier Firms); AF (All Firms); DAMM (Domestic Automotive Main Manufacturers); OAMM (Overseas Automotive Main Manufacturers); PC (Parent Company)

A significant difference has been found between local and foreign firms in of the following items: *own choice* (at 1 % level), *DAMM* (at 10 % level), *PC* (at 1 % level), and *foreign partner* (at 1 % level). There is also a highly significant difference between the direct suppliers and non-direct suppliers in terms of the *DAMM* item (at 1 % level) (chi-square test).

4.8.2 Innovation Activities

A total of 146 firms (88 %) stated that they conduct innovation activities (109 LF and 37 FF) (Table 4.32). The number of the direct suppliers performing innovation activities is 119 (90 %). That the number of the firms performing innovation activities (146) is higher than those performing R&D activities (133) shows that some firms conduct innovation activities without making R&D (13 firms). This finding is consistent with those of earlier studies: many firms in the late industrializing countries carry out innovation activities without making R&D (see Berger, 2005). Of 146 firms performing innovation activities, 125 (86 %) reported to conduct R&D, as well. Thus, it can be deduced that there is a strong relationship between the R&D and innovation activities of the firms. A significant difference has been found between local and foreign firms in terms of innovation activities (at 10 % level) (chi-square).

Cooperation Level with Partners in Innovation Activities

The firms performing innovation activities were asked to evaluate how intensely they cooperate with the external cooperated parties for these activities on a five-point Likert scale, involving a range from 1 (no cooperation) to 5 (very intense).

Table 4.32 Distribution of surveyed firms by innovation activities and cooperation level with partners

		LF	FF	DSF	AF	LF	FF	DSF
		%	%	%	%	*N*	*N*	*N*
Carry out Innovation Activities >—YES[a] (146 firms)		90.83*	82.22*	90.15	88.48			132
Innovation Partners-Intense Cooperation b								
DS	4≥	**37.61**	37.84	38.66	37.67	109	37	119
OS	4≥	36.70	32.43	33.61	35.62	109	37	119
DAMM	4≥	36.70	27.03	**40.34*****	34.25	109	37	119
OAMM	4≥	26.61	35.14	32.77***	28.77	109	37	119
PC	4≥	9.17***	**75.68*****	29.41	26.03	109	37	119
R&D institutions	4≥	21.10	8.11	19.33**	17.81	109	37	119
Private engineering and consultancy firms	4≥	22.02*	5.41*	18.49	17.81	109	37	119
Universities	4≥	15.60	16.22	17.65	15.75	109	37	119

Source: Author calculations based on survey results

Notes: ***, **, * denote significance level at the 1 %, 5 %, and 10 %, respectively ([a]chi-square, two-sided, b Mann–Whitney *U* test, two-sided). The item for innovation activities does not add up to 100 % because the questionnaire included further category called *No*, but has not been identified in the table. However, information on this category can be obtained by subtracting the stated proportion from 100 %. The items for innovation partners do not add up to 100 % because the questionnaire included further categories called *1—No Cooperation, 2—Little, 3—Middle* for partners, but have not been identified in the table. 4≥: Responses indicating the degree of intense as *Intense* (4) or *Very Intense* (5) are presented here.

Legend: LF (Local Firms); FF (Foreign Firms); DSF (Direct Supplier Firms); AF (All Firms); DS (Domestic Suppliers); OS (Overseas Suppliers); DAMM (Domestic Automotive Main Manufacturers); OAMM (Overseas Automotive Main Manufacturers); PC (Parent Company)

When the items are analyzed individually by exempting both 0 (no response) and 1 (no cooperation) choices, the median of five items is 4, the median of one item is 3.5, and the median of two items is 3. The items had the lowest mean value of 3 (*universities*) and the highest mean value of 4.11 (*parent company*). Therefore, the items that received the ratings of 4 or 5 were categorized as *intense* (4≥), those that received the rating of 3 in *medium* category (=3), and those that received the rating of 2 in *low* category (=2). In Table 4.32, the distributions of the firms whose answers for the items were in *intense* category (4≥) are examined.[36]

The analysis of the responses of all firms revealed that minimum 16 % (*universities*) and maximum 37 % (*DS*) of the firms conduct highly intense cooperation with more than one external party for innovation activities. The first three external parties with whom these firms cooperate are *DS* (38 %), *OS* (36 %), and *DAMM* (34 %). Here, it is observed that the cooperation in the innovation field is very

[36] Furthermore, when the items are examined together, on average 10 % of the firms chose *0 (no response)*, 35 % 1 (no cooperation), and 55 % a scale between 2 and 5 {25 % 2 (low) or 3 (medium), 30 % 4 (intense) or 5 (very intense)}.

intense with *DAMM* as it is with *DS* and *OS*.[37] Therefore, the geographical location of and the distance between supply/main industries do not matter much for these three categories. As a large part of the DAMM is composed of foreign-ownership or affiliates of MNCs, MNCs[38] play an important role in the cooperation of the supplier firms operating in the supply industry with respect to innovation activities. Relatively, the least-intensive cooperation is with *universities* (16 %).

Foreign firms seem to carry out the highest cooperation with *PC* (76 %). It is followed by *DS* (38 %) and *OAMM* (35 %). The least-intensively cooperated party is *private engineering and consultancy firms* with 5 %. Furthermore, the number of foreign firms that have extensive relations with *OAMM* is higher than that of foreign firms that have extensive relations with *DAMM*. These findings (more relations with *PC* and *OAMM*, fewer relations with R&D institutions and various private engineering-consulting institutions) seem to confirm our previous findings in terms of foreign supplier firms: they are generally the affiliates of MNCs in Turkey; they are the global suppliers of the AMMs; they have high production and technological capabilities; they realize basic innovation and R&D activities in their headquarters, and therefore they are dependent on parent companies.

The external party with which direct suppliers cooperate the most intensively is *DAMM* (40 %) as expected, and the least-cooperated party is *universities* (18 %).

Among these external parties, the local firms cooperate the most intensively with *DS* (38 %), *OS* (37 %), and *DAMM* (37 %), while the least-intensively cooperated ones are the affiliated *PC* (9 %) and *universities* (16 %).

A major finding here is that while most of the foreign firms (76 %) make very intensive cooperation with parent companies for innovation activities, only 10 % of the local firms make cooperation with parent companies. This is because the relations of local and foreign firms with parent companies are structurally different. That is to say, most of the local firms affiliated to the parent company (holding) are generally one of the group firms who are affiliated to this parent company in terms of management. At the same time, these group firms operate in different industries, and therefore each firm acts independently of other group firms in a variety of fields such as production, R&D, and innovation. The parent companies here serve as the supervisory, regulatory and the most authorized decision making body in terms of management. On the other hand, foreign firms are mostly the affiliates of the multinationals. The basic technological activities (R&D, innovation, design, etc.) of these firms are carried out at the main headquarters of the affiliated multinational. Therefore, the local firms are affiliated to the parent companies only in terms of management whereas the foreign firms are affiliated to the parent companies in terms of not only management but also any commercial activities such as R&D, production, and innovation.

[37] Here, the OS providing machinery-equipment and intermediate goods to the survey firms are concerned.

[38] DAMM, OAMM and foreign suppliers.

A significant difference has been found between local and foreign firms in terms of the items *PC* (at 1 % level) and *private engineering and consultancy firms* (at 10 % level). A significant difference between the direct suppliers and non-direct suppliers has also been found in terms of the items *DAMM* (at 1 % level), *OAMM* (at 1 % level), and *R&D institutions* (at 5 % level) (Mann–Whitney *U* test).

Technological Product and Process Innovation Activities[39] (Technological Output)

The firms were asked whether they had developed or put on the market any technological innovations in the production process, product, services, and logistic in the last three years.[40] Table 4.33 shows the distributions of the firms that said *Yes* to these questions concerning innovation activities. Similar distributions are observed among three groups of firms with respect to the innovation activities. Accordingly, the most important field in which the firms performed innovation activities is *production process* with 91 %. The other two fields are *product innovation* (73 %) and *supporting activities related to processes* (71 %). The fields in which the least innovation activities are performed are *service innovations* (45 %) and *innovations for logistics, delivery, and distribution* (42 %). This table shows that the firms perform more innovation activities in the *production process* and are more successful at making technological innovations than the non-technological innovation activities (e.g. logistics, distribution, and delivery). Another important finding is that foreign firms perform more innovation activities in logistics, distribution, and delivery than other firms.

A significant difference has been detected between local and foreign firms in terms of the item *innovations for logistics, delivery, and distribution* (at 10 % level), and between the direct suppliers and non-direct suppliers in terms of the item *innovation activities for product* (at 5 % level) (chi-square test).

---

[39] "**Technological product and process (TPP) innovation activities** are all those scientific, technological, organizational, financial and commercial steps which actually, or are intended to, lead to the implementation of technologically new or improved products or processes. **A technologically new product** is a product whose technological characteristics or intended uses differ significantly from those of previously produced products. **A technologically improved product** is an existing product whose performance has been significantly enhanced or upgraded. **Technological process innovation** is the adoption of technologically new or significantly improved production methods, including methods of product delivery" (see OECD, 1997, pp. 32–39).

[40] The questions and items are derived based on definitions from OECD's Oslo and Frascati manuals, and also technological innovation questionnaire form used by TurkStat in the innovation survey (OECD, 1997, 2002, 2005).

Table 4.33 Types of innovation activities (in the last 3-years)

	LF	FF	DSF	AF	LF	FF	DSF
Innovation activities-YES	%	%	%	%	*N*	*N*	*N*
New or significantly improved methods for production process(es)	**90.76**	**91.11**	**90.91**	90.85	119	45	132
New or significantly improved product(s)	72.27	75.56	77.27**	73.17	119	45	132
New or significantly improved supporting activities related to processes (maintenance systems, and operations for purchasing, accounting, and computing)	72.27	68.89	72.73	71.34	119	45	132
New or significantly improved service(s)	45.30	44.44	42.75	45.06	117	45	131
New or significantly improved logistics, delivery, and distribution methods	38.46*	51.11*	42.31	41.98	117	45	130

Source: Author calculations based on survey results
Notes: ***, **, * denote significance level at the 1 %, 5 %, and 10 %, respectively (chi-square, two-sided). Fisher's Exact test has been used if one or more cells have an expected count ≤5. The items do not add up to 100 % because the questionnaire included a further category called *No*, but have not been identified in the table.
Legend: LF (Local Firms); FF (Foreign Firms); DSF (Direct Supplier Firms); AF (All Firms)

Patents (Technological Output)

Table 4.34 provides information about the number of owned patents (NOPAT) and the newly-applied patents (NAPAT) by the firms. According to this, 107 of 154 firms (69 %) neither own any patents nor have made any patent applications (68 % for both LF and DSF and 75 % for FF). 22 % of the patent holder 47 firms have 1 to 5 patents, and 8 % have more than 5; the mean number of patents owned by these firms is 4.7, and the mean number of patent applications is 6.41. Neither between local and foreign firms nor between the direct suppliers and non-direct suppliers has a significant difference been found in terms of NOPAT and NAPAT (chi-square test).

The data in Table 4.34 raises the question of why the majority of the supplier firms in the sample did not take any patents. Although 81 % of these firms conduct R&D and 88 % conduct innovation activities, 69 % of them do not have any patents showing how critical the situation is. Two possible reasons behind this are as follows:

- First of all, the firms in the automotive industry may prefer to keep the new technologies they developed confidential in such times when the competition among the firms is fierce, and the technological innovations and developments are very rapid and essential. Since the patents are publicly available information, the firms avoid put their competitive power and their future assets at risk by not obtaining patents for the newly-developed technologies. The interviews with the executives of both AMMs and supplier firms revealed that the firms in the Turkish automotive industry are not willing to take the patents of the technologic products and processes they have developed. The most important reason for it, as specified by the firms, is that they did not want to reveal with the patent the

Table 4.34 Number of patents owned and applied

			Owned + Applied (NPAT)			
# of patents	Owned (NOPAT) *N*	Applied (NAPAT) *N*	LF (114 firms) %	FF (40 firms) %	DSF (125 firms) %	AF (154 firms) %
0	109	107	**67.54**	**75**	**68**	69.48
1	15	12	1.75	0	1.6	1.30
2	4	4	6.14	8	6.4	6.49
3–5	15	15	7.02	8	6.4	7.14
6–8	6	5	3.51	5	4.8	3.90
9–10	2	2	5.26	0	4.8	3.90
10+	3	6	8.77	5	8	7.79
Total	**154**	**151**	**100**	**100**	**100**	**100**

Source: Author calculations based on survey results
Notes: ***, **, * denote significance level at the 1 %, 5 %, and 10 %, respectively (chi-square, two-sided). Fisher's Exact test has been used if one or more cells have an expected count ≤5.
Legend: LF (Local Firms); FF (Foreign Firms); DSF (Direct Supplier Firms); AF (All Firms); NOPAT (Number of Owned Patents); NAPAT (Number of Applied Patents); NPAT (Number of Owned and Applied Patents)

confidential technologic information. After all, it develops with long researches and high costs in such an important field as automotive. They also consider it as a waste of time and sources to apply many countries to obtain a patent.

- Secondly, the fact that 69 % of the surveyed firms do not have patents, although a significant part of them perform R&D and innovation activities, indicates that they do not perform these activities at an advanced level. As we are going to mention in the Chap. 6 and conclusion section in detail, it has been observed that the domestic R&D departments of these firms generally work on partial modifications and improvements of the products, and their adjustment to the production line.[41] Also, foreign firms refrain from having patents (or they have a few patents) because they are the affiliates of MNCs, and therefore perform advanced level innovation and R&D activities to take patent at their main headquarters. Conversely, this might indicate that the firms are dependent on external sources for the technologies in which the possibility of taking patents for the automotive supply industry is high, and this dependency somewhat accounts for the low patent-taking tendency of the supplier firms.

[41] According to a study analysing the R&D activities of the firms with foreign capital operating in Turkish manufacture industry, the firms in automotive main industry trust their brands rather than patents to protect the innovations: See Erdil, Pamukçu, Erden, Göksidan, and Kepenek (2011).

4.8.3 Absorptive Capacity

Six alternative indicators were formed in order to obtain information on the absorptive capacity of the surveyed firms (Table 4.35):

- Share of engineers in total employment
- Share of white-collar personnel in total employment
- Share of R&D expenditures in turnover (R&D intensity)
- Share of export in turnover (export intensity)
- Number of patents obtained
- Sales per employee

The first two indicators show the quality level of the firms regarding human capital, the third indicator R&D potentials as a technological input, the fourth indicator export-oriented level (assuming that export-oriented firms have higher capabilities), the fifth indicator the number of patents obtained as an output of the technological capabilities, and the last indicator productivity levels. The absorptive capacities of the firms are reviewed through these indicators.

The data on the human capital structure show that foreign firms employ more highly-qualified personnel than local firms. The mean (median) rates of both the engineers (significant at 1 % level) and the white collar workers (significant at 5 %

Table 4.35 Alternative indicators of absorptive capacity

		LF	FF	DSF	AF
Share of engineers in TE (%)	*N*	**119**	**42**	**129**	**161**
	Mean	6.79***	11.33***	8.41	7.98
	Median	5.26	7.60	6.25	5.78
Share of white-collar personnel in TE (%)	*N*	**119**	**42**	**129**	**161**
	Mean	16.6**	20.9**	17.35	17.72
	Median	14.84	16.16	14.9	15.03
R&D intensity (%)	*N*	**104**	**35**	**111**	**139**
	Mean	2.55	2.91	2.77	2.64
	Median	1	2	1.17	1
Export intensity (%)	*N*	**120**	**45**	**132**	**165**
	Mean	34.53*	43.78*	35.30	36.83
	Median	26.50	40.00	25.00	30.00
# of Patents owned	*N*	**114**	**40**	**125**	**154**
	Mean	1.56	0.83	1.51	1.37
	Median	0	0	0	0
Sales per employee (TL)	*N*	**104**	**37**	**113**	**141**
	Mean	178,318**	318,480**	199,768	215,097
	Median	136,833	183,200	153,440	148,734

Source: Author calculations based on survey results

Notes: ***, **, * denote significance level at the 1 %, 5 %, and 10 %, respectively (two independent samples *t*-test, two-sided). Equal variances are assumed between groups.

Legend: LF (Local Firms); FF (Foreign Firms); DSF (Direct Supplier Firms); AF (All Firms); TE (Total Employment)

level) in the total employees are higher in foreign firms (*t*-test). Having a high amount of highly-skilled work force is an important indicator of technological capabilities because the most important requirement of operating with complex and advanced production technologies, and performing R&D and innovation activities is to have highly-skilled human capital.

As regards the rate of R&D expenditure of the firms in the total sales (R&D intensity), the foreign firms spend more on R&D activities than the local firms on average. Moreover, the median value indicates that at least 50 % of the foreign firms spend 2 % of their total sales on R&D whereas local firms spend 1 %. This indicates that foreign firms make more investment on R&D. It is also the result of having more highly-skilled human capital.

Foreign firms are mainly more export-oriented than local counterparts (significant at 10 % level) as we examined in Sect. 4.1.5 (*t*-test). Nearly half of the sales of the foreign firms on average is made to the overseas markets. In other words, at least 50 % of the foreign firms export 40 % of their total sales to overseas markets (median 40 %). The overseas markets require the capability of dealing with a more competitive market as well as the high technological capabilities pertaining to the products and production processes (e.g. quality control, low defect rate, high quality, testing, design capability). Therefore, it can be suggested that the foreign firms who make more production for the overseas markets possess higher amount of these technologies.

Highly-qualified human capital, innovation, and R&D activities are technological development inputs of the firms, and patents are the outputs of this development. Local firms seem to have more patents than the foreign firms on average. Nevertheless, this does not mean that the local firms are more successful at converting the technological innovations into patents than the foreign firms due to the reasons discussed in section "Patents (Technological Output)". In addition, one of the main reasons for the local firms' having a higher average of patents in the sample is that three of them have a high number of patents, and thus increase the average.

The average total sales per person (productivity) for the foreign firms are almost twice as high as the local firms (significant at 5 % level) (*t*-test). This indicates that foreign firms are more productive than the local firms. The most significant effect of the human capital quality, the level of R&D, and innovation activities, or the technological capability of the firms, shows itself in the productivity level, and the foreign firms are dramatically more capable than the local firms regards this indicator.

In sum, we can assume that foreign firms have higher absorptive capacity than the local firms according to these six indicators.

In terms of direct suppliers, all of them, but *the number of patents owned,* have higher mean rates than the local firms but lower mean rates than the foreign firms. This is significant as it shows that the absorptive capacity of the direct suppliers is lower than that of the foreign firms.

4.9 Performance and Future Projections

4.9.1 *Performance Increases: Improvements Related to Production Capability*

The firms were asked to evaluate the improvements (performance increases) achieved in terms of production capabilities in the last three years on a five-point Likert scale ranging from *1 = very increased* (deterioration) to 5 = very decreased (improvement). The items pertaining to the production capability are *defect rate,*[42] *duration of on-time delivery*, *cycle time,*[43] and *average costs*. When the items are examined individually, the median of four items is obtained as 4, the lowest mean value is 3.46 (*average costs*), and the highest mean value is 3.99 (*defect rate*). Therefore, the items evaluated at the level of 4 or 5 were handled in *decreased* category (4≥), the items evaluated at the level of 3 in *medium* category, and the items evaluated at the level of 1 and 2 were handled in *increased* category. Table 4.36 shows the distributions of the firms the responses of which fall in *decreased* category (4≥) (*improvement*).

60 % to 83 % of the firms stated that important improvements had been achieved with respect to these items in the last three years. The most important improvement seems to in the reduction of *defect rate* with 83 % (131 firms). This is followed by the improvements (reductions) in *duration of on-time delivery* with 68 % (108 firms), *cycle time* with 63 % (96 firms), and *average costs* with 60 % (94 firms). Apart from the *average costs* item, similar distributions are observed among the three groups of firms as to the other three items. Considerably more foreign firms (73 %) reported a reduction in the costs in the last three years than the local (55 %) and direct suppliers (62 %). This also indicates that foreign firms manufacture more efficiently.

A significant difference has been detected between local and foreign firms only in terms of the item *average costs* (at 1 % level). However, between the direct suppliers and non-direct suppliers, no significant difference has been found (Mann–Whitney *U* test).

[42] It is a standard measurement to measure the quality performances of the firms in automotive industry. It is generally measured as the parts per million (PPM). For example, this rate for a firm with 1,000 PPM means that 1,000 products per million manufactured or delivered by this firm are defective. These firms try to reduce this rate as much as possible.

[43] "Period required to complete one cycle of an operation; or to complete a function, job, or task from beginning to the end" (from http:\\businessdictionary.com).

Table 4.36 Distribution of surveyed firms by improvements in production capability (in the last 3-years)

		LF	FF	DSF	AF	LF	FF	DSF
Decreased (improvement)		%	%	%	%	*N*	*N*	*N*
Defect rate	4≥	**81.20**	**87.80**	**84.38**	82.91	117	41	128
Duration of on-time delivery	4≥	68.97	66.67	70.54	68.35	116	42	129
Cycle time[a]	4≥	62.93	65	62.99	63.46	116	40	127
Average costs	4≥	55.17***	73.17***	61.72	59.87	116	41	128

Source: Author calculations based on survey results
Notes: ***, **, * denote significance level at the 1 %, 5 %, and 10 %, respectively (Mann–Whitney *U* test, two-sided). The items do not add up to 100 % because the questionnaire included further categories called *1—Very Increased, 2—Increased, 3—Neither Increased Nor Decreased*, but have not been identified in the table. 4≥: Responses indicating the degree of improvements as *Decreased* (4) or *Very Decreased* (5) are presented here.
Legend: LF (Local Firms); FF (Foreign Firms); DSF (Direct Supplier Firms); AF (All Firms)
[a]"Period required to complete one cycle of an operation; or to complete a function, job, or task from start to finish" (from http:\\businessdictionary.com)

4.9.2 Reasons for Performance Increases (Improvements) in Production Capability

In this section, improvements in the production capability of the surveyed firms as a result of KTTs from customers are used as a proxy for the performance increases of the firms as Sects. 4.6.2 and 4.9.1.

The firms which had achieved improvement in their production capabilities and/or developed new technologies concerning the product or production process in the last years were asked to evaluate the items affecting their performances on a five-point Likert scale ranging from 1 (very ineffective) to 5 (very effective). When the items are evaluated individually, the median of six out of ten items is 4, the median of three is 3, and the median of only one is 2. The lowest mean value is 2.32 (*hiring skilled-specialist employees from other firms*), and the highest is 3.95 (*acquisition of new software, hardware, equipment and tools*). Therefore, the items evaluated at the level of 4 or 5 were handled in *effective* (4≥) category, the items evaluated at the level of 3 in *neither effective nor ineffective*, and the items evaluated at the level of 1 and 2 were handled in *ineffective*. Table 4.37 presents the distribution of the firms that responded to the items in *effective* category (4≥).

It was found out that the first five items were rated as *effective* by at least 72 % of the firms, and *acquisition of new software, hardware, equipment and tools* received the highest ranking with 83 % whereas *hiring skilled-specialist employees from other firms* received the lowest rating with 17 %.

Although *being a direct supplier of AMM* is ranked the second with 80 % in terms of all firms, it is ranked the first in terms of the foreign (81 %) and direct suppliers (88 %). That this is regarded as the most effective item by a large percentage of the direct suppliers points out the most important benefit of being a

Table 4.37 Reasons for performance increases (improvements) in production capability

		LF	FF	DSF	AF	LF	FF	DSF
Effective		%	%	%	%	*N*	*N*	*N*
Acquisition of new software, hardware, equipment and tools	4≥	**85.59***	76.74*	82.17*	83.23	118	43	129
Being a direct supplier of automotive main manufacturer	4≥	79.82	**80.95**	**88.46*****	80.13	114	42	130
Skill and capability improvement of employees through in-house training	4≥	80.51	75.00	79.23	79.01	118	44	130
Using of new production methods	4≥	80.34	68.18	74.42	77.02	117	44	129
Undertaking R&D activities	4≥	72.48	69.05	73.6*	71.52	109	42	125
Improvements in our suppliers' quality level	4≥	61.02	54.55	60.00	59.26	118	44	130
Knowledge and technology transfers from foreign manufacturers	4≥	45.37**	60.47**	47.97	49.67	108	43	123
Cooperation with universities and other institutions[a] in Turkey	4≥	39.42	35.14	35.34	38.30	104	37	116
Knowledge and technology transfers from domestic automotive main manufacturers	4≥	33.64	28.57	34.88***	32.24	110	42	129
Hiring skilled-specialist employees from other firms	4≥	17.31	15.79	16.81	16.90	104	38	119

Source: Author calculations based on survey results
Notes: ***, **, * denote significance level at the 1 %, 5 %, and 10 %, respectively (Mann–Whitney *U* test, two-sided). The items do not add up to 100 % because the questionnaire included further categories called *1–Very Ineffective, 2-Ineffective, 3-Neither Effective Nor Ineffective*, but have not been identified in the table. 4≥: Responses indicating the degree of effectiveness as *Effective* (4) or *Very Effective* (5) are presented here.
Legend: LF (Local Firms); FF (Foreign Firms); DSF (Direct Supplier Firms); AF (All Firms)
[a]Various research and consultancy institutions/firms, private or public research laboratories, etc.

direct supplier. Indeed, being a direct supplier affects the performances of the firms rather positively. The reasons of this situation might be as follows:

- They are to comply with continuously-updated quality improvement, cost and defect rate reduction plans by the AMMs,
- Continuous AMM monitoring and evaluation of the firms,
- Various KTTs and training activities concerning product and production process by the AMMs,
- Competition with other supplier firms in both the domestic and overseas markets,
- Strategic cooperation activities with the AMMs,
- Joint R&D and design activities or projects conducted for the development of product and production process with the AMMs.

A remarkable finding is that *the KTTs from foreign manufacturers* item (50 % for all firms) is considered to play a more important role than *KTTs from DAMM* item (32 % for all firms) in the performance increase among three groups of firms. This may be indicative of the fact that the supplier firms make more cooperation with the foreign firms (foreign ownership firms at domestic or overseas markets) in order to obtain new technologies concerning product and product development. As discussed earlier, the supplier firms procure new machinery, software, hardware, equipment and tools especially from the foreign firms, and foreign firms provide training, consulting and engineering services in this respect.

On the other hand, *KTTs from DAMM* is considered efficient in the ninth rank by 32 % of the firms. This might be interpreted positively or negatively as follows:

- The fact that it was chosen by one third of the firms can be regarded as positive in terms of indicating the KTTs to the firms, and their positive effects on the performances.
- On the other hand, it can be interpreted negatively since the rest of the firms did not choose this item; the AMMs are not keen on KTTs and are stricter in this respect and only make these transfers to certain suppliers. Also, the performance increase by these transfers is dependent on the technological capabilities, absorption, internalization, and comprehension level of the firms taking these transfers, and the supplier firms have less capability in this respect, which explains why this item was chosen by fewer firms.

Finally, *cooperation with universities and other institutions in Turkey* is in the eighth rank with 38 %. This is important in terms of pointing out that the cooperation between the firms and institutions, particularly with the universities, is still not at a satisfactory level.

A significant difference has been found between local and foreign firms in terms of the items *acquisition of new software, hardware, equipment and tools* (at 10 % level) and *KTTs from foreign manufacturers* (at 5 % level). A significant difference has also been detected between the direct suppliers and non-direct suppliers in terms of the items *acquisition of new software, hardware, equipment and tools* (at 10 % level), *being a direct supplier of AMM* (at 1 % level), *undertaking R&D activities* (at 10 % level), and *KTTs from DAMM* (at 1 % level) (Mann–Whitney *U* test).

4.9.3 Future Projections Related to Acquisition of New Technologies

Table 4.38 shows the distribution of the firms in terms of their future projections for new technology acquisition. Similar distributions are observed among three groups of the firms. The two most important items specified by 83 % of the firms are *get training* and *undertake R&D activities and/or improve existing R&D activities.* These are followed by *analyze rival products and processes* (60 %), *make joint R&D activities with AMM* (59 %), *get consultancy from universities* (32 %), *get consultancy from domestic* (29 %), *get consultancy from abroad* (25 %), *bring*

Table 4.38 Future projections related to acquisition of new technologies

	LF	FF	DSF	AF
	%	%	%	%
Get training	**86.55***	**75.56***	84.09	83.54
Undertake R&D activities and/or improve existing R&D activities	85.71*	**75.56***	**87.12*****	82.93
Analyze rival products and processes	63.87	51.11	59.09	60.37
Make joint R&D activities with automotive main manufacturers	58.82	60	64.39***	59.15
Get consultancy from Universities	35.29*	22.22*	31.82	31.71
Get consultancy from domestic	30.25	24.44	30.30	28.66
Get consultancy from abroad	25.21	24.44	26.52	25
Bring experts from abroad	17.65	15.56	18.94	17.07
Sign agreements to provide skill and knowledge transfer	17.65	11.11	13.64*	15.85
Hiring expert staff from automotive manufacturer(s)	6.72	2.22	6.82*	5.49
N	**119**	**45**	**132**	**164**

Source: Author calculations based on survey results
Notes: ***, **, * denote significance level at the 1 %, 5 %, and 10 %, respectively (chi-square, two-sided). Fisher's Exact test has been used if one or more cells have an expected count ≤5.
Legend: LF (Local Firms); FF (Foreign Firms); DSF (Direct Supplier Firms); AF (All Firms)

experts from abroad (17 %), and *sign agreements to provide skill and knowledge transfer* (16 %) items. *Hiring expert staff from AMM*, however, was chosen only by 5 % of the firms, and thus came the last in the ranking.

As can be seen in the table, the items other than *make joint R&D activities with AMM* were chosen by a higher percentage of the local firms. This indicates more efforts paid by the local firms to acquire new technologies, but it also reflects that foreign firms do not prefer these items since they make more technological production compared to local firms. This is verified by the fact that higher amount of foreign firms chose *make joint R&D activities with AMM* item, which requires high technology.

A significant difference has been detected between local and foreign firms in terms of the items *get training*, *undertake R&D activities and/or improve existing R&D activities*, and *get consultancy from universities* (each at 10 % level). A significant difference has also been found between the direct suppliers and non-direct suppliers in terms of the items *undertake R&D activities and/or improve existing R&D activities* (at 1 % level), *make joint R&D activities with AMM* (at 1 % level), and *sign agreements to provide skill and knowledge transfer* (at 10 % level) (chi-square test).

4.10 Summary

Major findings of this chapter can be summarized under nine headings in the following way:

- General characteristics of the surveyed firms
  - More than one quarter of the sample (27 %) is composed of foreign firms, and they are under the full control of foreign capital. Especially German firms dominate the foreign firms.
  - Most of the firms in the sample are medium (51 %) and large-sized (40 %) firms according to the number of employees. Moreover, the share of the firms with sales higher than TL 10M is 76 %. Foreign firms are large-sized by the number of employees and sales, and they are also younger (most of them were established after year 2000) than the local counterparts.
  - The large majority (80 %) of the firms in the sample are direct suppliers of AMMs in Turkey. The statistics related to firm-size (large average firm-size measured by the number of employees or by total sales) also confirms our expectations related to higher number of direct suppliers in the sample.
  - The large majority (92 %) of the firms are exporters. Generally, the statistics show that the number of the firms (14) that produce goods for only domestic market is very low in the sample; most of them probably produce goods for global production networks of which they are probably a part. However, foreign firms have higher export shares, and they are more export-oriented compared to the local firms. One of the explanations of this situation could be that they are incorporated into global production networks. This could also be the reflection of their high competition level because global markets are more competitive than domestic markets. Lastly, the major market of the 42 % is overseas countries.
  - More than half of the sample (52 %) manufactures high-tech (high-technology level) products, but the technology level of the foreign firms is higher than both the direct suppliers and local firms.
  - 21 % of the firms are part of a MNC, and 86 % of them have MNCs among their customers.
  - The statistics related to direct suppliers are generally following the same pattern for all firms, but they possess lower values than foreign firms, while higher values than local firms.

- KTTs related to the production process, product, financial, and training activities
  - It is found that various types of the transfers have *rarely* been provided to suppliers by their customers. However, transfers related to production process are provided more often than both product and financial transfers. These confirm that relationships between customers and suppliers are based on strategic production policies.

  - It is noteworthy that higher quality transfer types are being provided more to the foreign firms at higher rates compared to direct suppliers and local firms. Direct suppliers generally bear the same characteristics as local firms.
  - Financial transfers are less frequent among these transfers. Consumers generally do not make financial transfers to the suppliers, but if they do, they prefer making pre-financing of machinery and equipment and/or prepayment before delivery. Financial transfers are also mostly provided to foreign firms at higher rates.
  - Various training types and modes are being also provided to personnel of suppliers by customers. Training activities are mostly provided to *production/operation staff* in the mode of *off-the job training*. However, foreign firms get more training activities than local firms. This situation does not mean that foreign firms need more training because of the lower capability of their human capital; on the contrary, customers prefer to work mostly with foreign firms because of their high technological capabilities and absorptive capacity level. These kinds of activities are generally provided by customers on the basis of new product development process. Therefore, this confirms that strategic relationships between foreign firms and AMMs (customers) are very strong, and it reflects that it is necessary to have highly qualified personnel to utilize in these activities. The finding that more visual training is provided to local firms compared to foreign firms confirms this situation.

- Market and competition
  - Domestic market is the major market of the firms, and it is followed by the European countries (Germany, France, and England are the first three markets in EU). The third major market of the firms is the Middle East.
  - The main competitors of the local firms are in Turkey, and they also agree that domestic competition is very intense. On the other hand, the main competitors of the foreign firms are abroad, and they also agree that global competition is fierce. This means that foreign firms do not consider the domestic firms as their competitors since they are larger scale firms with superior technological capabilities.

- Technological capabilities
  - Firms seem to have high-tech production capabilities such as on-time delivery, quality control, design, testing, and analytical capability. However, foreign firms have better capabilities in most of the areas requiring higher technology and knowledge. This situation explains why they have more industrial quality certifies, and they conduct more technology agreements in terms of knowledge and technology acquisition, and development than local counterparts. On the other hand, direct suppliers have better capabilities as to the features required by AMMs for the suppliers and meet higher quality levels compared to other firms.
  - Firms have acquired technologies from many sources. However, the major source differs across the three groups of the firms. Indeed, DS is the most

important technology source for local firms, PC for foreign firms, and DAMM for direct suppliers. Universities are ranked the last as a technology source.
- Most of the firms (61 %) are in an intermediate position in terms of their design capabilities. In other words, design capabilities of the majority of firms are rather weak. However, direct suppliers have better design capability compared to foreign and local firms.
- The majority of the firms carry out co-design activities related to product with their customers, and also they are included in this process from the beginning. The most important partner in these activities is DAMM, which is followed by OAMM. This points out that the most important partner of the firms in co-design activities is AMMs domestically or overseas. On the other hand, the least cooperated partner is universities.

- Supply chain
  - The majority of the firms (86 %) produce goods for MNCs, and this shows that firms are a part of global production networks, and they are also capable to produce for MNCs.
  - The majority of the firms (83 %) carry out their total sales mainly to a sole customer, and 40 % of the firms become the sole supplier of their most important customers (AMMs).
  - The majority of the foreign firms (84 %) are subsidiaries of MNCs and global suppliers of the AMMs as expected. On the other hand, the minority of the local firms (36 %) are dependent on domestic firm/group/holding.
  - The main customers of the firms are AMMs at domestic (46 %) and overseas (19 %). This confirms that most of the firms in the sample are first-tier suppliers of AMMs. However, foreign firms (77 %) carry out more production for AMMs than both direct suppliers (73 %) and local firms (60 %) do.
  - Most two important input sources of the firms are DS and OS. They are followed by DAMM for local firms and PC for both foreign firms and direct suppliers.

- Cooperation
  - The firms engage in cooperation activities with various external parties, and the majority of them attach considerable importance to these activities to improve their technological and production capabilities. However, strategic collaborations between foreign direct suppliers and AMMs are the most strong, and foreign direct suppliers benefited from these activities much more than local direct suppliers did.
  - As a result of assistance received from customers, firms achieved very important developments in many areas such as quality control, production process, engineering, design, productivity, export, and profitability. However, local firms are affected more positively by these assistances. This indicates that local firms are not capable enough of absorbing this assistance.

- Direct suppliers
  - 80 % of the firms in the sample are the direct suppliers of the AMMs in Turkey.
  - Firms generally apply to the AMMs by their own efforts to be a direct supplier; however, local firms spend more effort in order to be a direct supplier.
  - Firms extensively benefit from the assistance of AMMs in fields of product, production process, and design. However, foreign direct suppliers benefit from these assistances more as a result of having higher technological capabilities and absorptive capacity.
- R&D, innovation activities and absorptive capacity
  - The rate of the firms that conduct innovation activities (88 %) is higher than that of the firms conducting R&D activities (81 %). The majority of the firms (60 %) prefer to conduct these R&D activities in a separated R&D department. AMMs are also an important factor affecting R&D activities of the firms.
  - Foreign firms make three times more R&D expenditures compared to local counterparts. Direct suppliers get more financial support for R&D from government or various institutions.
  - Firms conduct cooperation with external parties for innovation, and MNCs play an important role in this process. Direct suppliers conduct innovation activities more with AMMs, whereas foreign firms with parent company and local firms with domestic suppliers. However, the least cooperated external party is the universities.
  - Firms are more successful at making technological innovation activities especially in the field of production process; however, they do not prefer to get patents.
  - According to the indicators of the absorptive capacity of the surveyed firms, foreign firms have higher absorptive capacity than both the direct suppliers and local firms.
- Performance and future projections
  - The majority of the firms had achieved performance increase in production capabilities in the last three years. However, foreign firms performed more than both direct suppliers and local firms.
  - Being a direct supplier of AMM is regarded as the most influential factor by both the direct suppliers and foreign firms for the performance increase. By contrast, cooperation with the universities is not regarded by the firms as a strong influence for the performance increase.
  - Local firms are seen to make more efforts to acquire new technologies since they are involved in less technological production compared to foreign firms.

In conclusion, foreign firms are seen to dominate the local ones in terms of many indicators and absorptive capacity level (such as KTT, technological capabilities, size, export intensity, qualified human capital, producing high-tech and complex products, performing R&D and innovation activities, and productivity). As a result, foreign firms set up more extensive relations with their customers and benefited more from these. Furthermore, within the three groups of firms, direct supplier firms perform better than local firms as to the above-discussed indicators but not better than foreign firms.

Chapter 5
Survey Findings: Econometric Analysis

In this chapter, an econometric analysis is conducted by using gathered survey quantitative data to identify the determinant factors impacting different types of the KTTs provided by customers to suppliers at inter-firm level. To this end, the chapter focuses on various characteristics and activities of the suppliers discussed in Chap. 4 within nine main headings.[1] In the first two sections, econometric model and explanatory variables introduced in the regressions are examined. In the third section, Factor Analysis (FA) method used as a data reduction technique for the scales in the construction of the dependent variables is presented. Then, in the fourth and fifth sections, logit and ordinal logit estimation methods are examined; econometric findings are presented, and analyzed in detail. Lastly, findings are summarized.

5.1 Model

In order to analyze the determinant factors impacting various characteristics and activities of the surveyed suppliers and different types of the KTTs provided by customers to suppliers, logit and ordinal logit regression analyses are used depending on the nature of the dependent variable. The following topics are examined by applying logistic regression analysis to survey data:

- Factors affecting the chances of being a direct supplier of AMMs in Turkey,
- Determinants of R&D and product innovation activities of the suppliers (Almeida & Fernandes, 2008; De Negri & Turchi, 2007; Evenson & Westphal, 1995; Pamukçu, 2003),

[1] All analyses and estimations were performed by using STATA/SE 10.1 program.

A. Sönmez, *Multinational Companies, Knowledge and Technology Transfer*, Contributions to Management Science, DOI 10.1007/978-3-319-02033-4_5,

- Determinants of technology agreements, acquisition and sources (Basant, 1997; Braga & Willmore, 1991; Evenson & Westphal, 1995; Katrak, 1989, 1990, 1997),
- Determinants of the co-design activities related to products and the partners in cooperation for such activities.

Then, the following topics are examined by using ordinal regression analysis:

- Determinants of the KTTs provided by customers to suppliers at inter-firm level related to *production process*, *product*, *training activities*, and *financial assistance* (Berger, 2005; Giroud, 2003; Techakanont, 2002; Techakanont & Terdudomtham, 2004),
- Determinants of the technology capabilities of the suppliers in terms of production and design,
- Factors that affect the cooperation activities of the suppliers with other firms (Berger, 2005; Giroud, 2003; Techakanont, 2002; Techakanont & Terdudomtham, 2004; Wasti, Kozan, & Çınar, 2009; Wasti, Kozan, & Kuman, 2006),
- Determinants of the improvements in production capability as a result of assistance (support, advice, or knowledge transfer) received from customers (Berger, 2005; Giroud, 2003; Techakanont, 2002; Techakanont & Terdudomtham, 2004),
- Factors that affect the benefits of being a direct supplier,
- Determinants of the cooperation level with partners in innovation activities,
- Determinants of the performance increases/improvements in the last 3 years related to production capability.

The basic model to be estimated is as follows:

$$\begin{aligned} Y = {} & \beta_0 + \beta_1 FIRMAGE + \beta_2 FIRMSIZE + \beta_3 EXPINT + \beta_4 FOREIGN \\ & + \beta_5 DIRECTSUPP + \beta_6 LOCALGROUP + \beta_7 CLIENTMNC + \beta_8 RDDEPT \\ & + \beta_9 PRODUCTINNO + \beta_{10} SENGPER + \beta_{11} HIGHTECHPROD \\ & + \beta_{12} HIGHDESCAP + \beta_{13} ONESUPP + \beta_{14} SSUBCONT + \beta_{15} RDSUB + e \end{aligned}$$

Where, Y indicates the dependent variable which is binary (for logistic regression) or ordinal (for ordinal logistic regression). In the analyses, we estimate the several variations of the above model. The definitions of explanatory variables are defined in detailed below. The brief descriptions of the explanatory variables, descriptive statistics and correlation matrix are provided in Appendices K, L, and M, respectively.

5.2 Explanatory Variables

Based on the collected primary survey data at firm-level, 15 indicators related to various characteristics, activities, technological capabilities, and customer relationships of the supplier firms have been constructed to be used as explanatory variables in the regressions.[2] This section explains how to create these variables and their expected impacts on the dependent variables. These variables include the followings:

1. FIRMAGE, *age of the firm*, is the natural logarithm of the age of the supplier firm. It has been constructed by subtracting the establishment year of the supplier from 2010, which is the year when the survey was conducted (see Sect. 4.1.2). FIRMAGE may indicate its experience in the automotive industry in terms of production, and proxy the extent to which it might have trust-based relationships with its customers. The higher the trust between both partners, the less the transaction costs incurred and the higher the probability is to experience KTT-related activities with customers. They can easily establish relationships with customers or attempt to do so. Moreover, experienced suppliers may have more business networks and strategic relationships with customers compared to the newly established suppliers. On the other hand, more recently established suppliers may act more aggressively in contrast to the older suppliers, which may show signs of rigidity and inability to adapt to a changing environment. Moreover, they may have some disadvantages such as financial support, lack of information, market experience, and technological capabilities.
2. FIRMSIZE,[3] *size of the firm*, is measured as the logarithm of the total number of employees (see Sect. 4.1.3). FIRMSIZE may proxy a host of variables potentially affecting the KTTs, technological capabilities, R&D, and innovation activities of the suppliers. It may be an indicator that reflects intra-firm institutionalization, scale and scope economies in the production process, cost and availability of financial resources, and extent of the labor division within the firm. On the other hand, it may negatively affect the KTT-related activities of suppliers since large suppliers may be self-sufficient and demand less KTT from customers. However, increasingly, even the human and financial resources of the largest firms cannot be sufficient for conducting R&D and innovation activities, leading them to collaborate with other firms. In addition,

[2] Total number of the explanatory variables constructed from survey data was 27. In the result of the data analyses, it was detected that 12 of them were not appropriate to use in the analyzes because of high number of missing values, high correlation rates with other variables, yielding insignificant results, etc. so remaining 15 variables were used in the analyses. These 12 unused variables are TURNOVER, FOREIGNGROUP, PARTMNC, RDEXP, WORKTGT, RDINT, PRODUCTIVITY, INNO, RD, MC, TECHPROD, and DESCAPAB. Please see next footnotes for the description of these variables.

[3] It was preferred to use this variable instead of "the logarithm of the total sales value" variable (TURNOVER), since the latter includes high number of missing values.

it is also admitted that firm size may act as a proxy for production capabilities since it is related to the production capacity or scale of firm. Moreover, interviews with the AMMs showed that they tend to consider strong production capability as a necessary condition for the establishment of KTT-related activities with suppliers.

3. EXPINT, *export intensity*, is measured as the ratio of exports to total sales (see Sect. 4.1.5). In recent years, especially from 2000 onwards, AMMs as well as suppliers in Turkey have increased the proportion of their sales sold on global world markets. The intense competition prevailing abroad may oblige supplier firms to cooperate with their customers in order to improve the quality of their products, receive know-how pertaining to the production process in order to increase its efficiency and to carry out design-related activities. In some instances, these KTT-related activities may be conducted with the help of customers abroad. Therefore, the impact of this variable depends also on the position of supplier firms in the supply chain of the multinationals. Moreover, this variable may be an indicator of the level of the absorptive capacity of suppliers, since as discussed in Sect. 4.8.3, export markets require high technological capability to deal with a more competitive market.
4. FOREIGN,[4] *foreign firm*, is a dummy variable and shows that a supplier firm is owned by foreign agents (see Sect. 4.1.1). It takes the value of 1 if the share of firm equity owned by foreigners equals at least to 10 %, and 0 otherwise.[5] Foreign suppliers may possess a number of intangible proprietary assets that enable them to compete with domestic suppliers without any need of KTTs. On the other hand, these same proprietary assets may enable them to pursue advanced KTTs with their customers, which may operate in Turkey or abroad, especially if they are asked by their customers to manufacture technologically sophisticated parts and components. Conversely, these can serve as an obstacle to the foreign firm in the cooperation activities with other firms in the country in which it operates. If it is an affiliate of a multinational, it might transfer KTT from its parent firm (see section "Cooperation Level with Partners in Innovation Activities" in Chap. 4). Hence, the effect of this variable on the KTTs remains an empirical issue.
5. DIRECTSUPP, *direct supplier*, is a binary variable and takes the value of 1 if a firm declares being the *direct supplier* of at least one AMM operating in Turkey, and 0 otherwise (see Sects. 4.1.1 and 4.7). Direct supplier here means the first-tier suppliers working directly with the AMMs. This close relationship may be associated by the production of the relatively sophisticated

[4] It was found a high correlation rates between FOREIGN dummy variable and these two dummy variables constructed from survey: "Dependent on any overseas firm/group/holding" (FOREIGNGROUP) (0.88) and "being a part of MNC" (PARTMNC) (0.85). In addition, the correlation coefficient between these two last variables was equal to 0.83. Therefore, these two variables were not used in the regression analysis to avoid the problem of multicollinearity.

[5] In accordance with the previous chapters, we here also define foreign firms as firms if FS is at least 10 % or more in total capital of the firm.

part and components for customers and hence be more prone to the KTTs. By definition, the likelihood of lower-tier suppliers (such as second and third-tier) to benefit from such an advantage is extremely low. Having a good reputation in the sector, benefiting from a stable demand and being part of design activities in its early stages are other—potential—advantages of being a direct supplier which may exert a positive influence on KTT-related activities (see Sect. 4.7.3). It also shows that absorptive capacity of the supplier is high.

6. LOCALGROUP, *being a part of a group in Turkey*, is a dummy variable, and it informs us of whether a supplier firm is part of a larger group, a parent company or holding (see Sect. 4.5.1). Being part of a group can bring many advantages related to financial structure, R&D and innovation activities, market structure, business network, etc. Such a membership may be conducive to financial and training assistance and KTTs in case the supplier firm works for a manufacturer itself as part of the larger group, since this can reduce transaction costs and build trust between both firms. On the other hand, such a status may exert a negative effect on the KTTs since the role attributed to the supplier within its group may not be conducive to such relationships. This negative effect may also be the result of the formal technology transfer channels used by the group—i.e. technology licenses—which reduce the need for KTTs.
7. CLIENTMNC, *MNC among clients*, is also a binary variable indicating whether a supplier firm has a multinational among its customers (see Sect. 4.5.1). If this is the case, then multinationals can positively impact on the KTTs of suppliers by being more stringent on issue (such as delivery time, quality, and cost), and also by selecting among its suppliers those capable to act as co-designers. In this context, KTTs can take place from the same multinationals and/or various collaborations can be established with other firms. Of course, the position of suppliers in the supply chain of multinationals is also important for the final outcome. In summary, it may positively affect the technology, innovation, production and design capabilities of the suppliers. This assumption has been tested by introducing this binary variable in the regressions.
8. RDDEPT,[6] *possessing an R&D department*, is a binary variable indicating whether a supplier has a separate R&D department or not (see Sect. 4.8.1). It indicates that R&D activities are carried out by suppliers and they have also been made at an advanced level. These activities also show that supplier firm has a specific background related to production and especially in technology fields. This can increase the firms' self-sufficiency in the field of technology and reduce the need of KTTs from outside the firm. Moreover, this variable is an indicator of the level of the absorptive capacity of suppliers. A higher absorptive capacity may signal to AMMs the higher potential of a supplier in

[6] It was preferred to use this variable instead of "carrying out R&D activities" variable (RD), since the latter could not yield any significant results in the regression analyses. Although, we wanted to use "logarithm of the R&D expenditures" variable (RDEXP) and "share of the R&D expenditures in the total sales" variable (or R&D intensity (RDINT)) in the analyses, they included so many missing values and had high correlation rates with SENGPER variable.

such innovative activities (such as co-design, product quality improvement, and product development), and therefore foster cooperation between the two parties and give rise to the KTTs.

9. PRODUCTINNO,[7] *carrying out innovation activities related to product*, is also a dummy variable indicating whether a supplier carries out innovation activities related to product (see section "Technological Product and Process Innovation Activities (Technological Output)" in Chap. 4). This variable is an indicator of the level of the innovation activities in terms of technology and design capabilities of the suppliers. A higher product innovation activity may signal to AMMs the higher potential of a supplier in such activities. As an indicator, this variable may also play an important role related to strategic relationships in terms of new product development between AMMs and suppliers. Therefore, this can increase KTTs from AMMs to suppliers. Moreover, it could be an indicator of the absorptive capacity of the supplier firms.
10. SENGPER,[8] *share of engineers in total employment*, is measured as the ratio of engineers to total employment. This variable is an indicator of the high-skilled human capital level of the suppliers. Besides, a high amount of highly-skilled work force is an important indicator of advanced technological capabilities. This variable is also used as an alternative indicator to measure the absorptive capacity of suppliers because the most important requirement of operating with complex and developed production technologies and performing R&D and innovation activities is an advanced absorptive capacity at the firm-level (see Sect. 4.8.3).
11. HIGHTECHPROD,[9] *manufacturing high-tech products*, is a dummy variable indicating that the supplier manufactures high-tech products. Manufacturing technologically complex or high-tech products could be an indicator of the technological level of the supplier (see Sect. 4.1.6). After all, firms involved in high-tech production need to deal with developed production processes and innovation activities. Therefore, firms can lead KTTs to search for these activities. Moreover, manufacturing high-tech products may foster cooperation between AMMs and suppliers.
12. HIGHDESCAP,[10] *own a high-design capability*, is a dummy variable indicating that design capability of the supplier is quite high or developed (see Sect. 4.4.5). Design capability could be an important indicator for the supplier's

[7] It was preferred to use this variable instead of "carrying out innovation activities" variable (INNO), since the latter includes high number of missing values, has high correlation with RDDEPT, and could not yield any significant results in the regression analyses.

[8] Moreover, we want to use productivity variable, "logarithm of the total sales per employee" (PRODUCTIVITY) in the analyses; however, they include so many missing values and have high correlation rates with SENGPER.

[9] It was preferred to use this binary variable instead of "technological complexity level of the manufactured product" ordinal variable (TECHPROD) as (1) low, (2) medium, and (3) high, since the latter could not yield significant results in the regression analyses.

[10] It was preferred to use this binary variable instead of "design capability level of the supplier" ordinal variable (DESCAPAB) from 1 (low) to 4 (high) since the latter could not yield significant results in the regression analyses.

technological capability and innovation activities related to products. It may also play an important role in the supplier selection process, in the development of new products and in various cooperation activities with other firms (especially with AMMs). Moreover, to become a direct supplier, owning a high design capability is accepted as important criteria by AMMs (for details see Sect. 5.4.6). Lastly, to be capable of high design may foster cooperation and increase the probability of KTTs from customers.

13. ONESUPP,[11] *being the sole supplier of the most important customer*, is a dummy variable indicating that a supplier is the sole supplier of the most important customer per item purchased. For the reasons discussed in detail in Sects. 4.5.4 and 4.7, the effect of this variable is analyzed.
14. SSUBCONT,[12] *share of subcontracting contracts in total contracts*, is measured as the ratio of subcontracting *contracts* to total *contracts* (section 6, question 5 in the questionnaire, see Appendix G). These agreements are signed to perform some tasks in order to lower costs and share risks between suppliers and AMMs. It could also be indicative of the cooperation level between these firms in automotive supply industry since these activities are regularly performed with the same firms. Hence, these allow suppliers to further develop their skills and technological capabilities, and may give rise to KTTs to suppliers.
15. RDSUB, *R&D subsidy from government or various institutions*, is a dummy variable indicating that the supplier gets R&D support from various institutions (such as KOSGEB, TUBITAK, TEYDEB, TTGV, and EU) (see Sect. 4.8.1). It is expected that these kinds of subsidies show a positive effect on firms' R&D activities. Hence, the effect of this variable is analyzed only on *possessing an R&D department* (*RDDEPT*) dependent variable as an indicator of *performing R&D activities*.

Table 5.1 shows the summarized relationships between supplier characteristics and explanatory variables as a hypothetical based on literature and previous empirical studies. These variables may act as a proxy for supplier characteristics; in other words, they may be an indicator of the level of the supplier firm capabilities.

Summary statistics on the explanatory variables are provided in Appendices J and L. Moreover, a correlation matrix containing simple correlation coefficients between explanatory variables is given in Appendix M. As can be seen in the table, some of the correlations are significant, but they are below ±0.40 and even below 0.20. Therefore, multicollinearity between explanatory variables does not occur and does not affect the interpretations of the analysis results.

[11] It was preferred to use this variable (ONESUPP) instead of "main customer of the supplier firm" ordinal variable (MC) as 1 (DAMM), 2 (OAMM), 3 (DS) and 4 (OS) since the latter could not yield significant results in the analyses.

[12] It was preferred to use this variable instead of "logarithm of the duration of the working together with the most important customer" variable (WORKTGT) (section 6, question 4 in the questionnaire, see Appendix G) since the latter could not yield significant results and had high correlation with the FIRMAGE variable.

Table 5.1 Summary of the relationships between supplier characteristics and explanatory variables

Supplier characteristics	Explanatory variables
General characteristics	– *FIRMAGE* – *FIRMSIZE* – *FOREIGN* – *DIRECTSUPP* – *LOCALGROUP*
Production capability	– *EXPINT* – *RDDEPT* – *SENGPER* – *HIGHTECHPROD* – *HIGHDESCAP*
Innovative capability	– *RDDEPT* – *PRODUCTINNO* – *HIGHDESCAP*
Absorptive capacity	– *EXPINT* – *DIRECTSUPP* – *RDDEPT* – *PRODUCTINNO* – *HIGHDESCAP* – *SENGPER*
Buyer-supplier relationship	– *DIRECTSUPP* – *CLIENTMNC* – *ONESUPP* – *SSUBCONT*

5.3 Factor Analysis (FA)

Prior to the regression analyses, a number of interrelated items in the scales were combined. Therefore, before the regression analyses were run, some dependent variables were constructed from the scales to reduce the number of the related items (or the number of dependent variables) in such scales in the survey. For this aim, the items were subjected to FA or Principal Components Analysis (PCA).[13] By using the FA method as a data reduction technique, individual items in the scales were grouped or refined to form a smaller number of items.

In the literature, there are so many recommendations on sample size to run FA: a sample size from 100 to 300 observations (Froman, 2001; Hair, Black, Babin, Anderson, & Tatham, 2006; Meyers, Gamst, & Guarino, 2006), at least 150 observations (Tabachnick & Fidell, 2007) or 10 observations for each item in the scale

[13] These two terms (FA and PCA) are often used interchangeably by researchers in the literature but we prefer to use FA.

(Nunnally, 1978). In our study, a sample size of 165 is considered sufficient to conduct FA.

The suitability of data for the application of FA is assessed by both Kaiser-Meyer-Olkin Measure of Sampling Adequacy (hereinafter KMO measure) (Kaiser, 1970, 1974) and Bartlett's Test of Sphericity[14] (hereinafter Bartlett's chi-square) (Bartlett, 1954) statistics. For a good FA, KMO measure value should be at least 0.6 (Pallant, 2007) and Bartlett's chi-square should be statistically significant ($p < 0.05$) (Tabachnick & Fidell, 2007). These two statistics are provided at the end of each table of FA presented in each section. In the FA, PCA is used as an extraction method, and *Oblimin with Kaiser Normalization* is used as a rotation method.

The number of factors[15] was extracted using eigenvalue and loading factor of 0.4. By these methods, eigenvalues of the factors are analyzed, and the number of the factors is accepted according to their eigenvalues with more than one by using Kaiser's criterion (Hair, Babin, Money, & Samouel, 2003). In the selection of these items, a cut-off loading of 0.4 has been used to screen out items (Hair et al., 2006; Stevens, 2002); in other words, factor loadings less than 0.4 are removed from the scale to improve the scale. In addition, after conducting the FA, the reliability tests (Cronbach alpha tests) are again conducted to ensure that the new scales are statistically reliable (see section "Validity and Reliability Tests" in Chap. 3 for details), and reported together with the results of the FA in each section. Finally, in the computation of the dependent variables, a replacement method is applied to the items in each factor. Moreover, in the labeling of the factors, we looked for the highest loading items on each factor.

In the remaining part of this section, we analyze the dependent and explanatory variables to explore the association between them (identified in Sect. 5.2 and Table 5.1) on the basis of the theoretical discussions in the literature and previous empirical studies, thereby presenting the findings of the econometric estimations for examination.

5.4 Logistic Regression Analysis

In this section, logistic regression analysis, also called a *logit model*, is conducted in an attempt to identify factors impacting on various characteristics and activities of suppliers discussed in the previous chapter.[16] Some points about the logistic regression model are worth mentioning.

---

[14] In this test we want to reject the null hypothesis that the correlation matrix is an identity matrix.

[15] Actually PCA yields components but it is preferred to use *factor* term for referring to the group of related items.

[16] For details on the logistic models, see Amemiya (1981), Maddala (1983), McFadden (1984), Liao (1994), Long (2001), Greene (2008), and Verbeek (2008).

- All the dependent variables used in the analyses are binary variables (0 or 1).
- It is a nonlinear regression model and estimated by maximum likelihood (ML).
- Although ML estimation works well in small samples, it is not recommended with sample size less than 100. In the literature, there is not any consensus on the sufficient number of observation to conduct ML; however, it is considered that at least ten observations per parameter are sufficient to produce reasonable and robust results. Therefore, it was decided that the sample of 165 cases is enough to use in the ML estimation (Long, 1997).[17]
- Due to our cross-sectional data, a statistically significant relationship between explanatory and dependent variables will indicate an *association* between these variables rather than a causal relationship.
- The estimated coefficients obtained by logit model do not give any information about the change in the possibility of occurrence of dependent variable associated with the impact of a one unit change in the explanatory variable (marginal effect). Instead, they represent the change in the log-odds of the dependent variable associated with a unit change in the explanatory variable. For instance, it tells how the log of the odds in favour of *being a direct supplier*—compared to those which are not direct suppliers—change as *firm age* changes by a unit, given that all of the other explanatory variables in the model are held constant.[18]
- The formula used for the marginal effects differs from the formula used to estimate the coefficients, and it is a nonlinear function of all explanatory variables. In our study, it is not possible to use marginal effects because of many dummy variables included in the model.[19] However, the sign of a coefficient is the same as the sign of its marginal effect, and significance levels are very close to each other.[20] Therefore, we shall interpret the results over the signs of the coefficients. In other words, the results of the regression analyses are examined with an emphasis on the interpretation of the coefficients estimated.

---

[17] In addition, we remark that a limited number of the observations (165) that we have can be effective in the calculation of the standard errors of the coefficients. In other words, an explanatory variable that would be statistically significant in a situation of high number of observations can be insignificant at 10 % level in our study.

[18] When interpreting the findings concerning some of the variables, the coefficients showing the direction of the effects of an explanatory variable are calculated for a situation where other explanatory variables in the regression model are constant—*ceteris paribus*.

[19] Since the derivative used for marginal effect is not so appropriate to apply the dummy variables included to model as if they were continuous variable. In this case, the results are not so accurate and it might be optimistic to rely on these results (Greene, 2008).

[20] Coefficients also provide information about the relative marginal effect of two variables. In other words, if the coefficient of X variable is higher than that of Y variable, the marginal effects of X will be higher than that of Y. See Liao (1994), Greene (2008), and Verbeek (2008) about the computation of marginal effects and their relations with the coefficients.

5.4.1 Determinants of Being a Direct Supplier of AMMs in Turkey

Findings of the logistic regression analysis on the determinants of *being a direct supplier of at least one of the AMMs operating in Turkey* are given in Table 5.2. As discussed in Sects. 4.1.1 and 4.7, 132 firms out of 165 firms (80 %) stated that they are the direct suppliers of one or more AMMs in Turkey. The dependent variable takes the value of 1 if the firm is a direct supplier, and 0 otherwise. Explanatory variables have been included in the logit model in three stages to determine the problem of a possible multicollinearity: First, there are the eight variables (Model I); then three variables, SENGPER, HIGHTECHPROD and HIGHDESCAP, are added (Model II); and lastly ONESUPP and SSUBCONT variables are added to the model (Model III).

Three explanatory variables exert a positive and significant effect on the probability of *being a direct supplier of AMMs in Turkey*: FIRMAGE (significant at 10 % level), FIRMSIZE (significant at least at 5 % level) and SENGPER (significant at 5 % level). In other words, this finding shows that probability of being a second or third- tier supplier (versus being a direct supplier) increases with being a younger firm/with a decrease in total employment/with a decrease in the share of engineers in total employment, ceteris paribus.[21]

A significant (at 5 % level) and negative association exists between EXPINT and dependent variable, ceteris paribus. The finding may be explained by the fact that supplier firms establish intense relations with the firms operating abroad with the increase of export share in their turnover; therefore, the possibility for *being a direct supplier of domestic main industry* decreases (the main firm to which they supply directly may be located abroad), ceteris paribus.

Four dummy explanatory variables have also a statistically significant and positive effect on the probability of *being a direct supplier*: FOREIGN (at 5 % level), LOCALGROUP (at 5 % level), PRODUCTINNO (at least at 10 % level) and HIGHTECHPROD (at 10 % level). In other words, probability of *being a direct supplier of the AMMs* is higher for the firms/with foreign ownership,/dependent on a local group,/carrying out product innovation activities or/manufacturing high-tech products, ceteris paribus.[22] These findings may indicate that technological competence of foreign firms operating in the field of automotive supply industry is superior to local firms; this case can be explained by the advantages provided by the intangible assets of multinationals to their associates. Additionally, being involved in product innovation activities and manufacturing products of high technology

---

[21] In other words, for a one unit increase in FIRMAGE/FIRMSIZE/SENGPER, the log odds of being direct supplier increases by 0.59/0.893/0.076 given that all of the other variables in the model are held constant (see Table 5.2).

[22] The coefficients for dummy variables are interpreted in a slightly different way. For instance, being a foreign firm (versus non-being) increases the log odds of being direct supplier by 1.328 (see Table 5.2).

Table 5.2 Determinants of being a direct supplier: logistic regression analysis

	Direct supplier		
	Model I	Model II	Model III
FIRMAGE	0.524 (0.310)*	0.590 (0.358)*	0.347 (0.397)
FIRMSIZE	0.462 (0.281)	0.610 (0.296)**	0.893 (0.337)***
EXPINT	−0.018 (0.007)**	−0.020 (0.008)**	−0.020 (0.009)**
FOREIGN	1.328 (0.554)**	0.642 (0.588)	0.156 (0.667)
LOCALGROUP	0.920 (0.569)	0.921 (0.567)	1.255 (0.613)**
CLIENTMNC	0.052 (0.659)	−0.041 (0.768)	0.066 (0.845)
RDDEPT	0.011 (0.472)	−0.334 (0.576)	−0.386 (0.670)
PRODUCTINNO	0.793 (0.474)*	1.115 (0.524)**	1.472 (0.621)**
SENGPER		0.076 (0.040)*	0.100 (0.062)
HIGHTECHPROD		0.911 (0.491)*	1.037 (0.567)*
HIGHDESCAP		−0.128 (0.551)	0.193 (0.631)
ONESUPP			0.660 (0.546)
SSUBCONT			0.000 (0.009)
DIRECTSUPP			
RDSUB			
Obs. (N)	157	156	145
Likelihood	−64.536	−60.255	−51.569
Pseudo R2	0.157	0.211	0.260
McFadden's adj R2	0.040	0.054	0.059
Wald chi2 test	25.99 (0.0011)	29.61 (0.0018)	35.89 (0.0006)

Source: Author calculations based on survey results
Notes: ***, **, * denote significance level at 1 %, 5 % and 10 %, respectively. Robust standard errors of the coefficients are shown in parentheses. All models include intercepts significant at 1 % level but not shown in the results for space reasons. Two-tailed test. P-values are in brackets in the last rows.

indicate that some competences which should be present in direct supplier firms are available in the firms engaged in these activities according the viewpoint of AMMs. Higher possibility of firms affiliated to a local group to be the direct supplier is similarly the result of these firms having more extensive, technological and developed business networks.

5.4.2 Determinants of the Decision to Have an R&D Department

Table 5.3 presents the estimation results on the determinants of *the decision to have a separate R&D department*. The dependent variable takes the value of 1 if the firm has an R&D department, and 0 otherwise. As discussed in detail in Sect. 4.8.1, 133 firms out of 165 firms (81 %) stated that they perform R&D activities, and 99 of them (60 %) reported that they possess a separate R&D department. However, regression analyses generated by using the dependent variable of *performing R&D*

Table 5.3 Determinants of the decision to have an R&D department: logistic regression analysis

	R&D department		
	Model I	Model II	Model III
FIRMAGE	−0.169 (0.355)	0.036 (0.382)	0.312 (0.423)
FIRMSIZE	0.149 (0.227)	0.331 (0.252)	0.458 (0.268)*
EXPINT	0.003 (0.008)	0.005 (0.008)	0.007 (0.009)
FOREIGN	−0.059 (0.490)	−0.433 (0.552)	−0.454 (0.607)
LOCALGROUP	0.772 (0.427)*	0.851 (0.445)*	0.892* (0.504)*
CLIENTMNC	0.396 (0.609)	0.019 (0.713)	−0.549 (0.623)
RDSUB	1.941 (0.409)***	1.928 (0.450)***	1.939 (0.467)***
PRODUCTINNO	1.853 (0.431)***	1.998 (0.451)***	2.106 (0.498)***
DIRECTSUPP	−0.349 (0.488)	−0.529 (0.525)	−0.585 (0.608)
SENGPER		0.083 (0.036)**	0.121 (0.042)**
HIGHTECHPROD		0.090 (0.430)	0.178 (0.459)
HIGHDESCAP		1.043 (0.456)**	1.263 (0.472)***
ONESUPP			−0.312 (0.498)
SSUBCONT			−0.013 (0.008)
Obs. (N)	157	156	145
Likelihood	−77.969	−70.679	−61.535
Pseudo R2	0.263	0.326	0.374
McFadden's adj R2	0.168	0.202	0.222
Wald chi2 test	46.92 (0.0000)	47.15 (0.0000)	57.51 (0.0000)

Source: Author calculations based on survey results

Notes: ***, **, * denote significance level at 1 %, 5 % and 10 %, respectively. Robust standard errors of the coefficients are shown in parentheses. All models include intercepts significant at 1 % level but not shown in the results for space reasons. Two-tailed test. P-values are in brackets in the last rows.

activities did not yield significant results. Therefore, regression analyses regarding the issues determining *the possession of R&D department* were made. Possessing an R&D department bears great importance as it indicates that R&D activities are performed at an advanced level.

A significant and positive association was found between two explanatory variables, FIRMSIZE (at 10 % level) and SENGPER (at 5 % level), and the dependent variable of *possessing an R&D department*. It was observed that the higher the firm size or the share of engineers in total employment, the higher the probability of *owning an R&D department*. The literature review carried out by Cohen (2010) reveals that the possibility of *conducting R&D activities* throughout the manufacturing industry and in many sub-industries increases in line with the size of the firm. Greater possibility for firms having a greater share of engineers in the total employees to *possess an R&D department* when compared with the firms not having such a share may be associated with the relatively higher technological capability and absorption capacity of firms ranking in this category. Furthermore, the effect of these two explanatory variables on the possibility of *being a direct supplier* was also found positive and statistically significant (see Sect. 5.4.1); therefore, the relation between these variables can be interpreted similarly.

The estimated coefficients of the four dummy variables are also statistically significant and exert a positive effect on the probability of *owning an R&D department*: LOCALGROUP (at 10 % level), PRODUCTINNO (at 1 % level), HIGHDESCAP (at least at 5 % level), and RDSUB (at 1 % level). It should be noted that two dummy variables, LOCALGROUP and PRODUCTINNO, have also significant and positive effect on the probability of *being a direct supplier* (see Sect. 5.4.1). In other words, the probability of *being a direct supplier* or *owning an R&D department* is higher for the firms dependent on a local group and conducting product-innovation activities, ceteris paribus. Similarly, having a high design capability increases the probability of *owning an R&D department.* Moreover, R&D subsidies have a significantly positive effect on the firms' R&D activities.

In sum, these findings are important as they show that firms, which have high design capabilities, manufacture high quality and innovative technological products, *possess R&D departments.* Besides, they may indicate that firms operating in supply industry manufacture high quality products for AMMs in global production chains. Also, an important discovery regarding R&D incentives is that the incentives which were put into effect in 1990s but have become widespread since mid-2000s have positive effects on R&D activities of automotive supply industry.

5.4.3 Determinants of the Innovation Activities

The estimation results pertaining to the determinants of *the innovation activities of suppliers* are displayed in Table 5.4. The dependent variable takes the value of 1 if the firm conducts innovation activities, and 0 otherwise. As discussed in Sect. 4.8.2, 146 out of 165 firms (88 %) stated that they conduct innovation activities. Explanatory variables have been included in the logit model in three stages: First, there are the seven variables (Model I); then two variables, HIGHTECHPROD and HIGHDESCAP, are added (Model II); and lastly the CLIENTMNC variable is added to the model (Model III).

The impact of the three variables on the probability of carrying out *innovation activities* is statistically significant: While EXPINT (significant at least at 5 % level) and CLIENTMNC (significant at 10 % level) explanatory variables exert positive impact on this probability; FOREIGN variable (significant at least at 10 % level) exerts a negative impact. Either an increase in the share of export in total sales or presence of MNC among customers increases the possibility of conducting *innovation activities*, ceteris paribus. This indicates that firms manufacturing for export market or MNC customers meet both the required conditions (technology, design, quality, delivery, defect rate, etc.) regarding their products, re-design these products according to the conditions of the export markets, and carry out technological *innovation activities* needed for international competition. One disadvantage of *being a foreign-capital firm* is that foreign firms are less likely to make innovations at host country when compared to the local firms, ceteris paribus, which was

Table 5.4 Determinants of the innovation activities: logistic regression analysis

	Innovation activities		
	Model I	Model II	Model III
FIRMAGE	−0.480 (0.512)	−0.529 (0.546)	−0.572 (0.546)
FIRMSIZE	0.082 (0.212)	0.083 (0.210)	0.050 (0.217)
EXPINT	0.486 (0.227)**	0.510 (0.246)**	0.420 (0.249)*
FOREIGN	−1.514 (0.721)**	−1.436 (0.724)**	−1.465 (0.794)*
DIRECTSUPP	0.392 (0.722)	0.473 (0.726)	0.649 (0.728)
RDDEPT	0.744 (0.577)	0.792 (0.577)	0.822 (0.593)
SENGPER	0.460 (0.491)	0.472 (0.511)	0.205 (0.504)
HIGHTECHPROD		−0.561 (0.602)	−0.494 (0.634)
HIGHDESCAP		0.268 (0.594)	0.280 (0.619)
CLIENTMNC			1.245 (0.705)*
Obs. (N)	141	140	138
Likelihood	−45.773	−45.127	−42.888
Pseudo R2	0.118	0.128	0.134
McFadden's adj R2	0.036	0.065	0.088
Wald chi2 test	17.85 (0.0127)	20.30 (0.0162)	20.85 (0.0222)

Source: Author calculations based on survey results
Notes: ***, **, * denote significance level at 1 %, 5 % and 10 %, respectively. Robust standard errors of the coefficients are shown in parentheses. All models include intercepts significant at 1 % level but not shown in the results for space reasons. Two-tailed test. P-values are in brackets in the last rows.

confirmed by a series of studies on the developing countries.[23] This finding may be indicative of the fact that MNC affiliates, assuming that foreign firms are the affiliates of multinationals, are equipped with the technology and knowledge of the main firm. Thus, it may mean that they do not carry out innovation activities in the host countries, or they perform *incremental innovation* in order to adapt their products to local conditions. Standardization of consumer preferences and emergence of the *world products* phenomenon in the twenty-first century may affect the *innovation/R&D activities* of foreign firms negatively. Indeed, the results of the present study also show that FOREIGN variable exerts a negative, yet insignificant, impact on the probability of *possessing an R&D department* in Sect. 5.4.2 (see Table 5.3). These findings overlap with our survey findings (see section "Cooperation Level with Partners in Innovation Activities" in Chap. 4) that foreign firms conduct basic innovation and R&D activities in their headquarters.

[23] UNCTAD (2005) and references in there, see Pamukçu (2003) and Dayar and Pamukçu (2011) for Turkey.

Determinants of the Product Innovation Activities[24]

The results of the regression analysis on the determinants *of product innovation activities* are given in Table 5.5. The dependent variable takes the value of 1 if the firm said *Yes* to the question concerning the performance of product innovation activities, and 0 otherwise. As discussed in detail in section "Technological Product and Process Innovation Activities (Technological Output)" in Chap. 4, 120 (73 %) firms out of 164 firms stated that they perform product innovation activities. Explanatory variables have been included in the logit model in three stages: First, there are the seven variables (Model I); then two variables, HIGHTECHPROD and HIGHDESCAP, are added (Model II); and lastly the CLIENTMNC variable is added to the model (Model III).

EXPINT explanatory variable exerts a significant a positive impact on the probability of *performing product innovation activities* (significant at 5 % level) as well as on the probability of *innovation activities* (see Sect. 5.4.3). This finding, as interpreted in the possibility of *performing innovation activities*, indicates that firms manufacturing for global markets carry out various product innovation activities.

Table 5.5 Determinants of the product innovation activities: logistic regression analysis

	Product innovation activities		
	Model I	Model II	Model III
FIRMAGE	−0.042 (0.360)	0.033 (0.366)	0.153 (0.364)
FIRMSIZE	0.024 (0.145)	0.006 (0.147)	−0.011 (0.147)
EXPINT	0.344 (0.152)**	0.368 (0.153)**	0.341 (0.159)**
FOREIGN	0.230 (0.571)	0.201 (0.602)	0.332 (0.611)
DIRECTSUPP	0.804 (0.482)*	0.805 (0.497)	0.660 (0.507)
RDDEPT	0.857 (0.445)*	0.772 (0.445)*	0.671 (0.450)
SENGPER	−0.465 (0.300)	−0.475 (0.324)	−0.453 (0.338)
HIGHTECHPROD		0.135 (0.453)	0.269 (0.459)
HIGHDESCAP		0.265 (0.416)	0.187 (0.415)
CLIENTMNC			−0.149 (0.648)
Obs. (N)	140	139	137
Likelihood	−74.699	−73.000	−71.491
Pseudo R2	0.098	0.105	0.094
McFadden's adj R2	0.002	0.018	0.045
Wald chi2 test	15.30 (0.0323)	16.88 (0.0506)	14.68 (0.1440)

Source: Author calculations based on survey results

Notes: ***, **, * denote significance level at 1 %, 5 % and 10 %, respectively. Robust standard errors of the coefficients are shown in parentheses. All models include intercepts significant at 1 % level but not shown in the results for space reasons. Two-tailed test. P-values are in brackets in the last rows.

[24] We also tested the determinants of innovation activities related to production process, supporting activities, service, logistics, delivery and distribution (see Table 4.33); however, regression analyses did not yield significant results.

Two dummy variables, DIRECTSUPP and RDSUB (each one significant at 10 % level), also positively impact the possibility of *performing product innovation activities*. Being a direct supplier increases the possibility of conducting *product-innovation activities*, ceteris paribus. As explained previously (see Sect. 4.7), this finding is the result of the fact that the firms which are the direct suppliers of AMMs uses high technology, meeting the various product quality and manufacturing standards, and establishing stronger technological cooperation with the AMMs. Some firms receive R&D incentives. These firms' increasing likelihood of conducting *product innovation activities* when compared with the firms not receiving is important, for it shows that such incentives affect R&D activities positively (see Sect. 5.4.2) and foster the firms' *performing innovation activities*.

5.4.4 Determinants of the Technology Agreements (TA)

This section examines the determinants of *the technology agreements* (TA) of suppliers with other firms so as to develop the KTTs. Prior to the regression analysis, FA is conducted for technology agreements scale. The details of the FA including the reliability of the factors (Cronbach's alpha) are shown in Table 5.6. The results reveal that 11 types of *the technology agreements* were combined in two factors (groups). The first factor was labeled as *low-advanced technology agreements* (Factor 1), and the second one as *advanced technology agreements* (Factor 2).[25]

The estimation results on the determinants of *the technology agreements* performed by the supplier firms are presented in Table 5.7. Dependent variables take the value of 1 if the firm performs *technology agreements* with other firms, and 0 otherwise.[26]

Four explanatory variables exert a significantly positive influence on the probability of performing *low-advanced technology agreements* (listed under Factor 1 in Table 5.6): FIRMSIZE (significant at 5 %), DIRECTSUPP (significant at 5 %), RDDEPT (significant at 10 %), and SENGPER (significant at 5 %). On the other hand, four explanatory variables have a statistically significant impact at 10 % level on the probability of performing *advanced technology agreements* (listed under Factor 2 in Table 5.6): SENGPER, SSUBCONT, and PRODUCTINNO variables impact these kinds of agreements positively, while LOCALGROUP variable impacts them negatively.

These findings reveal that the possibility of making such kind of *technology agreements* regarding complex and advanced technology which includes the KTT increases according to the technology level of the firms (RDDEPT, SENGPER, PRODUCTINNO), the increase in the share of subcontracting agreements

---

[25] The reliability of the factors is also high, with a high Cronbach's alpha value over 0.7.

[26] As presented in detail in Sect. 5.3, a replacement method is applied to the items in each factor in the computation of the dependent variables for technology agreements (TA).

Table 5.6 Factor analysis of the technology agreements (TA)

Items	Factor 1 low-advanced TA	Factor 2 advanced TA	Cronbach's alpha (a)
1. Turn-key projects	0.559		0.737
2. Management contracts	0.671		
9. Technical assistance agreement	0.513		
3. Licensing agreement		0.505	0.793
4. Know-how agreement		0.610	
5. Purchasing of engineering services		0.469	
6. Joint venture agreement		0.653	
7. Purchasing of license and/or patent		0.632	
8. Agreements on personnel exchange		0.534	
10. International subcontracting agreements		0.443	
11. Agreements on the determination of product design and/or technical specifications			
N = 165, KMO measure = 0.650, Bartlett's chi-square = 203.32 (0.0000)			

Source: Author's calculations based on survey data
Notes: Table shows the factor loadings of each item above 0.4. The item 11 either does not fit well with the other items, or shows the lowest loading on factors. Therefore, it was removed from the analysis. Extraction method: principal component analysis; rotation method: Oblimin with Kaiser normalization. P-value is in bracket.
Legend: TA (Technology Agreements)

Table 5.7 Determinants of the technology agreements (TA): logistic regression analysis

	Technology agreements (TA)	
	Factor 1 (low-advanced TA)	Factor 2 (advanced TA)
FIRMAGE	−0.110 (0.327)	−0.369 (0.324)
FIRMSIZE	0.568 (0.260)**	0.128 (0.221)
EXPINT	0.007 (0.008)	0.007 (0.008)
FOREIGN	0.198 (0.540)	−0.305 (0.473)
DIRECTSUPP	1.139 (0.532)**	−0.278 (0.560)
LOCALGROUP	0.248 (0.451)	−0.677 (0.397)*
CLIENTMNC	−1.133 (0.727)	0.105 (0.599)
RDDEPT	0.809 (0.423)*	0.350 (0.472)
PRODUCTINNO	0.129 (0.454)	0.893 (0.530)*
SENGPER	0.109 (0.055)**	0.062 (0.036)*
HIGHTECHPROD	0.024 (0.409)	0.272 (0.403)
HIGHDESCAP	0.332 (0.424)	0.108 (0.409)
ONESUPP	−0.088 (0.415)	0.460 (0.394)
SSUBCONT	−0.002 (0.007)	0.012 (0.007)*
Obs. (N)	145	145
Likelihood	−76.935	−85.057
Pseudo R2	0.208	0.112
McFadden's adj R2	0.054	−0.045
Wald chi2 test	26.23 (0.0242)	16.17 (0.3028)

Source: Author calculations based on survey results
Notes: ***, **, * denote significance level at 1 %, 5 % and 10 %, respectively. Robust standard errors of the coefficients are shown in parentheses All models include intercepts significant at 1 % level but not shown in the results for space reasons. Two-tailed test. P-values are in brackets in the last rows.

(SSUBCONT), and the size of the firm (FIRMSIZE). Moreover, being a direct supplier (DIRECTSUPP) increases the possibility of making such kind of technology agreements when compared to those which are not direct suppliers. The negative effects of being a part of a local group (LOCALGROUP) may be that such kind of agreements are mostly realized by the group to which the firms are affiliated, or the processes which are subject to the agreements are mostly carried out by the affiliated main group firms rather than the third party firms. Still, the negative effects of being a part of the local group are debatable.

5.4.5 Determinants of the Sources of Technologies Acquired and/or Used

This section discusses the determinants of the sources of technologies acquired and/or used by the firms (see Sect. 4.4.4). FA is conducted for *the sources of technologies* scale. The results of the FA including the reliability of the factors are shown in Table 5.8, wherein seven types of the sources of technologies were combined in three factors (groups). The first factor was labeled as *AMMs* (Factor 1), and the second one as *lower-tier suppliers* (Factor 2), and the last one as *universities and other firms*[27] (Factor 3).

The estimation results on the determinants of the technology sources acquired and/or used by the suppliers are displayed in Table 5.9. Dependent variables take the value of 1 if a firm acquires the technology from that source, and 0 otherwise.

A significant relationship is found between FIRMAGE explanatory variable and *the AMMs* dependent variable (Factor 1) at 10 % level. A similar result is also obtained for the FIRMAGE variable on *the universities and other firms* (Factor 3) dependent variable (at 10 % level). Accordingly, operating in the automotive supply industry for a long time increases the possibility of obtaining technology resources from *the AMMs*, and decreases the possibility of obtaining these resources from both *the universities and other firms*. In other words, ceteris paribus, the more experience a firm has in the field of automotive supply industry, the higher the possibility of working with *the AMMs* (for technology acquisition) is and the lower the possibility of working with both *the universities and the other firms* is. While the positive effect of operating in the automotive supply industry for a long time on this possibility can be regarded as indicative of the relations and competence established with *the AMMs*, the negative effect can be perceived as an indicative of the efforts of mostly the new firms to acquire technology mainly from *the universities and other resources*.

FIRMSIZE exerts a significantly positive impact at 10 % level on the probability of acquiring technologies from *the AMMs* dependent variable (Factor 1). A similar result is also obtained for this variable on *the universities and other firms* (Factor 3)

[27] Private engineering and consultancy firms.

Table 5.8 Factor analysis for the sources of technologies

Items	Factor 1 AMMs	Factor 2 lower-tier suppliers	Factor 3 uni. and other firms	Cronbach's alpha (α)
1. DAMM	0.887			0.735
2. OAMM	0.869			
3. DS		0.738		0.813
4. OS		0.844		
6. Universities			0.736	0.546
7. Private engineering and consultancy firms			0.861	
5. PC				
N = 165, KMO measure = 0.570, Bartlett's chi-square = 238.22 (0.0000)				

Source: Author's calculations based on survey data

Notes: Table shows the factor loadings of each item above 0.4. The item 5 either does not fit well with the other items, or shows the lowest loading on factors. Therefore, it was removed from the analysis. Extraction method: principal component analysis; rotation method: Oblimin with Kaiser normalization. P-value is in bracket.

Legend: DAMM (Domestic Automotive Main Manufacturers); OAMM (Overseas Automotive Main Manufacturers); DS (Domestic Suppliers); OS (Overseas Suppliers); PC (Parent Company)

dependent variable (highly significant at 1 % level). This finding shows that the probability of acquiring the technologies from *the AMMs* or *universities and other firms* increases with *firm size*.

FOREIGN dummy variable exerts a significantly negative impact on the probability of acquiring technologies from *the lower-tier suppliers* (Factor 2) dependent variable (significant at 5 % level). In other words, firms with foreign capital reduce the probability of technology acquisition from *the lower-tier suppliers* compared to those without foreign capital, ceteris paribus. This dummy variable also reduces the probability of acquiring technologies from *the AMMs* (Factor 1) and *universities and other firms* (Factor 3) dependent variables. Moreover, it has a negative, yet insignificant, impact on *domestic* and *overseas dep*endent variables, as explained below. When we focus on the negative effect of being a foreign-capital firm, the negative effect points out that being a foreign-capital firm, ceteris paribus, affects technology acquisition from *domestic* and *overseas* sources negatively, which brings up the question of in what way foreign firms acquire technology. These findings may indicate that foreign-capital firms acquire technology from the multinationals they are affiliated to, which may point out that the foreign affiliate is dependent on the main firm in terms of the technology.

LOCALGROUP dummy variable exerts a significant impact at 10 % level on *the AMMs* dependent variable (Factor 1). A similar result is also obtained for this variable on *the universities and other firms* (Factor 3) dependent variable (significant at 10 % level). According to this finding, *being a part of a local group* affects negatively the probability of acquiring technologies from *the AMMs*, while it positively affects that from *the universities and other firms*. The preference of the firms affiliated to a local group for their technology supply may depend on the

Table 5.9 Determinants of the technology sources: logistic regression analysis

	Sources of the technologies used and/or acquired				
	Factor 1 (AMMs)[a]	Factor 2 (lower-tier suppliers)[b]	Factor 3 (uni. and other firms)[c]	Domestic[d]	Overseas[e]
FIRMAGE	0.485 (0.289)*	−0.111 (0.299)	−0.526 (0.310)*	−0.102 (0.316)	0.017 (0.303)
FIRMSIZE	0.443 (0.238)*	0.228 (0.232)	0.717 (0.239)***	0.197 (0.244)	0.477 (0.235)**
EXPINT	−0.006 (0.007)	−0.005 (0.008)	−0.005 (0.007)	−0.010 (0.008)	0.008 (0.007)
FOREIGN	−0.273 (0.466)	−0.966 (0.482)**	−0.558 (0.506)	−1.009 (0.497)**	−0.495 (0.480)
DIRECTSUPP	0.371 (0.526)	−0.644 (0.579)	−0.010 (0.560)	0.962 (0.543)*	−1.043 (0.548)*
LOCALGROUP	−0.702 (0.408)*	0.109 (0.408)	0.782 (0.411)*	−0.321 (0.444)	0.111 (0.412)
CLIENTMNC	0.564 (0.558)	0.195 (0.541)	0.468 (0.640)	1.196 (0.569)**	−0.098 (0.568)
RDDEPT	−0.436 (0.420)	0.023 (0.479)	0.460 (0.476)	0.057 (0.479)	−0.017 (0.457)
PRODUCTINNO	−0.159 (0.470)	−0.400 (0.531)	0.485 (0.516)	−0.777 (0.541)	−0.417 (0.504)
SENGPER	0.032 (0.023)	−0.043 (0.024)*	0.019 (0.025)	−0.016 (0.026)	0.035 (0.030)
HGHTECHPROD	0.106 (0.367)	0.169 (0.382)	0.395 (0.411)	0.537 (0.408)	0.668 (0.382)*
HGHDESCAP	−0.344 (0.368)	0.038 (0.394)	−0.094 (0.416)	−0.666 (0.410)	−0.087 (0.386)
ONESUPP	0.185 (0.377)	0.287 (0.397)	−0.617 (0.424)	−0.168 (0.412)	0.105 (0.389)
SSUBCONT	0.006 (0.006)	0.009 (0.006)	0.001 (0.006)	−0.003 (0.008)	0.005 (0.007)
Obs. (N)	145	145	145	145	145
Likelihood	−91.346	−87.458	−84.467	−79.462	−87.740
Pseudo R2	0.090	0.086	0.149	0.107	0.084
McFadden's adj R2	−0.059	−0.070	−0.002	−0.061	−0.073
Wald chi2 test	15.26 (0.3605)	15.06 (0.3742)	29.39 (0.0092)	19.07 (0.0045)	15.98 (0.0352)

Source: Author calculations based on survey results

Notes: ***, **, * denote significance level at 1 %, 5 % and 10 %, respectively. Robust standard errors of the coefficients are shown in parentheses All models include intercepts significant at 1 % level but not shown in the results for space reasons. Two-tailed test. P-values are in brackets in the last rows.

[a]Domestic and overseas automotive main manufacturers

[b]Domestic and overseas suppliers

[c]Universities, private engineering and consultancy firms

[d]Domestic automotive main manufacturers and domestic suppliers

[e]Overseas automotive main manufacturers and overseas suppliers

extent of the privileged relations with the parent company or the subsidiaries of the conglomerate, or they benefit from the network relations in the company.

SENGPER variable has a negative influence on the probability of technology acquisition from *the lower-tier suppliers* (Factor 2) (significant at 10 % level). This finding shows that an increase in the skilled human-capital level of the supplier gives rise to a decrease in the possibility of technology acquisition from *the lower-tier suppliers*. On the other hand, effects of this variable on the other dependent variables are positive, though statistically insignificant.

Technology acquisition resources were divided into two groups of *domestic*[28] and *overseas*,[29] and tested as dependent variables. The regression results confirm the findings of the first three columns. They demonstrate that (last two columns of Table 5.9—columns 4 and 5) three variables have a significant effect on the *domestic* technology resources: *being a foreign-capital firm* (FOREIGN), which affects the usage of domestic technology resources negatively (significant at 5 % level), *being a direct supplier* (DIRECTSUPP) (at 10 % level) and *presence of MNC among customers* (CLIENTMNC) (at 5 % level), which affect them positively (column 4). The three variables having significant effect on the *overseas* technology resources are: *firm size* (FIRMSIZE), *being a direct supplier* (DIRECTSUPP) and *manufacturing high technology products* (HIGHTECHPOD) (column 5). The variables of *firm size* and *manufacturing high technology products* affect overseas technology resources positively; *being a direct supplier* affects them negatively. Especially the finding specifying that *being a direct supplier* affects the usage of *domestic* technology resources positively but the usage of *overseas* technology resources negatively is important. This finding points out that the important resources of the technology acquired by the direct supplier firms are especially AMMs operating in Turkey (DAMM), and followed by domestic suppliers (DS).

5.4.6 Determinants of the Co-design Activities Related to Products

Findings of the regression results on the determinants of *co-design activities of the suppliers with the customers pertaining to the products* are presented in Table 5.10. As examined in detail in Sect. 4.4.6, 105 out of the 165 firms (64 %) stated that they perform *co-design activities with their customers as to the products*. The dependent variable takes the value of 1 if the firm performs co-design activities with their customers, and 0 otherwise. Explanatory variables have been included in the logit model in three stages: First, there are the eight variables (Model I); then PRODUCTINNO variable is added (Model II); and lastly the three more variables,

[28] DAMM and DS.

[29] OAMM and OS.

Table 5.10 Determinants of the co-product design activities: logistic regression analysis

	Co-design activities related to products		
	Model I	Model II	Model III
FIRMAGE	0.167 (0.289)	0.151 (0.284)	0.192 (0.287)
FIRMSIZE	0.405 (0.224)*	0.400 (0.229)*	0.395 (0.239)*
EXPINT	0.012 (0.007)*	0.011 (0.007)	0.012 (0.007)*
FOREIGN	0.292 (0.507)	0.216 (0.521)	0.163 (0.547)
DIRECTSUPP	1.028 (0.481)**	0.861 (0.495)*	0.884 (0.508)*
LOCALGROUP	−0.133 (0.403)	−0.153 (0.416)	−0.174 (0.419)
CLIENTMNC	−0.110 (0.519)	−0.240 (0.533)	−0.216 (0.535)
RDDEPT	0.951 (0.370)**	0.701 (0.394)*	0.636 (0.435)
PRODUCTINNO		0.981 (0.420)**	0.913 (0.432)**
SENGPER			0.001 (0.022)
HIGHTECHPROD			0.157 (0.410)
HIGHDESCAP			0.070 (0.396)
Obs. (N)	158	157	156
Likelihood	−89.641	−86.444	−85.468
Pseudo R2	0.137	0.160	0.161
McFadden's adj R2	0.050	0.062	0.033
Wald chi2 test	22.10 (0.0047)	26.93 (0.0014)	27.28 (0.0070)

Source: Author calculations based on survey results
Notes: ***, **, * denote significance level at 1 %, 5 % and 10 %, respectively. Robust standard errors of the coefficients are shown in parentheses. All models include intercepts significant at 1 % level but not shown in the results for space reasons. Two-tailed test. P values are in brackets in the last rows.

SENGPER, HIGHTECHPROD and HIGHDESCAP, are added to the model (Model III).

Findings show that five explanatory variables exert statistically significant and positive impact on the probability of *performing co-design activities with the customers as to the products*: FIRMSIZE (at 10 % level), EXPINT (at 10 % level), RDDEPT (at least 10 % level), PRODUCTINNO (at 5 % level) and DIRECTSUPP (at least at 10 % level). Enhancement of the possibility of *performing co-design activities related to product* by an increase in the export rate—ceteris paribus—indicates that the design of the products manufactured should be revised as the importance of the foreign markets increases in the turnover and this revision is realized in cooperation with the customers.[30] The positive effects of *possessing an R&D department* (or conducting R&D activities) and *performing product innovation activities* underline the requirement of having some technological threshold competences for joint product design. The positive effect of the *firm size* may refer to effect of "threshold scale" for joint product design. Consequently, the positive effect of *being a direct supplier* (being the supplier of at least one AMMs located in Turkey) on *the joint product design*

[30] Which customers are more cooperated in these activities is also analysed in the next section "Determinants of the Partners in Cooperation for the Co-design Activities Related to Products".

may refer to the fact that the firms (direct supplier firms) are engaged in more strategic relationships with the AMMs, or they get involved in the product development phase at the beginning in the design phase.

Determinants of the Partners in Cooperation for the Co-design Activities Related to Products

FA is conducted for *the partners in cooperation for the co-design activities related to product* scale. The results of the FA including the reliability of the factors are given in Table 5.11. As can be seen in the results, *partners in cooperation for the co-product design activities* were combined in three factors (groups). The first factor was labeled as *AMMs* (Factor 1), and the second one as *lower-tier suppliers* (Factor 2), and the last one as *universities and other firms*[31] (Factor 3).

Table 5.11 Factor analysis of the partners in cooperation for the co-product design activities

Items	Factor 1 AMMs	Factor 2 lower-tier suppliers	Factor 3 uni. and other firms	Cronbach's alpha (α)
1. DAMM	0.773			0.716
2. OAMM	0.843			
3. DS		0.831		0.695
4. OS		0.749		
6. Universities			0.815	0.706
7. Private engineering and consultancy firms			0.718	
5. PC				
N = 105, KMO measure = 0.634, Bartlett's chi-square = 236.86 (0.0000)				

Source: Author's calculations based on survey data

Notes: Table shows the factor loadings of each item above 0.4. The item 5 either does not fit well with the other items, or shows the lowest loading on factors. Therefore, it was removed from the analysis. Extraction method: principal component analysis; rotation method: Oblimin with Kaiser normalization. P-value is in bracket.

Legend: DAMM (Domestic Automotive Main Manufacturers); OAMM (Overseas Automotive Main Manufacturers); DS (Domestic Suppliers); OS (Overseas Suppliers); PC (Parent Company)

The findings of regression analysis on the determinants of *the partners in cooperation for the co-product design activities* are given in Table 5.12. Dependent variables take the value of 1 if the firm performs *co-design activities with the partner*, and 0 otherwise. The analyses in this section are performed on 105 firms that reported to perform *co-design activities with the customers* (partners).

The only variable that exerts a highly significant and positive impact on the probability of *performing cooperation with the AMMs for co-product design*

[31] Private engineering and consultancy firms.

Table 5.12 Determinants of the partners in cooperation for the co-product design activities: logistic regression analysis

	Partners in cooperation for co-product design activities		
	Factor 1 (AMMs)[a]	Factor 2 (lower-tier suppliers)[b]	Factor 3 (uni. and other firms)[c]
FIRMAGE	−0.716 (0.492)	−0.265 (0.434)	−1.178 (0.556)**
FIRMSIZE	0.016 (0.449)	−0.228 (0.235)	1.047 (0.357)***
EXPINT	0.022 (0.021)	0.010 (0.009)	0.001 (0.010)
FOREIGN	−0.578 (0.871)	0.202 (0.620)	−2.593 (0.909)***
DIRECTSUPP	3.343 (1.227)***	−1.093 (0.777)	−0.582 (0.891)
LOCALGROUP	0.217 (1.058)	0.598 (0.509)	0.344 (0.595)
CLIENTMNC	−0.639 (0.853)	−0.998 (0.822)	0.182 (1.078)
RDDEPT	−1.319 (1.285)	0.598 (0.569)	1.426 (0.821)*
PRODUCTINNO	0.596 (0.980)	−0.184 (0.655)	−0.534 (0.828)
SENGPER	0.006 (0.039)	−0.021 (0.030)	0.034 (0.040)
HIGHTECHPROD	−1.104 (1.181)	0.032 (0.486)	0.415 (0.635)
HIGHDESCAP	0.008 (0.843)	−0.605 (0.479)	0.254 (0.559)
Obs. (N)	100	100	100
Likelihood	−23.842	−61.463	−43.797
Pseudo R2	0.267	0.092	0.205
McFadden's adj R2	−0.133	−0.100	−0.031
Wald chi2 test	13.80 (0.0066)	10.04 (0.4285)	23.54 (0.0190)

Source: Author calculations based on survey results

Notes: ***, **, * denote significance level at 1 %, 5 % and 10 %, respectively. Robust standard errors of the coefficients are shown in parentheses. All models include intercepts significant at 1 % level but not shown in the results for space reasons. Two-tailed test. P-values are in brackets in the last rows.

[a]Domestic and overseas automotive main manufacturers

[b]Domestic and overseas suppliers

[c]Universities, private engineering and consultancy firms

activities dependent variable (Factor 1) is DIRECTSUPP (at 1 % level). In other words, *being a direct supplier firm* increases the likelihood of performing *co-product design activities with the AMMs*. This finding is an indicator of strategic alliances between the AMMs and direct suppliers. On the other hand, this variable has a negative impact on the other two dependent variables (Factor 2 and Factor 3); however, they are not statistically significant.

No variable that exerts a significant impact on the probability of performing *cooperation with lower-tier suppliers for co-product design activities* dependent variable was found (Factor 2).

There are four variables that exert significant impact on the probability of performing *cooperation with universities and other firms for co-product design activities* dependent variable (Factor 3): FIRMAGE (at 5 % level), FIRMSIZE (highly significant at 1 % level), FOREIGN (highly significant at 1 % level) and RDDEPT (at 10 % level). Findings show that, an increase in firm size or owning an R&D department increases the likelihood of performing *co-product design activities with universities and other firms*. On the other hand, being a firm with foreign ownership or being an older firm reduces this probability, ceteris paribus.

5.5 Ordinal Logistic Regression Analysis

This section firstly intends to analyze the determinants of the KTTs accruing from customers to their suppliers in Turkey automotive industry with ordinal logistic regression model based on the variables constructed from raw data. Surprisingly, there seems to be very few studies investigating the factors influencing the KTTs accruing from customers to their suppliers in the industry of emerging economies.[32] The quantitative study conducted in this section aims at filling the gap in this domain especially for Turkey. Secondly, we analyze the factors impacting on various characteristics and activities of suppliers discussed in the previous chapter. Although it shows similar characteristics with the logit model, a few points about the ordinal logistic model are worth mentioning.

- All KTT-related and other dependent variables constructed from the survey questionnaire are categorical but ordered variables—categorical variables with a sense of ordering.
- The estimated coefficient of an explanatory variable in this model inform us about the change in the log of odds of being in a higher level of the dependent variable, given that all the other variables in the model are held constant. For instance, Verbeek (2008) points out that for an ordinal dependent variable comprised of three categories with increasing intensity, a positive coefficient associated with an explanatory variable indicates that if this variable increases, the probability that the most intense category occurs will increase while the probability of the least intense category will decrease. The impact on the probability of the occurrence of the intermediate category is ambiguous since its probability might increase or decrease.[33]
- When using this regression model, the parallel regression assumption is maintained, meaning that the relationship between each pair of categories included in an explanatory variable does not change. This last point will be illustrated in section "Determinants of the KTTs Related to the Production Process", where the results of our econometric analysis are examined.

[32] Giroud (2003), Techakanont (2002), and Techakanont and Terdudomtham (2004).

[33] The probability of occurrence of the intermediate category may increase for some values of an explanatory variable and decrease for others: see Long (2001).

5.5.1 Determinants of the KTTs at Inter-firm Level

This section examines the impacts of the explanatory variables on the KTTs at inter-firm level accruing from customers to their suppliers in automotive industry in Turkey, especially those (1) related to *production processes*, (2) *product*, (3) *trainings*, and (4) *financial assistance* discussed in detail in Sect. 4.2.

Determinants of the KTTs Related to the Production Process

In this section, an econometric analysis is conducted to identify the factors impacting on KTTs related to *production process*. Prior to the analysis, FA is conducted on the 13 items that make up the *KTTs related to production process*. The details of the FA including the reliability of the factors are shown in Table 5.13. According to these results, 13 types of *the KTTs related to production process* were combined in three factors (groups). The first factor was labeled as production process transfers at *high-technological level* (HTL) (Factor 1), the second one as those at *medium-technological level* (MTL) (Factor 2), and last one as those at *low-technological level* (LTL) (Factor 3).[34]

The estimation results on the determinants of *KTTs related to production process* are presented in Table 5.14. The dependent variables conducted by FA and used in the regression analyses are in ordinal nature, and they take the value of 0 (*never*), 1 (*rarely*), and 2 (*frequently*) according to transfer degree. Explanatory variables have been included in the ordinal logit model in two stages. First, there are the 12 variables (Model I); then 2 variables, ONESUPP and SSUBCONT, are added (Model II).

A significantly negative association exists between FIRMAGE and the frequency of *production process-related KTTs at high-technology level* (Factor 1), which is significant at 10 %. In a similar way, there is also a significantly negative association between FIRMAGE and *those transfers at medium-technology level* (Factor 2) at 5 % level. Alternatively, a 1 year increase in the age of a supplier reduces by 0.545 points the log of odds of being in a higher level of *the KTTs at high-technology level* and reduces by 0.657 points those transfers at *medium-technology level*, i.e. in the *frequently* category compared to the combined *rarely and never* category or in the combined *frequently and rarely* categories compared to the *never* category.[35] In other words, *KTTs related to production processes at either high or medium-technology level* decrease with the level of experience of the

---

[34] The reliability of the factors is also high, with a high Cronbach's alpha value over 0.7.

[35] As discussed earlier, the value of the estimated coefficient implies that being a more experienced firm reduces the probability of more frequent KTTs at either high or medium technology level (*frequently*) while it increases the probability of non-use (*never*) of this type of KTT.

Table 5.13 Factor analysis of the KTTs related to the production process

Items	Factor 1 high-technology level	Factor 2 medium-technology level	Factor 3 low-technology level	Cronbach's alpha (α)
1. Assistance for R&D activities	0.861			0.779
2. Assistance for logistic management	0.468			
3. Provide documentations	0.620			
4. Know-how	0.603			
11. Assistance for design	0.739			
7. Customer sent its expert staff to stay at your plant for a certain period of time for assistance in solving problems in the production process		0.536		0.758
8. Assistance for business management		0.620		
9. Assistance for quality control methods		0.718		
12. Assistance for productivity-related problems		0.794		
13. Consumer assigned its expert staff in the establishment of production process of the plant		0.754		
5. Supply of machinery, tools and equipment			0.633	0.797
6. Supply of raw material			0.799	
10. Patent and/or license rights granted				
N = 162, KMO measure = 0.872, Bartlett's chi-square = 605.18 (0.0000)				

Source: Author's calculations based on survey data
Notes: Table shows the factor loadings of each item above 0.4. The item 10 either does not fit well with the other items, or shows the lowest loading on factors. Therefore, it was removed from the analysis. Extraction method: principal component analysis; rotation method: Oblimin with Kaiser normalization. P-value is in bracket.

suppliers, ceteris paribus. This may simply indicate that many experienced suppliers in the Turkish automotive industry do possess a high level of production capabilities, which makes it unnecessary to make frequent *high* and *medium-technology level production process transfers* from their customers.

A significantly positive association exists between FIRMSIZE and the frequency of *production process-related KTTs at medium-technology level* (Factor 2), which is significant at the 10 % level. In other words, an increase in FIRMSIZE positively affects the likelihood of occurrence of *production process transfers at medium-technological level*, ceteris paribus. Alternatively, we would say that, for a one unit increase in FIRMSIZE, the odds of the *frequently* category of production process related KTTs at *medium-technology level* versus the *rarely* and *never* categories of those KTTs are 0.482 times greater, given that the other variables in the model are

Table 5.14 Determinants of the KTTs related to production process: ordinal logistic regression analysis

	Factor 1 (high-technological level)		Factor 2 (medium-technological level)		Factor 3 (low-technological level)	
	Model I	Model II	Model I	Model II	Model I	Model II
FIRMAGE	−0.545 (0.328)*	−0.440 (0.322)	−0.559 (0.282)**	−0.657 (0.313)**	−0.216 (0.315)	−0.144 (0.330)
FIRMSIZE	0.320 (0.209)	0.292 (0.227)	0.482 (0.268)*	0.392 (0.268)	0.122 (0.213)	−0.025 (0.228)
EXPINT	−0.010 (0.006)	−0.009 (0.007)	−0.014 (0.007)*	−0.011 (0.008)	−0.019 (0.006)***	−0.015 (0.007)**
FOREIGN	−0.807 (0.469)*	−0.751 (0.476)	−0.736 (0.456)	−0.606 (0.446)	0.692 (0.496)	0.733 (0.516)
DIRECTSUPP	0.550 (0.448)	0.673 (0.477)	0.448 (0.438)	0.583 (0.483)	0.346 (0.477)	0.519 (0.491)
LOCALGROUP	−0.802 (0.411)*	−0.772 (0.419)*	−0.422 (0.423)	−0.542 (0.438)	−0.289 (0.416)	−0.286 (0.424)
CLIENTMNC	0.387 (0.465)	0.382 (0.490)	0.517 (0.512)	0.555 (0.516)	−1.000 (0.594)*	−1.065 (0.602)*
RDDEPT	0.735 (0.418)*	0.578 (0.449)	0.473 (0.434)	0.656 (0.448)	0.646 (0.469)	0.533 (0.492)
PRODUCTINNO	−0.044 (0.421)	0.091 (0.440)	0.165 (0.421)	0.003 (0.442)	−0.182 (0.477)	−0.236 (0.518)
SENGPER	−0.018 (0.025)	−0.029 (0.025)	0.011 (0.018)	0.002 (0.019)	−0.017 (0.022)	−0.022 (0.026)
HIGHTECHPROD	0.502 (0.343)	0.531 (0.353)	−0.105 (0.360)	−0.250 (0.363)	0.021 (0.370)	0.006 (0.386)
HIGHDESCAP	−1.240 (0.362)***	−1.299 (0.392)***	−0.379 (0.345)	−0.470 (0.366)	−1.095 (0.385)***	−1.001 (0.392)**
ONE.SUPP		0.337 (0.357)		−0.100 (0.384)		0.684 (0.390)*
SSUBCONT		0.004 (0.007)		0.011 (0.005)**		0.005 (0.008)
Obs. (N)	156	145	156	145	156	145
Likelihood	−126.245	−117.901	−127.539	−119.456	−94.021	−86.620
Pseudo R2	0.088	0.093	0.059	0.068	0.110	0.119
McFadden's adj R2	−0.014	−0.030	−0.044	−0.057	−0.013	−0.033
Wald chi2 test	21.94 (0.0382)	26.25 (0.0240)	21.92 (0.0385)	25.40 (0.0309)	19.45 (0.0781)	21.11 (0.0988)

Source: Author calculations based on survey results

Notes: ***, **, * denote significance level at 1 %, 5 % and 10 %, respectively. Robust standard errors of the coefficients are shown in parentheses. All models include intercepts significant at 1 % level but not shown in the results for space reasons. Two-tailed test. P-values are in brackets in the last rows.

held constant. Because of the parallel regression assumption, the same increase, 0.482 times, is found between the combined categories of *frequently* and *rarely* production process related KTTs at *medium-technology level* and the *never* category of those transfers. This points out that the probability of benefit from these *process-related KTTs is high* for the suppliers that reached a certain scale.

A significant and negative association exists between EXPINT and the frequency of *production process-related KTTs at medium-technology level* (Factor 2), which is significant at 10 % level. A similar effect is also observed for KTTs at *low-technology level* (Factor 3) dependent variable at least at 5 % level. In other words, an increase in EXPINT reduces the probability of frequent occurrence of *production process transfers at either medium* or *low-technological level*, ceteris paribus.

A negative association was found between FOREIGN dummy variable and *production process-related KTTs* at *high-technology level* (Factor 1) (significant at 10 % level). The presence of the foreign capital in the firm reduces the probability of frequent occurrence of *production process transfers* at *high-technological level*, ceteris paribus, compared to local firms. Alternatively, for the FOREIGN variable, for a one unit increase in FOREIGN, i.e., going from 0 to 1, the odds of *frequently* provided *production process-related KTTs* at *high-technology level* versus the combined *rarely* and *never* categories are 0.807 times lower, given that all of the other variables in the model are held constant. Likewise, because of the parallel regression assumption, the odds of the combined *frequently* and *rarely* categories versus *never* is 0.807 times lower, given that all of the other variables in the model are held constant. This finding may indicate that foreign-capital firms do not need high amount of domestic transfers at *high-technology level* related to the production process since they manufacture products of advanced technology. It may also indicate that the KTTs have already been provided by the parent company (MNC) located abroad before the investment is made in Turkey and/or while the manufacturing process is in progress after the investment (assumed that suppliers with foreign ownership are the affiliates of multinationals in abroad). This situation confirms that foreign ownership firms import technologies from their parent companies rather than the AMMs operating in Turkey.[36] On the other hand, this dummy variable negatively impacts the probability of frequent occurrence of *production process transfers* at *medium-technological level*, while it positively impacts those at *low-technological level*. However, neither of them is statistically significant.

There is a significant and negative association between LOCALGROUP dummy variable and *production process-related KTTs* at *high-technology level* dependent variable (Factor 1) (significant at 10 % level). In other words, *being a part of a local group* exerts a negative impact on the probability of frequent occurrence of *production process transfers* at *high-technological level*, ceteris paribus. A finding

[36] This finding consistent with the findings in the report prepared for YASED (see Erdil, Pamukçu, Erden, Göksidan, & Kepenek, 2011).

similar to the above-mentioned one was obtained for foreign firms the headquarters of which are located abroad and operating in the field of automotive supply industry in Turkey. Therefore, the negative relation between the variable of LOCALGROUP and these transfers may be interpreted in a similar way—the sole difference here is the substitution of the foreign firm with the local firm. This dummy variable also impacts on negatively the probability of frequent occurrence of *production process transfers* at either *the medium* or *low-technological level*, though not significantly.

CLIENTMNC dummy variable has a significant and negative influence on the *production process-related KTTs* at *low-technological level* (Factor 3) at 10 % level given that all the other variables in the model are held constant. This means that the presence of at least one MNC among the customers reduces the probability of frequent occurrence of *production process transfers* at *low-technological level*, ceteris paribus. This finding indicates that MNCs do not provide KTTs at *low-technology level* regarding the production process of the supplier firms operating with MNCs. The most important reason is that transfers on such level may be performed by MNCs which are the customers of the suppliers. On the other hand, this dummy variable positively impacts the probability of frequent occurrence of *production process transfers* at *either high* or *low-technological level*, though not significantly.

A positive and significant association was found between RDDEPT dummy variable and *production process-related KTTs* at *the high-technological level* (Factor 1) at 10 % level. The presence of an R&D department in the firm increases the probability of occurrence of *production process transfers* at *the high-technological level*, ceteris paribus. The presence of an R&D department in supplier firms indicates that these firms conduct R&D activities actively, and have at least technological background and competence. Therefore, it might be concluded that making transfers at *high-technology level* regarding the production process requires an advanced level of technology. This dummy variable also has a positive impact on the probability of frequent occurrence of *production process transfers* at either *medium* or *low-technological level*, though not significant.

HIGHDESCAP dummy variable negatively impacts on the probability of occurrence of *production process transfers* at both (1) *high-technological level* (highly significant at 1 % level) (Factor 1) and (2) *low-technological level* (significant at least at 5 % level) (Factor 3). In other words, firms that have high design capability negatively affect the probability of occurrence of *production process transfers* at either *high* or *low-technological level*, ceteris paribus. Since the high design capability requires significantly advanced technological manufacturing capabilities, it may be assumed that supplier firms having this characteristic do not need to acquire these transfers from their customers. This dummy variable also impacts the probability of frequent occurrence of *production process transfers* negatively at *medium-technological level*, but not at a statistically significant level.

A positive association was found between ONESUPP dummy variable and *production process related KTTs* at *low-technological level* (Factor 3) at 10 % level. This means that probability of occurrence of *production process transfers* at *low-technological level* is higher for the firms that are the sole suppliers of the customers per item purchased, ceteris paribus. It is worth underlining that KTTs are provided to the firms by the customers, of whom they are the sole supplier.

SSUBCONT has a significant and positive impact on *the production process related KTTs* at *medium-technological level* (Factor 2) at 5 % level. That is, an increase in SSUBCONT variable positively impacts the probability of occurrence of *production process related KTTs* at *medium-technological level.* This dummy variable also impacts the probability of frequent occurrence of *production process transfers* positively at either *high* or *low-technological level, though* statistically insignificant.

There are four explanatory variables that do not have any significant effects on the *production process related-KTTs* provided by customers to suppliers. These are DIRECTSUPP, PRODUCTINNO, SENGPER and HIGHTECHPROD variables. It is known that *being a direct supplier* of AMM has superiorities in many areas including the KTTs. Similarly, conducting product innovation activities, having high level of skilled employees, and the capability of manufacturing high technology products show that the firm is superior in terms of technological competence. It is interesting that these superiorities are not reflected on the KTTs related with the production processes in automotive supply industry.

Determinants of the KTTs Related to the Product

In this section, an econometric analysis is conducted to identify factors impacting on *KTTs related to product.* Prior to the regression analysis, FA is conducted for the product transfers scale. The results of the FA are presented in Table 5.15. As can be seen in the table, types of the *KTTs related to product* were combined in only one factor. The factor was labeled as *product related-KTTs* (Factor 1).

The estimation results on the determinants of *KTTs related to product* are presented in Table 5.16. The dependent variable conducted by FA and used in the regression analysis is in ordinal nature, and they take the value of 0 (*never*), 1 (*rarely*), and 2 (*frequently*) according to transfer degree.

RDDEPT (at 10 % level), PRODUCTINNO (at 5 % level) and ONESUPP (at 5 % level) dummy variables have a significant and positive impact on the probability of frequent occurrence of *product related-KTTs* dependent variable. This means that each one of the variables, *presence of R&D department/conducting product-innovation activities/being sole supplier of the customer* increases the probability of occurrence *product related KTTs* provided by customers to suppliers, holding all other variables constant.

HIGHDESCAP dummy variable exerts a significant and negative impact on the probability of frequent occurrence of *product related-KTTs* dependent variable at 5 % level. Put differently, *being a firm with high design capability* impacts the likelihood of occurrence of *product related-KTTs* negatively, ceteris paribus. This situation may point out that firms with high design capability do not need *product related-KTTs*.

Table 5.15 Factor analysis of the KTTs related to the product

Items	Factor 1 product KTT
1. Assistance related to product designs	0.720
2. Joint operation related to product	0.831
3. Joint design activity related to product	0.789
4. Product specifications	0.681
5. Technical specifications, original design or technical drawings for products	0.642
N = 163, KMO measure = 0.787, Bartlett's chi-square = 226.45 (0.0000)	

Source: Author's calculations based on survey data
Notes: Extraction method: principal component analysis; rotation method: Oblimin with Kaiser normalization. P-value is in bracket. See Appendix G for the Cronbach's alpha value of the scale.

Table 5.16 Determinants of the KTTs related to product: ordinal logistic regression analysis

	Product related KTTs
FIRMAGE	−0.023 (0.338)
FIRMSIZE	0.154 (0.181)
EXPINT	−0.003 (0.008)
FOREIGN	−0.082 (0.524)
DIRECTSUPP	−0.086 (0.566)
LOCALGROUP	−0.516 (0.477)
CLIENTMNC	0.190 (0.690)
RDDEPT	1.104 (0.572)*
PRODUCTINNO	1.147 (0.554)**
SENGPER	−0.033 (0.021)
HIGHTECHPROD	0.483 (0.439)
HIGHDESCAP	−0.951 (0.466)**
ONESUPP	0.960 (0.453)**
SSUBCONT	−0.009 (0.010)
Obs. (N)	126
Likelihood	−89.968
Pseudo R2	0.047
McFadden's adj R2	−0.029
Wald chi2 test	23.93 (0.0467)

Source: Author calculations based on survey results
Notes: ***, **, * denote significance level at 1 %, 5 % and 10 %, respectively. Robust standard errors of the coefficients are shown in parentheses. All models include intercepts significant at 1 % level but not shown in the results for space reasons. Two-tailed test. P-values are in brackets in the last rows.

There are ten explanatory variables that do not have any significant effects on the probability of occurrence of *product related-KTTs* dependent variable. It is quite interesting that, of these variables, especially DIRECTSUPP, FIRMSIZE, EXPINT, SENGPER and HIGHTECHPROD ones do not have any impact on *product related-KTTs*. The superiorities of *being a direct supplier* were discussed earlier. It is also known that large firms have many advantages compared to medium and small-sized firms. The other three variables also show that firms are superior in terms of technological competence and competitiveness in global markets. Interestingly, these superiorities of the firms are not reflected on *product related-KTTs* in the automotive supply industry.

Determinants of the Training Activities Provided by Customers

In this section, an econometric analysis is conducted in to identify factors impacting on *training activities* provided by the customers to the suppliers. Evidently, training activities provided by the cooperating customers could be an indicator of the KTTs from customers. As discussed in detail in Sect. 4.2.4, training could be in the mode of visits to customer plant, seminars, courses, and applied training given by the expert staff of the customers. These kinds of activities enable the exchange of tacit knowledge especially from customers to suppliers. In addition, they increase the technological capabilities of the suppliers.

Prior to the regression analysis, FA is conducted for the training scale. The details of the FA are presented in Table 5.17. According to these results, *KTTs through training* were combined in one factor. The factor was labeled as *training* (Factor 1).

The findings of the regression analysis on the determinants of *KTTs through training* are given in Table 5.18. The dependent variable conducted by FA and used in the regression analysis is in ordinal nature, and they take the value of 0 (*never*), 1 (*rarely*), and 2 (*frequently*) according to transfer degree.

Both DIRECTSUPP (at 5 % level) and FOREIGN (at 10 % level) dummy variables have a significantly positive influence on the probability of frequent occurrence of *training activities*. That is, either *being a direct supplier* or *being a foreign firm* positively impacts the likelihood of frequent occurrence *of training activities* provided to suppliers, ceteris paribus. These findings are important as they show that the likelihood of benefit from such activities is much higher for the direct suppliers or foreign firms than for other firms. This also may suggest that customers choose amongst their direct suppliers or foreign suppliers in order to provide the afore-mentioned *training activities*. *Being a direct supplier* especially suggest that

Table 5.17 Factor analysis of the KTTs through training

Items	Factor 1 training
1. Training on technologies used in production	0.851
2. Training of production/operation staff (engineers, technicians etc.)	0.870
3. Training of management staff	0.879
N = 164, KMO measure = 0.722, Bartlett's chi-square = 189.92 (0.0000)	

Source: Author's calculations based on survey data

Notes: Extraction method: principal component analysis; rotation method: Oblimin with Kaiser normalization. P-value is in bracket. See Appendix G for the Cronbach's alpha value of the scale.

Table 5.18 Determinants of the KTTs through training: ordinal logistic regression analysis

	Training
FIRMAGE	0.402 (0.313)
FIRMSIZE	0.112 (0.154)
EXPINT	0.004 (0.007)
FOREIGN	0.877 (0.463)*
DIRECTSUPP	1.094 (0.508)**
LOCALGROUP	0.004 (0.434)
CLIENTMNC	−0.097 (0.537)
RDDEPT	0.360 (0.491)
PRODUCTINNO	−0.256 (0.546)
SENGPER	−0.033 (0.032)
HIGHTECHPROD	−0.457 (0.380)
HIGHDESCAP	−0.311 (0.372)
ONESUPP	0.628 (0.398)
SSUBCONT	0.011 (0.006)*
Obs. (N)	127
Likelihood	−124.971
Pseudo R2	0.091
McFadden's adj R2	−0.025
Wald chi2 test	25.06 (0.0339)

Source: Author calculations based on survey results
Notes: ***, **, * denote significance level at 1 %, 5 % and 10 %, respectively. Robust standard errors of the coefficients are shown in parentheses. All models include intercepts significant at 1 % level but not shown in the results for space reasons. Two-tailed test. P-values are in brackets in the last rows.

close relationships between suppliers and customers increase the probability of frequent *trainings* given to their personnel. Put differently, this finding indicates that *not being the first-tier supplier* reduces the probability of occurrence of *KTTs through training* provided by customers.

SSUBCONT dummy variable exerts a statistically significant and positive impact on the probability of frequent use of *training activities* (significant at 10 % level). An increase in SSUBCONT variable impacts the probability of frequent use of *training activities* provided by customers to suppliers positively.

Other explanatory variables do not have any statistical impact on *the training activities* dependent variable.

Determinants of the Financial Transfers Provided by Customers

Other assistance that is provided by the customers to the suppliers is direct or indirect *financial transfers* (*supports*). Although these assistances are not directly connected to the KTTs arising from customers, clearly the effective use of these transfers should be ensured. In addition, these could be an indicator of the strategic

Table 5.19 Factor analysis of the financial transfers

Items	Factor 1 for strategic relationship	Factor 2 for procurement	Cronbach's alpha (α)
1. Unilateral financial aid	0.774		0.771
2. Loans with low interest rates	0.741		
3. Contribution to risk capital	0.685		
4. Pre-financing of machinery, equipment and tools		0.623	0.812
5. Prepayment for orders before delivery		0.650	
N = 161, KMO measure = 0.696, Bartlett's chi-square = 95.12 (0.0000)			

Source: Author's calculations based on survey data
Notes: Extraction method: principal component analysis; rotation method: Oblimin with Kaiser normalization. P-value is in bracket.

relationships between suppliers and customers related to production, so they also may indicate KTTs. Lastly, suppliers can continue to operate or they make R&D activities related to products easier through financial support provided by customers.

Prior to the regression analysis, FA is conducted for *financial transfers* scale. The details of the FA including the reliability of the factors are shown in Table 5.19. According to these results, *financial transfers* were combined in two factors. These two factors were labeled as financial transfers for *strategic relationship* (Factor 1) and those for *procurement* (Factor 2).[37]

The estimation results on the determinants of the *financial transfers* are presented in Table 5.20. The dependent variables conducted by FA and used in the regression analyses are in ordinal nature, and they take the value of 0 (*never*), 1 (*rarely*), and 2 (*frequently*) according to transfer degree.

It should be specified at this point that 151 (93 %) of the firms never benefit from the *unilateral financial aid* of the customers and 11 firms (7 %) rarely benefit from these *aids*. The number of the firms that stated they had never benefited from the *loans with low interest rates* is 152 (93 %) (nine firms expressed that they rarely benefit from the loans; two firms stated that they frequently benefit from the loans). Similarly, 157 firms (96 %) responded by selecting *never* to the question of how often the customers contributed to *risk capital* (five suppliers specified that they rarely benefited, and one supplier declared that it frequently benefited). The inequality in the distribution of the responses given for these questions is greater than the other questions in the survey.

FIRMSIZE exerts a significant positive impact on the frequency of provision of *financial transfers related to strategic relationship* at 5 % level (Factor 1). This finding means that the likelihood of the provision of *financial transfers* increases with the firm size. It may be due to the fact that being a larger supplier with

[37] The reliability of the factors is also high, with a high Cronbach's alpha value over 0.7.

Table 5.20 Determinants of the financial transfers: ordinal logistic regression analysis

	Financial transfers	
	Factor 1 (for strategic relationship)	Factor 2 (for procurement)
FIRMAGE	−0.626 (0.493)	0.003 (0.282)
FIRMSIZE	0.700 (0.348)**	0.252 (0.205)
EXPINT	−0.003 (0.009)	0.002 (0.006)
FOREIGN	−2.048 (0.806)**	0.310 (0.435)
DIRECTSUPP	−0.432 (0.668)	0.126 (0.425)
LOCALGROUP	−0.091 (0.546)	0.329 (0.357)
CLIENTMNC	0.433 (0.823)	−0.230 (0.447)
RDDEPT	0.202 (0.586)	0.224 (0.380)
PRODUCTINNO	−0.811 (0.601)	0.524 (0.436)
SENGPER	0.048 (0.042)	−0.029 (0.024)
HIGHTECHPROD	0.835 (0.607)	0.032 (0.357)
HIGHDESCAP	−0.614 (0.532)	−0.827 (0.353)**
ONESUPP	−0.313 (0.533)	0.683 (0.352)*
SSUBCONT	−0.002 (0.008)	0.001 (0.007)
Obs. (N)	144	145
Likelihood	−54.846	−147.285
Pseudo R2	0.123	0.068
McFadden's adj R2	−0.132	−0.033
Wald chi2 test	12.30 (0.5820)	23.41 (0.0539)

Source: Author calculations based on survey results

Notes: ***, **, * denote significance level at 1 %, 5 % and 10 %, respectively. Robust standard errors of the coefficients are shown in parentheses. All models include intercepts significant at 1 % level but not shown in the results for space reasons. Two-tailed test. P-values are in brackets in the last rows.

significant production capabilities may reduce the risks associated with the reimbursement of the loan granted. However, it can be considered as a reflection of the shortcomings in the capital markets.

FOREIGN dummy variable negatively impacts the probability of occurrence of *financial transfers related to strategic relationship* (Factor 1), which is significant at 5 % level. *Presence of foreign ownership* in the firm impacts the probability of occurrence of these transfers in a negative way, ceteris paribus. This finding indicates that foreign-capital firms do not require *financial transfers* on high levels since they have stronger financial structures. It may also indicate that the foreign-capital firms receive more *financial assistance* from the foreign parent company located abroad.

HIGHDESCAP exerts a significant negative impact on the frequency of provision of *financial transfers related to procurement* at 5 % level (Factor 2). In other words, *having a high design capability* reduces the probability of provision of those transfers, ceteris paribus. This situation points out that firms with high design capability do not need these kinds of *financial assistance*.

A significantly positive association exists between ONESUPP and the frequency of provision of *financial transfers related to procurement* (Factor 2) dependent variable (significant at 10 % level). The probability of the provision of those transfers is higher for the firms that are the sole supplier of the customers per item purchased, ceteris paribus. This situation is important since it indicates that such transfers are provided by customers to their sole suppliers.

5.5.2 Determinants of the Technology Capabilities of the Suppliers

This section examines *the technological capabilities* of the supplier firms for production activities when compared with other rivalry or leader firms in the industry. Prior to the regression analysis, FA is conducted for the *technological capabilities* scale. The details of the FA including the reliability of the factors are shown in Table 5.21. Here, ten types of the *technological capabilities* were combined in two factors. The first factor was labeled as *technological capabilities*

Table 5.21 Factor analysis of the technology capabilities

Items	Factor 1 design and R&D capabilities	Factor 2 production capabilities	Cronbach's alpha (α)
1. R&D capability	0.733		0.824
3. Expertness on CAD-CAM-CAE	0.604		
4. Co-designer capability	0.650		
7. Design capability	0.858		
8. Product improvement capability	0.822		
5. Quality control capability		0.887	0.766
6. Testing and analytical capability		0.820	
10. On-time delivery		0.642	
2. Reach to lower prices			
9. Automation level in production process			
N = 165, KMO measure = 0.715, Bartlett's chi-square = 424.46 (0.0000)			

Source: Author's calculations based on survey data

Notes: Table shows the factor loadings of each item above 0.4. The items 2 and 9 either do not fit well with the other items, they show the lowest loading on factors. Therefore, they were removed from the analysis. Extraction method: principal component analysis; rotation method: Oblimin with Kaiser normalization. P-value is in bracket.

Table 5.22 Determinants of the technology capabilities related to production, design and R&D: ordinal logistic regression analysis

	Technology capabilities related to production, design and R&D			
	Factor 1 (design and R&D capabilities)		Factor 2 (production capabilities)	
FIRMAGE	−0.030 (0.273)	−0.192 (0.286)	−0.185 (0.343)	−0.399 (0.335)
FIRMSIZE	0.302 (0.223)	0.305 (0.227)	0.152 (0.208)	0.149 (0.226)
EXPINT	0.003 (0.006)	0.005 (0.007)	−0.004 (0.006)	−0.003 (0.007)
FOREIGN	−0.353 (0.374)	−0.321 (0.387)	−0.382 (0.464)	−0.312 (0.490)
DIRECTSUPP	0.516 (0.546)	0.570 (0.591)	0.461 (0.559)	0.364 (0.580)
LOCALGROUP	−0.978 (0.393)**	−1.052 (0.410)**	0.530 (0.397)	0.584 (0.424)
CLIENTMNC	1.072 (0.479)**	1.198 (0.494)**	0.731 (0.539)	0.691 (0.594)
RDDEPT	0.819 (0.351)**	1.091 (0.393)***	−0.596 (0.449)	−0.586 (0.503)
PRODUCTINNO	0.716 (0.378)*	0.641 (0.417)	0.151 (0.461)	0.053 (0.488)
SENGPER	0.025 (0.021)	0.019 (0.022)	0.025 (0.024)	0.036 (0.026)
HIGHTECHPROD	0.155 (0.338)	0.152 (0.343)	−0.143 (0.375)	−0.088 (0.392)
HIGHDESCAP	0.574 (0.349)*	0.408 (0.360)	0.448 (0.414)	0.661 (0.446)
ONESUPP	−0.763 (0.334)**	−0.791 (0.350)**	−0.732 (0.378)*	−0.797 (0.379)**
SSUBCONT		0.007 (0.007)		0.010 (0.008)
Obs. (N)	152	144	153	145
Likelihood	−175.314	−165.478	−114.078	−105.519
Pseudo R2	0.119	0.130	0.061	0.084
McFadden's adj R2	0.034	0.035	−0.071	−0.063
Wald chi2 test	61.66 (0.0000)	61.64 (0.0000)	13.92 (0.3796)	19.35 (0.1519)

Source: Author calculations based on survey results

Notes: ***, **, * denote significance level at 1 %, 5 % and 10 %, respectively. Robust standard errors of the coefficients are shown in parentheses. All models include intercepts significant at 1 % level but not shown in the results for space reasons. Two-tailed test. P-values are in brackets in the last rows.

related to design and R&D (Factor 1), and second one was labeled as those related to *production* (Factor 2).[38]

The estimation results on the determinants of the *technological capabilities of the suppliers related to design and R&D* and *production* are presented in Table 5.22. The dependent variables conducted by FA and used in the regression analyses are in ordinal nature, and they take the value between 1 (low) and 5 (high) on a five-point Likert scale according to their specified capability level. Explanatory variables have been included in the ordinal logit model in two stages. First, there are the 13 variables in column 1; then SSUBCONT variable is added to column 2.

Four dummy variables exert positive and significant impact on *the technological capability level of the suppliers related to design and R&D* (Factor 1) dependent variable: CLIENTMNC (at 5 % level), RDDEPT (at 5 % level), PRODUCTINNO (at 10 % level) and HIGHDESCAP (at 10 % level). *The presence of MNC among*

[38] The reliability of the factors is also high, with a high Cronbach's alpha value over 0.7.

customers/having an R&D department/conducting product innovation activities/ having high design capability positively affects the likelihood of the increases of technological capability levels of the suppliers in *design and R&D*, ceteris paribus. The impacts of each of these four dummy variables on the technological capability level of the suppliers related to *production* (Factor 2) dependent variable were analyzed. RDDEPT variable was found to have a negative impact, while other three were found to have a positive impact. However, none of them was statistically significant. It is a reasonable finding that these three dummy variables positively affect more advanced competences such as *design and R&D* rather than capabilities regarding *production* such as logistics, delivery on time, quality control, and testing capability. In sum, these findings indicate that KTTs are provided to suppliers by cooperated multinational customers, and particularly the suppliers with developed technological capabilities benefit from these transfers.

A negative association was found between LOCALGROUP and the technological capability level of the suppliers related to *design and R&D* (Factor 1) dependent variable (significant at 5 % level). *Being dependent on any local group-firm-holding* affects negatively the likelihood of the increases of technological capability levels related to *design and R&D*, ceteris paribus. On the other hand, the effect of this variable on *the technological capability level of the suppliers related to production* (Factor 2) is positive, though statistically insignificant. The underlying factors behind this finding should be examined in another study.

ONESUPP dummy variable also has a negative impact on *the technological capability level of the suppliers-related to design and R&D* (Factor 1) and *-related to production* (Factor 2) dependent variables (significant at 5 % level). In other words, *being the sole supplier* of the most important customer per item purchased affects negatively the likelihood of the increases of technological capability levels—*related to design and R&D*, as well as—*related to production*, ceteris paribus.

The factors underlying behind these findings that LOCALGROUP and ONESUPP dummy variables negatively affect the technological capability levels of the suppliers should be further examined in another study.

5.5.3 Determinants of the Inter-firm Cooperation (Why Suppliers Cooperate with Other Firms)

This section examines the determinants of the motivations of suppliers for *cooperating with other firms*. Firstly, FA is conducted for the inter-firm cooperation activities scale. The results of the FA are shown in Table 5.23, wherein nine types of the motivations for *cooperating with other firms* were combined in one factor. The factor was labeled as *cooperation with other firms* (Factor 1).

The estimation results on the determinants of the motivations of the supplier activities in *cooperation with other firms* are given in Table 5.24. The dependent variable constructed by FA and used in the regression analysis is in ordinal nature, and they take the value between 1 (*very unimportant*) and 5 (*very important*) on a five-point Likert scale according to importance degree.

Table 5.23 Factor analysis of the inter-firm cooperation

	Factor 1
Items	Cooperation with other firms
1. Carrying out R&D activities	0.802
2. Know-how transfer	0.723
3. Reducing/sharing production costs/risks	0.759
4. Replacing technologically phased out products with the new ones	0.794
5. Establishing long-term strategic partnership	0.731
6. Improving product quality	0.840
7. Opening up global markets	0.785
8. Entering new technology fields	0.863
9. Learning about new technologies	0.840
N = 165, KMO measure = 0.896, Bartlett's chi-square = 10 28.82 (0.0000)	

Source: Author's calculations based on survey data

Notes: Extraction method: principal component analysis; rotation method: Oblimin with Kaiser normalization. P-value is in bracket. See Appendix G for the Cronbach's alpha value of the scale.

Table 5.24 Determinants of the cooperation with other firms: ordinal logistic regression analysis

	Cooperation with other firms
FIRMAGE	−0.215 (0.301)
FIRMSIZE	0.249 (0.198)
EXPINT	0.003 (0.007)
FOREIGN	−1.020 (0.392)***
DIRECTSUPP	0.132 (0.524)
LOCALGROUP	−0.077 (0.346)
CLIENTMNC	−0.134 (0.484)
RDDEPT	1.008 (0.369)***
PRODUCTINNO	−0.298 (0.426)
SENGPER	−0.010 (0.039)
HIGHTECHPROD	−0.271 (0.347)
HIGHDESCAP	−0.073 (0.350)
ONESUPP	−0.279 (0.346)
SSUBCONT	0.012 (0.010)
Obs. (N)	139
Likelihood	−170.736
Pseudo R2	0.059
McFadden's adj R2	−0.040
Wald chi2 test	29.50 (0.0089)

Source: Author calculations based on survey results

Notes: ***, **, * denote significance level at 1 %, 5 % and 10 %, respectively. Robust standard errors of the coefficients are shown in parentheses. All models include intercepts significant at 1 % level but not shown in the results for space reasons. Two-tailed test. P-values are in brackets in the last rows.

The coefficient associated with FOREIGN dummy variable is negative and highly significant at the 1 % level. This demonstrates that the likelihood of involvement of the foreign suppliers in *cooperation activities with other firms*

(Factor 1) is lower compared to local firms, ceteris paribus. In other words, *being a foreign ownership firm* exerts a negative impact on the degree of importance attached by suppliers to the motives for *inter-firm collaboration activities* explained in detail in Table 5.23. The possible explanation of the negative effect is that foreign suppliers operating in the automotive supply industry in Turkey do not feel obliged to cooperate with other domestic firms in order to carry out the activities specified in Table 5.23 unlike the local firms. It is probably because these firms are technologically more capable than local firms, and they acquire the information needed for the activities such as KTTs, product development, and R&D from parent MNCs located in the headquarters abroad.

A highly significant and positive association exists between RDDEPT and *cooperation activities with other firms* at 1 % level. In other words, *presence of an R&D department* (RDDEPT) positively affects the degree of importance attached by suppliers to *all inter-firm collaboration motives* given in Table 5.23, ceteris paribus. This positive effect shows the importance of innovation capabilities beyond production capabilities for inter-firm collaborations to occur, and also points to the role played by the absorptive capacity in this process. Moreover, this finding that a supplier has an R&D competence shows a positive signal to enter into cooperation for other firms.

5.5.4 Determinants of the Improvements in Production Capability as a Result of Assistance[39] Received from Customers

This section analyzes the determinants of *the improvements (increases) in the production capability of suppliers*, caused by the assistance that has been provided by their customers for the last 3 years. Prior to the regression analysis, FA is conducted for *the improvements in production capability* scale. The results of the FA are presented in Table 5.25, wherein ten types of the improvements were combined in one factor. The factor was labeled as *improvements in production capability* (Factor 1).

The estimation results on the determinants of *the improvements (increases) in the production capability of suppliers during the last 3 years* are presented in Table 5.26. The dependent variable conducted by FA and used in the regression analysis is in ordinal nature, and they take the value between 1 (very decreased) and 5 (very increased) on a five-point Likert scale according to improvement degree.

LOCALGROUP dummy variable has a significantly negative impact on the degree of *improvements in the production capabilities of the suppliers* (significant at 10 % level). *Depending on any local group/firm/holding* reduce the probability of

[39] It refers to every support, advice, knowledge and technology transfers.

Table 5.25 Factor analysis of the improvements in production capability

Items	Factor 1 Improvements in production capability
1. Productivity	0.774
2. Profitability	0.703
3. Design capability (design new products and processes)	0.784
4. Engineering capability related to product	0.835
5. Improve production process and/or capability to find solutions for production problems	0.862
6. Improvement in quality control methods	0.844
7. Improvement in business management	0.812
8. Performance on export	0.683
9. Reach to lower output prices	0.748
10. Learning about new technologies and production processes	0.845
N = 165, KMO measure = 0.920, Bartlett's chi-square = 1180.92 (0.0000)	

Source: Author's calculations based on survey data
Notes: Extraction method: principal component analysis; rotation method: Oblimin with Kaiser normalization. P-value is in bracket. See Appendix G for the Cronbach's alpha value of the scale.

Table 5.26 Determinants of the improvements in production capability: ordinal logistic regression analysis

	Improvements in production capability
FIRMAGE	0.279 (0.272)
FIRMSIZE	0.067 (0.224)
EXPINT	0.009 (0.007)
FOREIGN	−0.430 (0.510)
DIRECTSUPP	0.078 (0.489)
LOCALGROUP	−0.661 (0.362)*
CLIENTMNC	−0.536 (0.380)
RDDEPT	−0.398 (0.430)
PRODUCTINNO	0.596 (0.437)
SENGPER	0.005 (0.040)
HIGHTECHPROD	0.181 (0.357)
HIGHDESCAP	0.637 (0.354)*
ONESUPP	0.266 (0.355)
SSUBCONT	
Obs. (N)	147
Likelihood	−130.653
Pseudo R2	0.054
McFadden's adj R2	−0.062
Wald chi2 test	19.24 (0.1158)

Source: Author calculations based on survey results
Notes: ***, **, * denote significance level at 1 %, 5 % and 10 %, respectively. Robust standard errors of the coefficients are shown in parentheses. All models include intercepts significant at 1 % level but not shown in the results for space reasons. Two-tailed test. P-values are in brackets in the last rows

increases in *improvement degree of the suppliers in certain fields related to production capability* (ten items stated in Table 5.25), ceteris paribus.

HIGHDESCAP exerts a positive impact on the degree of *improvements in production capabilities of the suppliers* (significant at 10 % level). In other words, *having a high design capability* affects in a positive way the probability of increases *in improvement degree of the suppliers in certain fields* stated in Table 5.25 in a positive way. It highlights that firms *having an advanced design capability* internalize such assistances, which increases their production capabilities.

5.5.5 Determinants of the Benefits of Being a Direct Supplier

This section analyzes the determinants of *the benefits of being a direct supplier of the AMM in Turkey* based on suppliers' experience in this field. Prior to the regression analysis, FA is conducted for *the benefits of being a direct supplier* scale. The details of the FA including the reliability of the factors are presented in Table 5.27. The results led to 14 types of the benefits combined in two factors: benefits related to *production technology* (Factor 1), and benefits related to *design and business relationships* (Factor 2).[40]

The estimation results on the determinants *of the benefits received from being a direct supplier of at least one of the AMMs in Turkey* are presented in Table 5.28. As discussed in Sects. 4.1.1 and 4.7, 132 firms out of 165 firms (80 %) stated that they are the direct suppliers of one or more AMMs in Turkey; therefore, the analyses in this section are performed on these 132 direct supplier firms. The dependent variables conducted by FA and used in the regression analyses are in ordinal nature, and they take the value between 1 (very unimportant) and 5 (very important) on a five-point Likert scale according to importance degree.

A negative and significant association was found between FOREIGN dummy variable and *the benefits related to production technology* dependent variable (Factor 1) at 10 % level. The ceteris paribus effect of *being foreign firm* is estimated to be negative on the probability of increases in importance degree of the benefits (see Factor 1) received from *being a direct supplier*. This variable also has a negative, though statistically insignificant, effect on the *benefits related to design and business relationships* dependent variable (Factor 2). One possible explanation of this negative effect is that foreign direct suppliers do not regard *the benefits of being a direct supplier* (items specified in Table 5.27) as very important—compared to local direct supplier firms—since their capabilities are highly advanced in terms of *production technology*, *design and business relationships*, and they probably acquire these benefits from the parent multinationals located in the headquarters abroad.

[40] The reliability of the factors is also high, with a high Cronbach's alpha value over 0.7.

Table 5.27 Factor analysis of the benefits pertaining to being a direct supplier

Items	Factor 1 production technology	Factor 2 design and business relationship	Cronbach's alpha (α)
4. Learning/improving new production processes	0.815		0.856
5. Learning/improving new quality control methods	0.640		
7. Improvement in testing and analyzing capabilities/techniques	0.465		
11. Imp roving in ability and knowledge about identifying and finding solutions to problems encountered in the production	0.619		
12. Increase in productivity	0.827		
13. Reducing risks involved with making decision to invest in new technology and/or machines	0.645		
14. Learning about new technologies	0.662		
1. Developing new business relationships with other firms		0.600	0.732
2. Good reputation and familiarness in sector		0.871	
3. Stable trade		0.666	
6. Better understand the customers' blueprint and technical specifications		0.688	
8. Attendance to early stage in the design activities and developing prototype		0.822	
9. Increase in design capability		0.811	
10. Access to international developed markets		0.441	
N = 132, KMO measure = 0.853, Bartlett's chi-square = 738.38 (0.0000)			

Source: Author's calculations based on survey data
Notes: Extraction method: principal component analysis; rotation method: Oblimin with Kaiser normalization. P-value is in bracket

SENGPER explanatory variable exerts a significant and negative impact on the *benefits related to production technology* (Factor 1) dependent variable at 5 % level. An increase in SENGPER reduces the probability of increases in the importance degree of *the benefits* stated in Table 5.27 under the title of Factor 1, ceteris paribus. This variable also has a negative, but insignificant, effect on *the benefits related to design and business relationships* (Factor 2) dependent variable. The main reason for the negative effect is that direct supplier firms which have highly skilled employees do not attach as much importance to *the benefits of being a direct supplier* (specified in Table 5.27) as the firms lacking highly skilled employees.

HIGHDESCAP dummy variable exerts a highly significant and positive impact on the degree of importance of *the benefits related to design and business relationships* (Factor 2) dependent variable at 1 % level. That is, the ceteris paribus effect of *having a high-design capability* positively affects the likelihood of increases in the

Table 5.28 Determinants of the benefits of being a direct supplier: ordinal logistic regression analysis

	Benefits of being a direct supplier	
	Factor 1 (production technology)	Factor 2 (design and business relationships)
FIRMAGE	0.228 (0.331)	−0.220 (0.308)
FIRMSIZE	0.192 (0.291)	0.142 (0.243)
EXPINT	0.007 (0.008)	0.000 (0.008)
FOREIGN	−0.846 (0.463)*	−0.405 (0.513)
DIRECTSUPP		
LOCALGROUP	0.582 (0.461)	0.070 (0.469)
CLIENTMNC	−0.669 (0.520)	0.202 (0.584)
RDDEPT	−0.375 (0.518)	−0.243 (0.594)
PRODUCTINNO	0.097 (0.463)	−0.568 (0.697)
SENGPER	−0.045 (0.022)**	−0.061 (0.040)
HIGHTECHPROD	0.632 (0.394)	0.345 (0.429)
HIGHDESCAP	0.136 (0.420)	1.400 (0.444)***
ONESUPP	0.284 (0.396)	−0.043 (0.407)
SSUBCONT	0.015 (0.007)**	0.024 (0.010)**
Obs. (N)	118	118
Likelihood	−100.510	−81.389
Pseudo R2	0.107	0.125
McFadden's adj R2	−0.035	−0.036
Wald chi2 test	30.40 (0.0041)	24 (0.0311)

Source: Author calculations based on survey results
Notes: ***, **, * denote significance level at 1 %, 5 % and 10 %, respectively. Robust standard errors of the coefficients are shown in parentheses. All models include intercepts significant at 1 % level but not shown in the results for space reasons. Two-tailed test. P-values are in brackets in the last rows

improvement degree of *the benefits related to design and business relationships* stated in Table 5.27 under the title of Factor 2. This dummy variable also has a positive, though statistically insignificant, effect on *the benefits related to production technology* (Factor 1) dependent variable. This finding reflects that direct supplier firms with advanced design capability consider *the benefits* specified in Table 5.27 as highly important based on their experience in being a direct supplier.

SSUBCONT explanatory variable positively affects the degree of importance of both *the benefits related to production technology* (Factor 1) and those *related to design and business relationships* (Factor 2) dependent variables (each one significant at 5 % level). In other words, an increase in SSUBCONT increases the probability of increases in importance degree of *the benefits* stated in Table 5.27 under the titles of both Factor 1 and Factor 2. This means that direct supplier firms signed subcontracting agreements with the AMMs to perform some tasks, and as they assess *the benefits of being a direct supplier* as very important. The main reason for the positive effect is that cooperation level between these firms in automotive supply industry allows suppliers to further develop their skills and technological capabilities, and may give rise to the KTTs to suppliers. Therefore, they regard *the benefits of being a direct supplier* (items specified in Table 5.27) as very important.

5.5.6 Determinants of the Cooperation Level with Partners in Innovation Activities

This section examines the determinants of *cooperation level of the suppliers with external partners in innovation activities.* Prior to the regression analyses, FA is conducted for the partners in *innovation activities* scale. The results of the FA including the reliability of the factors are shown in Table 5.29. Here, eight types of the partners in *innovation activities* were combined in three factors (groups). The first factor was labeled as *AMMs* (Factor 1), and the second one as *lower-tier suppliers* (Factor 2), and the last one as *universities and other firms*[41] (Factor 3).

The estimation results on the determinants *of cooperation level of the suppliers with external partners in innovation activities* are presented in Table 5.30. As discussed in Sect. 4.8.2, 19 firms out of 165 firms (12 %) stated that they do not conduct *innovation activities*; therefore, the analyses in this section are performed on 146 innovative firms. The dependent variables conducted by FA and used in the regression analyses are in ordinal nature, and they take the value between 1 (no cooperation) and 5 (very intense) on a five-point Likert scale according to the intensity level of the cooperation. Explanatory variables have been included in the ordinal logit model in two stages. First, there are the 12 variables (Model I); then 2 variables, ONESUPP and SSUBCONT, are added to the model (Model II).

Being a direct supplier, DIRECTSUPP, exerts a highly significant and positive impact on *the cooperation level of the suppliers in innovation activities with both the AMMs* (Factor 1) (at 1 % level), and *the universities and other firms* (Factor 3) (at 10 % level) dependent variables. On the other hand, this variable affects negatively the *cooperation level of the suppliers with lower-tier suppliers* (Factor 2) (at 10 % level). Being a direct supplier of AMM increases the probability of intensity in *cooperation level of the suppliers with either the AMMs* or *universities and other firms*, while it reduces the probability of intensity in *cooperation level of the suppliers with the lower-tier suppliers.* These findings are important since they indicate that *direct suppliers and AMMs actively cooperate for innovation.*

A significantly negative relation exists between FIRMAGE and *the cooperation level of the suppliers with the universities and other firms* (Factor 3) *related to innovation activities* dependent variable (at 10 % level). The probability of increases in cooperation level of the suppliers with *the universities and other firms* reduces with firm experience and maturity, ceteris paribus. In other words, *to cooperate with the universities and other firms in innovation activities* is more important for younger firms than for older ones.

---

[41] Private engineering and consultancy firms, R&D institutions.

Table 5.29 Factor analysis of the cooperation level with partners in innovation activities

Items	Factor 1 AMMs	Factor 2 lower-tier suppliers	Factor 3 uni. and other firms	Cronbach's alpha (α)
1. DAMM	0.811			0.772
2. OAMM	0.917			
3. DS		0.836		0.691
4. OS		0.870		
6. R&D institutions			0.725	0.667
7. Universities			0.748	
8. Private engineering and consultancy firms			0.773	
5. PC				
N = 146, KMO measure = 0.614, Bartlett's chi-square = 249.33 (0.0000)				

Source: Author's calculations based on survey data

Notes: Table shows the factor loadings of each item above 0.4. The item 5 either does not fit well with the other items, or it shows the lowest loading on factors. Therefore, it was removed from the analysis. Extraction method: principal component analysis; rotation method: Oblimin with Kaiser normalization. P-value is in bracket

Legend: DAMM (Domestic Automotive Main Manufacturers); OAMM (Overseas Automotive Main Manufacturers); DS (Domestic Suppliers); OS (Overseas Suppliers); PC (Parent Company)

There is a significantly negative relation between CLIENTMNC dummy variable and *cooperation level in innovation activities with the lower-tier suppliers* (Factor 2) dependent variable at 10 % level. *Presence of at least one MNC among customers* reduces the probability of increases in *cooperation level of the suppliers with the lower-tier suppliers*, ceteris paribus. This variable also exerts a positive, though insignificant, impact on the probability of increases in *the level of suppliers' cooperation with the universities and other firms* (Factor 3), while it affects negatively that with *the AMMs* (Factor 1).

Both RDDEPT (highly significant at 1 % level) and HIGHTECHPROD (significant at least at 10 % level) dummy variables exert positive and significant impact on *the cooperation level of the suppliers in innovation activities with the universities and other firms* (Factor 3) dependent variable. Either *the presence of an R&D department* or *manufacturing high-tech products* increases the probability of intensity in *cooperation level of the suppliers with the universities and other firms*, ceteris paribus.

PRODUCTINNO has a significant and positive influence *on the cooperation level of the suppliers in innovation activities with the AMMs* (Factor 1) at least at 5 % level. *Conducting product related innovation activities* increases the probability of intensity in *cooperation level of the suppliers in innovation activities with AMMs*, ceteris paribus. This finding indicates that there are strategic relationships in terms of product innovation activities between suppliers and AMMs, and also this confirms that innovation capability of the suppliers plays an important role in this close cooperation.

SENGPER exerts also a statistically significant positive impact on *the cooperation level of the suppliers in innovation activities with the AMMs* (Factor 1) at least

Table 5.30 Determinants of the cooperation level with partners in innovation activities: ordinal logistic regression analysis

	Factor 1 (AMMs)[a]		Factor 2 (lower-tier suppliers)[b]		Factor 3 (uni. and other firms)[c]	
	Model I	Model II	Model I	Model II	Model I	Model II
FIRMAGE	−0.033 (0.260)	0.010 (0.258)	0.064 (0.310)	0.027 (0.327)	−0.400 (0.281)	−0.540 (0.297)*
FIRMSIZE	0.238 (0.186)	0.152 (0.187)	−0.054 (0.182)	0.041 (0.198)	0.146 (0.189)	0.231 (0.208)
EXPINT	−0.001 (0.007)	−0.001 (0.007)	−0.002 (0.006)	−0.005 (0.007)	0.000 (0.006)	0.001 (0.007)
FOREIGN	−0.167 (0.446)	−0.014 (0.463)	0.309 (0.525)	0.241 (0.543)	−0.450 (0.443)	−0.3 14 (0.422)
DIRECTSUPP	2.243 (0.586)***	2.435 (0.630)***	−0.358 (0.375)	−0.756 (0.394)*	0.828 (0.468)*	0.900 (0.587)
LOCALGROUP	−0.372 (0.407)	−0.556 (0.422)	−0.112 (0.409)	−0.146 (0.436)	0.135 (0.386)	0.032 (0.394)
CLIENTMNC	−0.207 (0.522)	−0.108 (0.653)	−0.750 (0.416)*	−0.766 (0.491)	0.175 (0.577)	0.105 (0.666)
RDDEPT	−0.167 (0.426)	−0.062 (0.485)	0.252 (0.321)	0.226 (0.367)	1.169 (0.404)***	1.277 (0.456)***
PRODUCTINNO	1.418 (0.549)***	1.247 (0.560)**	0.018 (0.423)	0.154 (0.464)	−0.091 (0.463)	−0.073 (0.525)
SENGPER	0.049 (0.013)***	0.039 (0.017)**	−0.013 (0.017)	−0.009 (0.020)	−0.020 (0.021)	−0.021 (0.021)
HIGHTECHPROD	0.294 (0.365)	0.287 (0.368)	0.343 (0.337)	0.214 (0.353)	0.850 (0.380)**	0.691 (0.406)*
HIGHDESCAP	−0.115 (0.338)	−0.304 (0.412)	0.617 (0.355)*	0.675 (0.382)*	0.619 (0.360)*	0.675 (0.411)
ONESUPP		0.506 (0.404)		−0.067 (0.382)		−0.489 (0.451)
SSUBCONT		0.001 (0.008)		−0.001 (0.005)		0.006 (0.008)
Obs. (N)	134	125	132	123	132	122
Likelihood	−189.849	−175.786	−191.994	−175.724	−168.775	−155.098
Pseudo R2	0.111	0.116	0.026	0.033	0.084	0.093
McFadden's adj R2	0.036	0.025	−0.055	−0.066	−0.003	−0.012
Wald chi2 test	50.53 (0.0000)	49.50 (0.0000)	15.19 (0.2310)	18.37 (0.1904)	31.87 (0.0015)	32.19 (0.0038)

Source: Author calculations based on survey results

Notes: ***, **, * denote significance level at 1 %, 5 % and 10 % respectively. Robust standard errors of the coefficients are shown in parentheses. All models include intercepts significant at 1 % level but not shown in the results for space reasons. Two-tailed test. P-values are in brackets in the last rows

[a]Domestic and overseas automotive main manufacturers

[b]Domestic and overseas suppliers

[c]Universities, private engineering and consultancy firms

at 5 % level. An increase in SENGPER affects the probability of increases in *cooperation level of the suppliers in innovation activities with the AMMs*, ceteris paribus, in a positive way. Conversely, this variable exerts a negative, yet statistically insignificant, effect on the other two dependent variables: *the cooperation level of the suppliers in innovation activities* with *the lower-tier suppliers* (*Factor* 2) and that with *the universities and other firms* (Factor 3).

HIGHDESCAP dummy variable positively affects both of the dependent variables at 10 % level: (1) *the cooperation level of the suppliers in innovation activities* with *the lower-tier suppliers* (Factor 2) and (2) with *the universities and other firms* (Factor 3) (significant at 10 % level). In other words, *the presence of high design capability in a firm* increases the probability of intensity in *cooperation level of the suppliers in innovation activities* with both *the lower-tier suppliers* and *the universities and other firms*, ceteris paribus. Conversely, this variable exerts a negative, yet not significant, impact on *the cooperation level of the suppliers with the AMMs* (Factor 1).

5.5.7 Determinants of the Performance Increases in Production Capability

In this section, the determinants of *the developments achieved in terms of production capabilities of the suppliers in the last 3 years* are investigated. Prior to the regression analyses, FA is conducted. Table 5.31 displays the results of FA for the performance increase related to *production capability* scale. As can be seen in the table, the items pertaining to the *production capability* were combined in only one factor. The factor was labeled as *performance increase* (Factor 1).

The findings of the regression analysis on the determinants *of the improvements in production capability of the suppliers in the last 3 years*, which were accepted as a performance indicator, are presented in Table 5.32. The dependent variables conducted by FA and used in the regression analyses are in ordinal nature, and they take the value between 1 (very increased; deterioration) and 5 (very decreased; improvement) on a five-point Likert scale according to development level in

Table 5.31 Factor analysis of the performance increases (improvements)

	Factor 1
Items	Performance increase
1. Cycle time	0.655
2. Defect rate	0.679
3. Average costs	0.600
4. Duration of on-time delivery	0.772
N = 165, KMO measure = 0.662, Bartlett's chi-square = 64.43 (0.0000)	

Source: Author's calculations based on survey data
Notes: Extraction method: principal component analysis; rotation method: Oblimin with Kaiser normalization. P-value is in bracket. See Appendix G for the Cronbach's alpha value of the scale

Table 5.32 Determinants of the performance increases (improvements) related to production capability: ordinal logistic regression analysis

	Performance increase		
	Model I	Model II	Model III
FIRMAGE	−0.031 (0.350)	−0.049 (0.371)	−0.007 (0.375)
FIRMSIZE	−0.002 (0.244)	0.002 (0.245)	0.112 (0.272)
EXPINT	−0.011 (0.007)*	−0.011 (0.007)	−0.007 (0.007)
FOREIGN	1.365 (0.573)**	1.392 (0.563)**	1.259 (0.573)**
DIRECTSUPP	0.016 (0.621)	0.031 (0.616)	−0.066 (0.628)
LOCALGROUP	−0.483 (0.386)	−0.461 (0.391)	−0.466 (0.402)
CLIENTMNC	0.567 (0.603)	0.533 (0.600)	0.284 (0.629)
RDDEPT	0.608 (0.441)	0.596 (0.457)	0.485 (0.502)
PRODUCTINNO	−0.018 (0.469)	0.025 (0.482)	0.145 (0.513)
SENGPER	−0.026 (0.034)	−0.024 (0.033)	−0.012 (0.033)
HIGHTECHPROD		−0.210 (0.361)	−0.293 (0.383)
HIGHDESCAP		0.138 (0.381)	0.284 (0.385)
ONESUPP			0.000 (0.408)
SSUBCONT			−0.001 (0.006)
Obs. (N)	152	151	142
Likelihood	−135.251	−134.584	−122.963
Pseudo R2	0.053	0.055	0.050
McFadden's adj R2	−0.046	−0.058	−0.090
Wald chi2 test	14.32 (0.1589)	16.81 (0.1570)	15.80 (0.3259)

Source: Author calculations based on survey results

Notes: ***, **, * denote significance level at 1 %, 5 % and 10 %, respectively. Robust standard errors of the coefficients are shown in parentheses. All models include intercepts significant at 1 % level but not shown in the results for space reasons. Two-tailed test. P-values are in brackets in the last rows

production capability. Explanatory variables have been included in the ordinal model in three stages to determine the problem of a possible multicollinearity: There are first ten variables in column 1 (Model I); then two variables, HIGHTECHPROD and HIGHDESCAP, are added (Model II); and lastly the ONESUPP and SSUBCONT variables are added to the model (Model III).

EXPINT explanatory variable exerts a significantly negative impact on the *performance increases of the suppliers*' dependent variable (significant at 10 % level). That is, an increase in EXPINT reduces the probability of *performance increase (improvement) of the suppliers related to production capability* (Model I), ceteris paribus. This variable also exerts a negative, yet insignificant, impact on the other two models (Model I and II). EXPINT variable's negative effect (i.e. increase) on *the performance levels of the firms regarding the indicators* in Table 5.31, ceteris paribus, may indicate that there are qualitative differences (more complex and quality products) between the products presented by exporter suppliers to world markets and the products of the firms manufacturing mainly for domestic market.

Besides, it may indicate logistical problems to deliver on time the products manufactured mainly for world markets—increased distance between the importing country and Turkey may increase the seriousness of this problem. At the same time, the possible problems at the customs of both Turkey and export country, and problems arising from import regulations in the importing country may extend the delivery period of the products. Naturally, these kinds of problems are not in question for the firms manufacturing mainly for domestic market. Therefore, the increasing effect of the *export intensity* on defective product rate, average costs, delivery duration in time and cycle time may be depending on these qualitative differences.[42] Naturally, these are the speculative remarks, but they are as yet the most reasonable.

The coefficients of the FOREIGN variable are positive and statistically significant at 5 % level for all models. *The presence of foreign ownership in the firm* exerts a positive effect on the probability of *performance increases* (*improvement*) *of the suppliers related to production capability indicators* stated in Table 5.31, ceteris paribus. As discussed earlier, higher technological capabilities of the foreign-capital firms compared to the local firms may be interpreted with a greater possession of both tangible and intangible assets by the foreign-capital firms. It appears that foreign firms are superior to local firms in terms of many technological indicators and absorptive capacity (more export intensity, manufacture technologically more advanced and complex products, productivity, having high qualified personnel, performing more R&D activities, etc.) (see Sects. 4.1, 4.4, 4.8, and 4.10). Superiorities in these technological levels are predictive of higher efficiency for the foreign firms compared to the local firms (current studies point out that the efficiencies of local firms in terms of labor/total factor productivity is generally lower than the foreign firms operating in the same industry). If the differences in efficiency levels are reflected on the indicators in Table 5.31, the result in Table 5.32 can be obtained.

Other explanatory variables do not exert any statistically significant impact on the probability of *performance increases* (*improvement*) *of the suppliers*. It is quite interesting that, of these variables, especially CLIENTMNC, RDDEPT, EXPINT, FIRMAGE, DIRECTSUPP, and ONESUPP explanatory variables do not have any impact on the performance increases of the suppliers. It is interesting and no obvious to interpret that these competences of the firms not reflected on performance increases of the suppliers.

[42] Whether these interpretations reflect the truth or not and also if they reflect truth, the reasons thereof may be comprehended as a consequence of semi-structured interviews to be conducted with supplier firms within the scope of another study. In the literature review conducted for Turkey, no such study was found. Furthermore, we did not come up with any coding error in the data used herein.

5.6 Summary

Recapitulative tables for findings obtained in this chapter in terms of logistic and ordinal logistic regression analyses are presented in Tables 5.33 and 5.34. They contain signs of coefficients statistically significant at least at the 10 % level and that are associated with explanatory variables that impact on the dependent variables analyzed such as *being a direct supplier*, *possessing an R&D department*, *carrying out innovation activities*, *sources of the technologies*, *co-design activities*, different *types of KTTs—related to production process*, *product*, *training*, *and financial transfers*, *technological capabilities*, *benefits of being a direct supplier*, *cooperation activities*, and *performance increases*.

The effects of some important features of the suppliers that they were used as explanatory variables in the regression analyses on *being a direct supplier* and *possessing an R&D department* are summarized in Table 5.33. A larger number of variables positively influence both *being a direct supplier* and *possessing an R&D department*. Four variables, (1) *firm size*, (2) *being a part of a local group*, (3) *conduct product innovation activities*, and (4) *share of engineers in total employees*, are the common variables that positively impact both dependent variables. On the other hand, *being an experienced firm*, *being a foreign firm* and *manufacturing high technology products* positively impact the likelihood of *being a direct supplier*. The finding related to foreign ownership points out the difficulty of *being a direct supplier* of the firms without foreign partners. At the same time, it can be considered as a reflection of the dependence on foreign technology in automotive supply industry. Therefore, the dependence on foreign technology is not only unique to the automotive main industry. The only variable that exerts a negative impact on *being a direct supplier* is the *export intensity*. *R&D incentives* and *high design capability* variables affect the likelihood of *possessing an R&D department* positively. This finding indicates the positive role of *R&D incentives* in resolving the problem of insufficient number of suppliers that have R&D competences in automotive supply industry often voiced by AMMs[43] in Turkey.

The findings related to factors affecting *innovation* and *product innovation activities* of the suppliers are summarized in Table 5.33. *Export intensity* and *owning MNC among customers* exert a positive impact on the probability of *carrying out innovation activities*, while *being a foreign firm* exerts a negative impact on these activities. The negative impact of *the foreign ownership* indicates that the likelihood of *innovation activities* in the foreign firms is less compared to local firms, ceteris paribus. This finding does not point out any weakness in the innovation competences of the foreign firms. On the contrary, MNCs that have intangible assets have transferred these to their affiliates operating abroad; therefore, foreign firms in the host country (assumed that they are the affiliates of MNCs) are engaged less with these activities than do local firms. There are three variables that exert positive impact *on product innovation activities*: (1) *export intensity*,

[43] See Erdil et al. (2011).

Table 5.33 Recapitulative table of logistic regression analyses

	Direct Supplier	R&D Department	Innovation Activities	Product Innovation Activities	Technology Agreements (TA)	Sources of the Technologies	Co-Design Activities related to Products	Partners in Cooperation for Co-Design Activities related to Products
FIRMAGE	+					+ AMMs - Universities and other Firms		- Universities and other Firms
FIRMSIZE	+	+			+ LATA	+ AMMs + Universities and other Firms + Overseas*	+	+ Universities and other Firms
EXPINT	-		+	+			+	
FOREIGN	+		-			- Lower-Tier Suppliers - Domestic*		- Universities and other Firms
DIRECTSUPP				+	+ LATA	+ Domestic* - Overseas*	+	+ AMMs
LOCALGROUP	+	+			- ATA	- AMMs + Universities and other Firms		
CLIENTMNC			+			+ Domestic*		
RDDEPT				+	+ LATA		+	+ Universities and other Firms
PRODUCTINNO	+	+			+ ATA		+	
SENGPER	+	+			+ LATA + ATA	- Lower-Tier Suppliers		
HIGHTECHPROD	+					+ Overseas*		
HIGHDESCAP		+						
ONESUPP								
SSUBCONT					+ ATA			
RDSUB		+						

Source: Author calculations based on survey results

Notes: (+) Positive Effect and (-): **Negative Effect**. All are statistically significant at least at the 10% level. Two-tailed test.

Legend: AMMs (Automotive Main Manufacturers) *Domestic: Domestic AMMs and Domestic Suppliers. *Overseas: Overseas AMMs and Overseas Suppliers; LATA (Low-Advanced Technology Agreements); ATA (Advanced Technology Agreements)

Table 5.34 Recapitulative table of ordinal logistic regression analyses

	Knowledge and Technology Transfers				Technology Capabilities	Cooperation with other Firms	Improvements in Production Capability	Benefits of being a Direct Supplier	Partners in Innovation Activities	Performance Increases (Improvements)
	Production Process	Product	Training	Financial						
FIRMAGE	- HTL - MTL								- Universities and other Firms	
FIRMSIZE	+ MTL			+ Strategic Relationship						
EXPINT	- MTL - LTL									-
FOREIGN	- HTL		+	- Strategic Relationship		-		-Production Technology		+
DIRECTSUPP			+						+ AMMs - Lower-Tier Supplierrs + Universities and other Firms	
LOCALGROUP	- HTL				- Design and R&D		-			
CLIENTMNC	- LTL				+ Design and R&D				- Lower-Tier Supplierrs	
RDDEPT	+ HTL	+			+ Design and R&D	+			+ Universities and other Firms	
PRODUCTINNO		+			+ Design and R&D				+ AMMs	
SENGPER								-Production Technology	+ AMMs	
HIGHTECHPROD									+ Universities and other Firms	
HIGHDESCAP	- HTL - LTL	-		- Procurement	+ Design and R&D		+	+ Design and Business Relationship	+ Lower-Tier Supplierrs + Universities and other Firms	
ONESUPP	+ LTL	+		+ Procurement	- Design and R&D - Production					
SSUBCONT	+ MTL		+					+ Production Technology + Design and Business Relationship		

Source: Author calculations based on survey results

Notes: (+) Positive Effect and (-): Negative Effect. All are statistically significant at least at the 10% level. Two-tailed test. "Improvements" word in column 7 is used as a proxy for performance increases.

Legend: HTL (High-Technological Level); MTL (Medium-Technological Level); LTL (Low-Technological Level); AMMs (Automotive Main Manufacturers)

(2) *being a direct supplier*, and (3) *possessing an R&D department*. The most striking finding is that the positive impact of the *export intensity* is on both *innovation* and *product innovation activities* of the suppliers. This shows that the suppliers manufacturing for world markets are engaged in *innovation activities* to develop the quality of the products and also to compete with other firms in these markets. In other words, an increase in the share of the manufacturing conducted for the world markets in the turnover has forced the suppliers to build their innovation capabilities.

A larger number of variables positively influence the likelihood of performing *technology agreements* (TA) (Table 5.33). The factors affecting *the low-advanced level technology agreements* (LATA) positively are *firm size, being a direct supplier, possessing an R&D department* and *share of engineers in total employees*. On the other hand, the factors affecting the likelihood of performing *advanced technology agreements* (ATA) positively are the share of *subcontracting agreements in total sales, carry-out product innovation activities* and *share of engineers in total employees. Being a part of a local group* exerts a negative impact on such agreements. These findings indicate that technological capabilities and absorptive capacity of the suppliers play an important role in performing *technology agreements. Technology agreements* may be seen as an important channel in terms of the KTTs especially by the firms seeking to improve their competences in search of technology.

A large number of variables have impact on *the acquiring of technologies from various sources* (Table 5.33). *Firm age* and *firm size* are the two important variables that positively affect the acquisition of technologies from *the AMMs. Firm size* and *being a part of a local group* positively affect the likelihood of acquiring these technologies from *the universities and other firms*. Moreover, *firm size* and *manufacturing high technological product* variables have a positive impact on *the acquisition of technologies* from *overseas*[44] sources, while *being a direct supplier* and *having MNC among customers* have also a positive impact on that from *domestic*[45] sources. The variables that negatively affect *the acquiring of technologies from various sources* are *firm age* (on universities and other firms), *foreign ownership* (on both lower-tier suppliers and domestic sources), *being a direct supplier* (on overseas sources), *being a part of a local group* (on AMMs), and *share of engineer personnel in total employees* (on lower-tier suppliers).

The factors affecting *the co-design activities related to products* and *the partners cooperated in these activities* are summarized in Table 5.33. *Firm size* and *possessing an R&D department* variables positively influence the likelihood of both *performing co-design activities related to products* and *performing these activities with the universities and other firms*. Moreover, *conducting product innovation activities* and *export intensity* have also a positive impact on *the co-design activities of the suppliers related to products*. These findings indicate that

[44] OAMM and OS.

[45] DAMM and DS.

engaging in R&D and *product innovation activities*, an increase *in the number of employees* and *in firm's total sales* conducted to world markets trigger *cooperation activities of the suppliers in the field of product design*. These findings point out the advantages benefited by large-sized firms such as technological, managerial, financial, and human resources benefits. The *export intensity* may give an idea about the place of the suppliers in the MNCs' global production networks. The most striking finding is that *being a direct supplier* exerts a positive impact on both *the co-design activities of the suppliers related to products* and *performing these activities with the AMMs*. This last finding indicates the superiority of the direct suppliers (or first-tier suppliers) in the field of co-product design with *the AMMs*. The last finding is that *being an experienced firm* and *foreign ownership firm* negatively affect *the co-design activities performed with universities and other firms*.

The findings in Table 5.34 concerning the KTTs provided by customers to suppliers reveal the following observations:

- A larger number of variables influence *the production process–related KTTs* negatively than *the transfers related to product*. *Age of the firm, export intensity, foreign ownership, being a part of a local group, having MNC among customers*, and *high design capability* reduce the possibility of such *production process transfers* at *high* (*HTL*), *medium* (*MTL*) *or low* (*LTL*) *technology levels*. By contrast, *firm size, engaging in R&D activities, being sole supplier of the most important customer*, and *share of the subcontracting agreements in total sales* exert a positive impact on such transfers. This finding about the negative impacts of the variables on *KTTs for production process* is probably related to the already developed technological capabilities of the suppliers.
- The variables that positively affect *the product-related KTTs* are *being the sole supplier of the most important customer, engaging in R&D*, and *product innovation activities*. The variable *owning a high design capability* negatively affects these transfers or reduces the possibility of such transfers probably because of the developed product design capabilities of the suppliers.
- *Being a direct supplier of AMMs, share of subcontracting agreements in total sales* and *being a foreign firm* are the most important features that affect the frequency of *training activities* provided by customers. These *training activities* may be an important channel for the KTTs.
- With respect to *financial transfers*, four explanatory variables exert a significant effect: (1) *Firm size*, (2) *foreign ownership*, (3) *high design capability*, and (4) *being the sole supplier of the most important customer*. *Firm size* impacts the possibility of *financial transfers for strategic relationship* positively, whereas *foreign ownership* influences these transfers negatively. Moreover, *high design capability* negatively impacts on *financial transfers for procurement*, whereas *being sole supplier* positively influences such transfers.
- One of the most striking findings is that *being a direct supplier firm* does not have any significant impact on *financial, product*, and *production process transfers*. It should be taken into account that a majority of the firms participating in

our survey are direct suppliers (first-tier suppliers), which may possibly have affected these results.

In our study, *technological capabilities of the suppliers* measured in different dimensions (such as design, R&D, and production) have been accepted as a performance indicator. There are four variables that positively affect the likelihood of the increases of *design and R&D capabilities*: (1) *possessing an R&D department*, (2) *carrying out product innovation activities*, (3) *high design capability*, and (4) *having MNC among customers* (see Table 5.34). The first three findings already indicate the developed competences and high absorptive capacity of the suppliers; therefore, it is expected that the suppliers that have these characteristics have high technological capabilities in terms of such competences. The last finding may also indicate that suppliers manufacturing for multinationals operating in or outside of Turkey get these *KTTs related to design and R&D* from them.[46] We do not have any information about the share of multinationals in the total sales of the suppliers, and therefore we have to base this interpretation on the dummy variable included in the regression model. In addition, *being a part of a local group* exerts a negative impact on *the technological capabilities related to design and R&D*, *being a sole supplier* of the most important customer exerts a negative impact on such capabilities related to both (1) *design and R&D*, and (2) *production.*

In order for *the KTTs to be provided to suppliers by AMMs*, a very important factor—perhaps a prerequisite—is *the cooperation activities of the suppliers with other firms*. The main finding related to factors affecting the establishment of *cooperation activities with other firms* is that explanatory variables *being foreign ownership firm* reduce the probability of frequent occurrence of these activities, while *possessing an R&D department* increases such activities (see Table 5.34). R&D activity of the suppliers which is an important indicator of many competences increases the possibility of cooperation in order to obtain KTTs and to develop technology level. Other variables do not exert any statistically significant impact on the probability of *cooperation activities*.

The findings related to factors affecting *improvements in production capability of the suppliers for the last 3 years* accepted as another performance indicator are summarized in Table 5.34. *Being a part of a local group* reduces the probability of increases in improvements of the suppliers in certain fields (Table 5.25) related to production capability, while a *high design capability* exerts a positive impact on this probability.

The factors positively affecting the likelihood of increases *in improvement of benefits received from being a direct supplier of AMM* are *being a foreign firm*, *share of engineer personnel in total employees*, *high design capability* and *share of subcontracting agreements in total sales* (see Table 5.34). *Share of engineers* and *foreign ownership* variables exert negative impact on *the benefits related to production technology*. On the other hand, *share of subcontracting agreements*

[46] This finding may also give an idea about the place of the suppliers manufacturing for multinational located abroad in the global production chain of the multinational.

positively impacts on both *production technology* and *design and business relationship* benefits. *High design capability* variable exerts a positive impact only on *the benefits related to design and business relationships.*

An analysis of the findings in Table 5.34 in terms of *external partners in innovation activities* leads to the following observations. There are three variables that positively affect the likelihood of *the cooperation of the suppliers with the AMMs*: *being a direct supplier*, *conduct product innovation activities*, and *share of engineers in total employment.* The variables that positively affect such *cooperation activities with the universities and other firms* are *being a direct supplier*, *possessing an R&D department*, *manufacturing high technology product* and *high design capability*, whereas *firm age* variable exerts a negative impact on these activities with the same partners. On the other hand, *being a direct supplier* and *having a MNC among customers* exert negative impact on *the cooperation with the lower-tier suppliers*, while *high design capability* affects these activities with such partners positively.

Table 5.34 summarizes the findings related to factors affecting *performance increases* (*improvements*) *of the suppliers.* The performance increases are measured by indicators that determine the cooperation activities of the AMMs with suppliers such as cycle time, defect rate, average cost, and duration of on-time delivery. The findings show that *being foreign ownership firm* increases the probability of *performance increases* compared to local firms—ceteris paribus, while increase in *export intensity* results in a decrease in *performance increases.* The possible reasons for the negative effect of *export intensity* were already made in the relevant section.

Finally, the findings presented in Tables 5.33 and 5.34 in terms of explanatory variables used in the regressions bring to the fore five variables that affect the variety of factors positively: (1) *firm size*, (2) *being a direct supplier*, (3) *possessing an R&D department*, (4) *conducting product innovation activities*, and (5) *share of engineers in total employment.* These are assumed as important indicators of absorptive capacity of the suppliers except *firm size.* The most prominent variable among these is the *possessing an R&D department* or *engaging in R&D activities. Engaging in R&D activities*—that is an important indicator of competence—positively affects (a) *high level KTTs related to product and production process*, (b) *technological capabilities in terms of design and R&D*, (c) *cooperation activities with other firms*, (d) *co-design activities related to products*, (e) *product innovation activities*, (f) *carrying out low-advanced level technology agreements*, and (g) *cooperation with universities in both innovation and co-design activities related to products.* These draw attention to the importance of R&D competences for those suppliers which desire to be a co-designer, to carry out innovation activities, to obtain KTTs, to establish partnerships, and to develop their technological capabilities. Engaging in R&D activities is also considered to be an important factor by the AMMs in order to select their suppliers and to decide to collaborate with them (see Sect. 6.5). In conclusion, *possessing an R&D department* (or performing R&D activities) is found more important than any other factors.

Chapter 6
Findings and Analysis of the Interviews

This chapter evaluates the qualitative data obtained from the semi-structured in-depth analyses of the interviews conducted with 19 top-executives of the 11 AMMs operating in Turkey (DAMM), based on major topics and interview sub-sections. This chapter is composed of seven sections. In the first section, general information on the AMMs interviewed is presented. In the next three sections, we analyze the characteristics of the AMMs operating in Turkey; their cooperation activities with multinationals (global AMMs) as their foreign partners; R&D and technology-related activities and channels of the transfers realized by multinationals to the AMMs at intra-firm level. In the fifth and sixth sections, we detail the relations of the AMMs with their direct suppliers operating in Turkey, and analyze and evaluate the channels and determinants of the transfers at inter-firm level realized by the AMMs to their direct suppliers through backward linkages. The seventh section contains the results of the SWOT analysis of the industry in terms of the transfers based on the findings obtained in the framework of the study. In the analyses, similar views put together for the sake of keeping the identities of the AMMs interviewed confidential, and the findings are enriched with the viewpoints considered significant, as well. The interviews made with the top-executives of the AMMs took 32 h in total, and the text content of 447 pages were decoded following the interviews. The findings obtained from the interview analyses are also used for the quantitative survey results on supply industry attained in Chaps. 4 and 5 in a complementary way. Since there is no other similar study conducted in the industry on this subject as far as we know, no comparison of the findings could be possible. The findings tried to be put forward herein are the first findings concerning the industry and therefore are of major significance. In addition, they could contribute to more extensive, similar studies yet to follow, setting an example for them.

A. Sönmez, *Multinational Companies, Knowledge and Technology Transfer*, Contributions to Management Science, DOI 10.1007/978-3-319-02033-4_6,

6.1 Introduction

The global AMMs (or OAMM)[1] are the MNCs who invest in many countries on a global scale, who develop business relationships and establish partnerships, and who carry out manufacturing activities at the plants founded in the host countries under their own licenses as a result of those partnerships.[2] For this reason, we prefer to use the expressions *global AMM* and *MNC* interchangeably or to bear the same meaning hereafter.

The 11 AMMs we have made an interview with could be evaluated under three distinct foundation and management structures: (1) affiliates of multinationals (FS = 100 %), (2) capital partnerships in the form of joint-ventures based on the different foreign shares, established between a local group and multinational (29 % $<$ FS $<$ 86 %), and (3) independent local assemblers with no foreign share but performing manufacturing under multinational licenses[3] (see Appendix H). It was observed during the interviews that even the AMMs, with no apparent licensed manufacturing, perform manufacturing of their engines and various powertrain[4] parts locally under foreign licenses of the global AMMs as their foreign partner or import them directly.

It has been determined that the know-how, design and R&D headquarters of the AMMs with foreign-capital (FS $\geq$ 29 %) interviewed are located in the central countries of MNC who are their partners (Japan, USA, Germany, France, Italy, etc.).[5] It has been found out that it is these central headquarters who make the decisions on and strategies about manufacturing and the future, who develop issues of each and every kind of basic design, planning, R&D activities, manufacturing process supervision, operations management systems, etc., and who are the final decision making organs in these respects. Generally speaking, it is observed that the AMMs in Turkey are informed and notified of the decisions only after those decisions are made, except in highly unusual cases. Such kind of management style is also applicable to multinationals' other affiliates and plants around the world. In other words, the AMMs in the form of joint-venture capital partnerships are considered a 100 % MNC in management and technological sense, although their capital structure is different. It has been determined as a result of the interviews that there is also no difference between a MNC as a foreign partner and the

[1] See Sect. 3.7.3 for the decription of OAMM.

[2] Whereas some global AMMs specialize only in the personal car product range, some others specialize in both personal car and commercial vehicle product ranges.

[3] It is determined that these companies have several license agreements with various multinationals abroad in order to produce and/or assemble some imported systems/parts (especially powertrain parts) in their vehicles.

[4] Such parts (gearboxes, differentials, brakes, suspensions, axles, etc.) are imported from the partner MNC in general.

[5] For example, Japan is the headquarters for *Toyota* and *Anadolu-Isuzu*, Germany is the headquarters for both *M.A.N Türkiye* and *Mercedes-Benz Türk*, and France is the headquarter for *Renault*.

AMMs with joint-venture capital ownership in the real sense. For example, the company TOFAŞ defines himself as FIAT, FORD-OTOSAN as FORD, OYAK RENAULT as RENAULT and ANADOLU ISUZU as ISUZU. In this context, it is observed that each and every kind of cooperation activity between the partners (MNCs and AMMs in Turkey) are being performed mutually, that the manufacturing and management systems of multinationals abroad are being implemented at the domestic AMMs, and that there is no obstacle in front of any type of the KTTs conducted by multinationals to AMMs in Turkey. In addition, it is observed that the local engineers of the AMMs and foreign engineers of multinationals work continuously as a mixed team on many projects at home or abroad. In sum, the knowledge, experience, technology, culture and know-how owned by multinationals are shared in different degrees with the AMMs in Turkey as their partners (groups such as OYAK, OTOSAN, TOFAŞ, ANADOLU, and KOÇ), in consequence of the partnership agreements and cooperation.

Figure 6.1 shows the shares of all AMMs operating in Turkey (15 ea.), and among them, shares of 11 AMMs who were interviewed, in total production and sales during 2010 year, based on foreign ownership structure. As seen from Fig. 6.1b, the share of the AMMs interviewed in total production is 89 % while their share in total sales is 91 %. The ratio of the AMMs with domestic-capital (FS = 0 %), with whom an interview could not be performed, in total sales (total production) is 2 % (2 %), while the said ratio of the AMMs with majority-foreign ownership (50 % < FS < 86 %) is 5 % (7 %) and the ratio of AMMs with fully foreign-ownership (FS = 100 %) is 2 % (1 %), respectively. Therefore, we could assume comfortably that the sample interviewed represents the universe (the automotive main industry of Turkey) to a great extent (91 %) (see section "Representativeness and Reliability of the Sample" in Chap. 3 and Table 3.10 for details).

We will analyze the AMMs in accordance with the previous chapters based on the following three distinct foreign ownership structures:

1. Foreign ownership AMMs if FS is at least 10 % or more in total capital (FS ≥ 10 %),
2. Joint-ventures with different foreign shares (29 % < FS < 86 %) and with full foreign ownership (FS = 100 %) (affiliate of MNCs),
3. Minority (29 % < FS < 42 %) and majority foreign ownership AMMs (50 % < FS).

When Fig. 6.1 is evaluated based on the above-mentioned foreign ownership structures with respect to all AMMs (Fig. 6.1i):

- It is observed that the share of all the AMMs with foreign ownership according to the first category in the total production and sales is 97 % and 92 % respectively, while shares of the AMMs with domestic-capital (FS = 0 %) are only 3 and 8 %.
- When the AMMs are compared according to the second category, it is observed that the share of the AMMs in the form of joint-venture capital partnership (seven companies) in total production is 88 %, approximately ten times higher

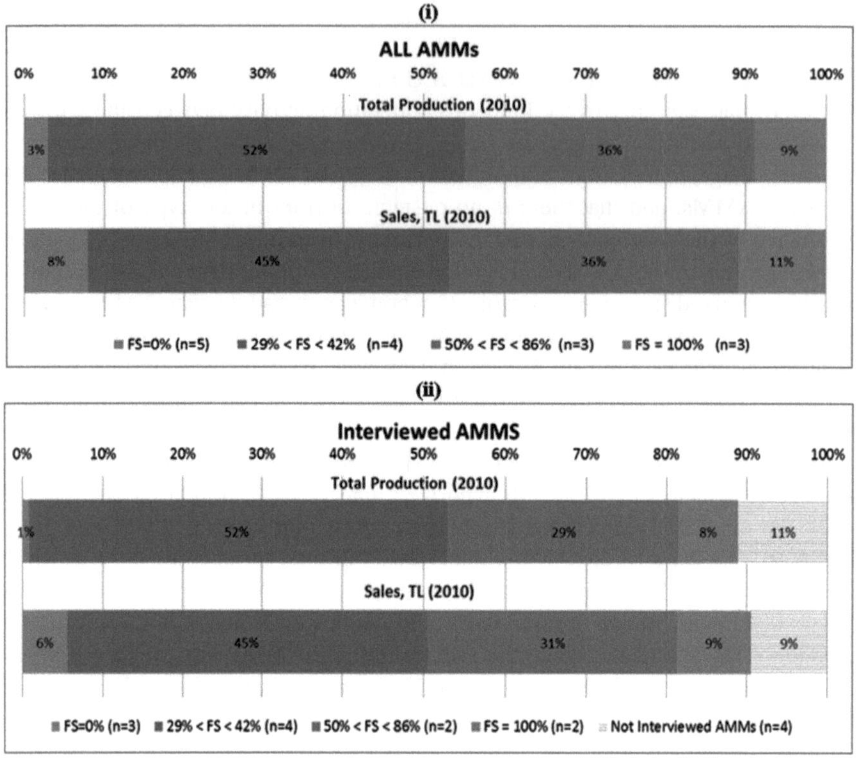

Fig. 6.1 Share of all and interviewed AMMs in total production and sales in 2010 by different foreign ownership. *Source*: Author calculations based on OSD statistics

than those of the AMMs with 100 % foreign capital (three companies) (9 %), and their share in total sales is 81 %, approximately seven times higher than those of the AMMs with 100 % foreign capital (11 %).

- Furthermore, when we make the same evaluations according to the third category, we observe that the AMMs have nearly equal shares [AMMs with minority (majority) foreign ownership is 45 % (52 %) in production, and AMMs with minority (majority) foreign ownership is 45 % (47 %) in total sales]. These statistics reveal that a major part (92 %) of the production activities in the Turkish automotive main industry is realized by the multinational affiliates or partners, and underline the fact that multinationals play a significant role in this respect. In other words, they point out that the Turkish automotive main industry is under the control and domination of the multinationals. Although we will analyze later in detail, it is understood from the information obtained from the interviews that the AMMs operating in Turkey are a part of multinationals and that they are included in their global production networks, and that those multinationals export their products initially to all EU countries, and to the countries in Eastern Europe, the Middle East, the Far East, Asia, North Africa, as well as the Balkan countries and USA, utilizing Turkey as a regional production base.

6.2 General Characteristics of the AMMs Interviewed

6.2.1 General Information

The information on the 11 AMMs interviewed could be summarized as follows:

Capital Structure: When their capital structure is reviewed, it is seen that: three of them are domestic and two are 100 % foreign-capital companies. The other six, however, have a joint-venture capital ownership structure, where foreign capital remains as minority (29 % < FS < 42 %) (four companies) and majority (FS ≥ 50 %) (two companies).

Scale: The AMMs interviewed are large-scale companies. The total number of their employees ranges between 500 and 8,400 people. The average company size in terms of the number of employees is above 3,000 people (3,390). See Appendix H for the capital of the companies, their total production capacities during the year 2011 and total production and sales figures during the year 2010.

Management: When we look into the management structure of the AMMs, we observe that the management and organizational structure of the joint-venture partnerships where foreign capital remains as minority are composed basically of the local managers appointed by the domestic partner, while it is observed that the MNC as the foreign partner is represented with a relatively small number of foreign managers (two or three people) within those structures. However, in the companies of a joint-venture capital partnership that are fully foreign companies or that have foreign capital in majority, the top management is established by the MNC. Individuals appointed to the said management positions are in general foreigners, but could be locals as well. Foreign managers are appointed generally from the central country where the MNC headquarters is located.

Vehicle Type: When the types of the vehicles manufactured by the companies are reviewed, it is observed that seven of them manufacture light commercial vehicles and/or buses, two manufacture personal cars and light commercial vehicles, one manufactures only personal cars and one manufactures only tractors[6] (see Appendix H for details). It has been understood from the interviews that the AMMs in Turkey have focused on and specialized in especially the manufacturing of buses and light commercial vehicles. In other words, we could say that Turkey is a commercial vehicle production base. Trucks, light trucks, pick-ups and buses in

[6] Two of the AMMs with domestic-capital manufacture also armoured vehicles and tanks in cooperation with *Aselsan*, *Roketsan* and *MKE*, under the scope of the defence industry. Such types of vehicles are designed and developed domestically and manufactured by obtaining relevant licenses. It has been determined that the AMMs have accumulated significant design, R&D experience and know-how via such projects.

various models and versions, addressing to all segments, are manufactured and sold to markets abroad.

Market: It is detected that on average 70–80 % of the vehicles manufactured by the AMMs are exported. Therefore, the companies are giving special attention to overseas markets rather than the domestic market. On average, half of the exports of the companies are to the EU countries (Portugal, France, Italy, Germany, Spain and England), while the rest to Eastern Europe, the Middle East, Asia, North Africa, Saudi Arabia and USA.

R&D: All of the AMMs interviewed expressed that they conduct R&D activities. Except one (TOYOTA), all of the AMMs have an R&D center in the framework of the R&D Law no. 5746 accepted in 2008. The number of full-time R&D personnel of the AMMs ranges between 55 and 960, while the average number of R&D personnel is 287. When the said value is compared to the average number of employees, it is understood that 8.5 % of the employees are R&D personnel. During the interviews, the managers stated that they consider it more appropriate to share the information on the ratio of their R&D expenses to their total sales, rather than let their R&D budget known, for confidentiality purposes. In view of that, it has been determined that the ratio of the annual R&D expenses of the companies to their total sales ranges between 2 and 5 %, being on average 3.5 %. When compared to the average R&D expenses in Turkey, it is clear that those rates are quite high. When the companies were asked about the reasons for the situation, they indicated that the ratios had been higher than normal levels for the last couple of years and that the most important reason for that case was the higher R&D investment amounts at the new product development phase and during establishment of an R&D center under the scope of R&D Law no. 5746, despite the global financial crisis. On the other hand, when the companies were asked about their estimations for the upcoming years, their responses were that those ratios would remain higher for the next couple of years and that it was planned that they would remain stable at 2–3 % in the medium term. It is observed that, in consequence of the AMMs' having specialized in the commercial vehicles, the ratio of the number of R&D personnel and R&D expenses of the AMMs manufacturing buses and commercial vehicles to their total sales are higher than those of the AMMs manufacturing personal cars.

6.2.2 Production Structure

It has been determined that across the automotive main industry, the most technology- and knowledge-intensive production based mainly on know-how and electronics is evident in the personal cars, while the most labor-intensive production is evident in the buses. In general, the basic vision and mission of the AMMs in Turkey is to supply the highest-quality systems and parts utilized during the manufacturing process, first from domestic sources, then from various sources abroad, at lowest prices and in a timely manner, and assemble these afterwards.

It has been discovered that the AMMs that are fully domestic or that have minority foreign ownership do not actually engage in manufacturing but rather are generally involved in assembly, design and R&D activities aimed at adaptive improvements basically of non-basic scientific nature.[7] On the other hand, the AMMs where foreign capital is in majority are basically assembly facilities and do not perform R&D or design activities—such activities are conducted at the MNC headquarters as the main centers, but they are in strategic cooperation with the said centers in terms of all the decisions taken on the manufacturing process.

According to the information gathered from the interviews, the production, design and R&D activities of the AMMs could be summarized in three different fashions, independent of the foreign capital structure:

1. Importing vehicles as a whole from the facilities of the MNC located abroad, as their foreign partner (as official distributors of the MNC): In this production structure, it has been observed that the AMMs that have a joint-venture capital structure, are allowed by their partner MNCs to perform small-scale arrangements, adaptive changes and parts modifications on the vehicles when required, in terms of compliance with the market and domestic highway conditions (not involving any production activities), at the plants of the AMMs located at home.
2. Manufacturing completely identical copies of the vehicles produced abroad, through localization process in Turkey, by means of utilizing the styles, concepts, designs and projects developed abroad by the foreign partner MNCs: In this production structure, the AMMs do not contribute to the vehicles manufactured in terms of design, R&D and advanced level of technology; the manufacturing process is conducted totally under the foreign licenses and at home. In other words, all basic activities (such as design, R&D, development, and test) depend on the MNC headquarters, and Turkey is utilized only as a production base. The technology, electronics, powertrain group based on know-how (transmission, drive shafts, and differentials), engines, steering wheel systems, smart control units containing embedded software, and safety parts or systems used for the manufacturing are all imported from abroad. Localization efforts are, on the other hand, performed mostly on various systems and components such as seats, sheet-metal parts, plastic parts, tires, lighting (headlights/tail lights etc.), windshields and windows, and internal trim materials, in cooperation with domestic suppliers. Although the rate of localization according to the value of the vehicle varies depending on vehicle types being manufactured and AMMs, it ranges from 40 to 50 % for the personal cars and 60 to 70 % for the commercial vehicles (see Sect. 6.2.4). As to the localization efforts and processes of the AMMs, it has been observed that they could conduct adaptive, small-scale improvement and development activities at their R&D centers in Turkey. In this respect, the responsibility for the vehicles manufactured for the Turkish market lies with the R&D unit located in Turkey.

[7] Two of the AMMs have underlined the fact that some of the critical and major parts which cannot be procured from the suppliers are manufactured in-house.

3. Planning, designing, developing, testing and manufacturing the vehicle and its systems from scratch, totally by the local engineers of the AMMs with domestic-capital (FS = 0 %) manufacturing buses and commercial vehicles[8]: In this production structure, it is observed that the most critical parts of the vehicle are produced on the co-design and/or co-development basis, together with the direct suppliers located at home or abroad, adhering to the basic design and style. In this production structure, however, the basic components and parts of the vehicles (such as engine, and powertrain) are also imported from abroad. When we look at the origins of the parts, we could say that, in general, the powertrain parts are imported from the multinational headquarters located abroad while system parts (such as safety systems, and steering box) are as a whole imported from the global-mega suppliers and then mounted on the vehicle. Other critical sub-parts are procured from the other direct suppliers at home or abroad, on the basis of co-design or co-development, while less complex parts are generally procured from the domestic suppliers. As to the foreign-capital AMMs interviewed, they have underlined the fact that, in the case of the style, concept and design of the vehicle being totally local, in order for the vehicle to be manufactured in the facilities located in Turkey and then exported to the world markets, it is imperative that those vehicles be approved by the partner MNC, since the vehicles to be manufactured will bear the trademark of the multinational who is the foreign partner. When it comes to the intellectual property rights concerning the design, it is observed that those rights vary depending on the business relations between the partners and financial structure, but in general those rights belong to the partner who has developed the design. On the other hand, there has been no such problem for the AMMs with domestic-capital, since they have been using their own trademark. In this scope, it has been determined that the design which has been developed in this way is subjected to various tests by the foreign experts and engineers at the headquarters of the relevant multinational, and that it is authorized to be manufactured and exported to other countries if deemed appropriate as a result of those tests.[9] As to the AMMs with domestic-capital, the approval is granted by the domestic AMM if the style, design and engineering of the vehicle are local; or by the partner multinational if it is not.

Among the AMMs interviewed, there are companies who fall within only one of the production structures we have summarized above, as well as the AMMs falling within two or all three of them. That being said, it could be stated that the AMMs in the personal car category operate in production structures 1 and 2, while the AMMs

[8] The manufacturing process from scratch is composed of the following phases: conceptual design, detailed design, design verification, developing, prototype, and tests for quality and durability, pre-production tests, serial production and post-production tests.

[9] One of the multinationals has the only partner company in the bus class located in Turkey, manufacturing buses under their own trademark only in Turkey together with their local partner and exports them to the world.

in the commercial category operate in production structures 2 and 3. One of the most significant findings is that 90 % of the R&D, innovation and technology-related activities of the AMMs in Turkey are focused on the vehicles falling in the bus and light commercial vehicle class, produced in the third (3) manufacturing structure.

The most critical operations conducted by the AMMs in relation with the production process are grouped under five main topics:

1. **Deciding on from which suppliers and how the parts to be used in the production process will be procured**: The suppliers with whom the company will work together are determined according to the criteria concerning the components to be required, such as quantity, design, cost, quality, reliability, durability, delivery, and defect rate.
2. **Material technologies**: Continuous optimization of the manufacturing-related structure; researching and developing higher-quality, more comfortable, lighter and more robust technological materials.
3. **Torque-related processes**: Assembling and torqueing the materials supplied accurately and properly; using the tools correctly; and making and checking their calibrations and performing their maintenance and service activities in a complete manner. These processes are considered a highly critical operation by the AMMs.
4. **Welding**: Workmanship of the welding works of the parts (especially in the car-body); welding quality; performing welding works accurately and uniformly each and every time. It is seen that such welding work is performed by robots on the production line of some AMMs with foreign-capital who manufacture personal cars. On the other hand, it is observed that majority of domestic- and foreign-capital AMMs who manufacture buses and commercial vehicles do not employ any robot technologies. It has been determined that the assembly and welding work at the production facilities of the MNCs abroad are performed totally by robots, and that the quality of the workmanship takes precedence, since the use of robot technologies is not feasible due to the fact that the quantities produced in the domestic plants remain at very low levels.
5. **Check-outs, test and measurements**: Conducting each and every kind of test and making measurements on both the parts used and the final products before they come out of the production line, and approving them and making the product ready for delivery to the final user.

6.2.3 R&D, Innovation, and Technology Activities

Another finding of the study is that the majority of the AMMs continuously perform R&D, innovation and technology-related activities on both the products and the production processes, with a view to producing more comfortable, safer, higher-quality and less expensive vehicles than those produced by their rivals.

We could summarize the R&D activities of the AMMs in three classes based on the information obtained:

1. Improvement and development activities on the products which could be industrialized in the short term (3–5 years).
2. R&D activities which could be transformed into products and therefore be commercialized in the medium term (minimum 5 years).
3. Basic scientific R&D activities and research on the products which could be industrialized in the long term (10–15 years), conducted by the R&D units, private research laboratories and headquarters of the multinational as foreign partner.

The most important finding is that the AMMs in Turkey do not engage in the basic R&D projects falling in the third category. It has been determined that the most significant R&D activities have been concentrated on the first category, followed partly by the second category. It has also been observed that the R&D activities of the AMMs include manufacturing of the new products at low- and mid-tech level, and re-designing, improving and developing the former parts in terms of quality and targeted use. It has also been determined that as problems arise in the field of the topics studied under the scope of such R&D projects, the need to co-operate with MNCs, universities and supply industry increases, as well.

It has been discovered that the AMMs with foreign-capital benefit, to the utmost degree, from the advantages of being included in the global production network of the multinationals, i.e. being a global company, when it comes to such R&D activities. In this scope, strategic cooperation is established between the R&D centers of the multinationals abroad and the R&D centers located in Turkey, the local and foreign engineers work as a mixed team on various projects at home or abroad, and all kinds of support can be provided by the multinationals for resolution of the problems faced with. It could be said that the common goal of the companies is to give birth to new and different products involving intensive know-how and technology, suitable for the market concerned, to improve the existing products, and to make new designs.

Some of the AMMs have stated that after they have established an R&D center in Turkey, certain departments (such as design, parts development, and new products design), which are currently included in the structure of the multinational headquarters abroad are planned to be shifted by multinationals to this center in Turkey. In addition, it has been discovered that two of the AMMs with a foreign partner, operating in Turkey, have a separate R&D center affiliated to the R&D center of the multinational as foreign partner. However, these centers at home are not fully-equipped R&D centers but rather work together with the multinational headquarters on the improvement projects for the vehicle to be manufactured at home.

The findings on the R&D, innovation and technology-related activities of the AMMs, obtained in consequence of the interviews, may be summarized as follows:

- Trying to implement to full extent the MNC systems, and trying to improve and develop them continuously, on the issues of product, quality, occupational safety, production processes and business management and under the supervision of such processes,

- Conducting R&D and improvement activities for the vehicles to be produced and sold locally in Turkey, on issues such as compatibility with the domestic market and highway conditions,
- To conduct R&D activities on innovative vehicle technologies and composite, alternative, new materials. It has been determined that the AMMs in Turkey have a serious accumulation of know-how on such issues,
- To perform co-design and co-development activities through joint projects at the new product development phases, by establishing strategic partnerships with the partner MNC, global-mega suppliers and domestic direct suppliers,
- To develop and enhance the engineering and designing skills of the R&D personnel continuously, to employ advanced computer software in analyses, to perform virtual 3D design and mathematical modeling activities and to produce engineering drawings based on dynamics and statics,
- To perform optimization activities concerning the powertrain,
- To perform partial design verification, virtual verification, physical verification and virtual crash test activities,
- To conduct studies on vehicle manufacturing that comply with the EU regulations, on designing and safety-related activities, and on products or product lines,
- To work on issues concerning improvement of exhaust-emission tests,
- To perform activities on obtaining patents.

6.2.4 Domestic Input Rate and Major Sources of Inputs

It has been determined, according to the analyses of the interviews, that the rate of the domestic inputs[10] used in the vehicles produced at home has been gradually increasing from personal car production to bus production. Based on the information obtained, the rate of the domestic inputs increases up to 40–50 % in the production of personal cars, 50–60 % in the production of trucks and pick-up trucks, and 70 % in the production of buses. Since more parts are used in the production of buses, their domestic value added is higher.

Regardless of foreign-capital ownership and the type of the vehicle produced, system parts (such as powertrain, engines, safety systems, steering wheel systems, electronic hardware containing embedded software, and brakes) are imported from the multinational headquarters basically located in Japan, Germany, USA, Italy, France and England and/or from global-mega suppliers. The market accommodating the global-mega suppliers is an oligopoly market composed of leading players who manufacture such system parts as stated above, supplying many multinationals around the world with the parts concerned (see Sect. 3.7.2). Less complex and labor-intensive sub-parts are generally procured from the direct suppliers located at home. The inputs procured from domestic sources are mostly such products as iron

[10] Raw materials, intermediate goods, etc.

and steel, sheet-metal parts, windshield and personal's window glasses, plastic parts, seats, glove compartment, tires, rubber parts, plastic tubes, signal lamps, headlights and tail lights.

While the average rate of the inputs that are procured from the domestic sources and used for the production in terms of parts' quantity is nearly 85 %, the rate of the domestic inputs in the total cost drops to 40–50 %. The most important reason for this is that the imported parts and systems are the basic products composing the vehicle and increase the cost of the inputs due to the fact that they are more expensive, since they are predominantly electronic, technological and know-how inputs. Also considering the fact that the parts and components procured from the direct suppliers in Turkey contain imported inputs (indirect import), the rate of the domestic inputs in the total vehicle production drops even further. Moreover, given that the average rate of the domestic direct suppliers of the AMMs (66 %) is twice the average rate of the overseas direct suppliers (33 %) (see Sect. 6.5.2 and Table 6.1), the average cost per supplier for the AMMs becomes twofold higher when it comes to overseas suppliers. In other words, although a smaller number of products in terms of parts' quantity are procured from a smaller number of overseas suppliers, the total cost paid for them is almost equal to the cost paid to the domestic suppliers.

6.2.5 Patent Policy

It has been observed that AMMs have gradually increased their efforts to obtain patents in recent years and that therefore they attach great importance to the training of the R&D personnel. It has been determined that, until the last 4–5 years, the AMMs did not make much effort to increase the number of their patents; but have been trying to increase the number by promoting obtaining patents via R&D centers established especially in the framework of the R&D Law no. 5746. The existence of an R&D center established under the scope of the law encourages obtaining patents has been indicated as the most important reason for such a situation. Therefore, the AMMs who were not interested in obtaining a patent to protect their products before are now identifying an increase in the number of their patents as their target in order to protect their rights and encourage innovation. Nevertheless, there exist different opinions among the AMMs with regard to obtaining patents. For some companies, increasing the number of patents in their possession constitutes an extremely important goal and indicator, and therefore their R&D personnel are encouraged to get patents by producing projects in the subject fields considered important by the company, and are rewarded to this end. Some companies, however, do not prefer to obtain a patent even if they do have innovations and developments specific to them. It has been stated that the general opinion and target of those companies, as a matter of fact, is obtaining patents for the technologies, innovations, products and studies which would bring higher commercial values and which need to be protected, rather than obtaining a patent for every minor innovation.

6.3 KTTs Through Partners in Strategic Collaborations

It has been observed that the AMMs with both domestic and joint-venture capital structure have established formal or informal strategic collaborations and contacts both at home and abroad with many partners, first of all with their partner MNC. Mutual relationships with such strategic partners have enabled KTTs to AMMs operating in Turkey on certain issues (such as production, production processes, supervision, quality, business management, and training to the AMMs), contributing greatly to the know-how of the AMM itself.

The partners with whom such collaborations have been established are listed as follows, according to their order of significance:

1. **MNC as foreign partner (global AMMs)**: Most critical technologies and know-how employed for the production are provided from the multinational headquarters. It has been determined, for example, that highly extensive engineering support is being obtained concerning the issues (such as production, production processes, quality control, and supervision). It is observed that various training and support activities aiming at implementing the self-proven *World Class Manufacturing* (WCM) systems[11] successfully (see Sect. 6.4.2), which are being implemented by the multinationals across their worldwide plants and also at the plants located in Turkey ensure noteworthy KTTs. It is understood that successful and continuous implementation of such systems requires a well thought-out planning, joint study, discipline, engineering, culture and vision. For this reason, serious training on the establishment and successful implementation of such systems are required and it may take many years.
2. **Licensor MNC**: KTTs are provided via joint agreements and projects executed with the licensor multinationals. This kind of activities is generally observed for the AMMs with domestic capital that produce some parts under multinational license at home.
3. **Global-mega suppliers**: It has been determined that during design, design verification, product development and production processes, KTTs are provided to the local engineers through joint projects, consultancy and support. The mega-suppliers (who are also multinationals) have a considerable amount of know-how on the powertrain, engines, safety system parts and electronic parts of high-tech nature, and they design, develop and manufacture these parts at their own R&D centers and headquarters abroad. Especially during the processes involving new parts development, the engineers of the domestic AMMs and engineers of the above-mentioned companies work together as a mixed team under the scope of the co-design and co-development activities. Therefore, various global-scale partnerships and collaborations among the companies

[11] Such as Toyota Production System (TPS).

emerge and pre-competition collaboration projects and/or joint projects are executed.[12]

4. **Overseas direct suppliers**: The parts and systems, new products, technologies, manufacturing methods and know-how that cannot be procured domestically are provided from these companies. During the procurement process, several collaboration, co-design and co-development activities are executed and strategic collaborations are established also with those companies.
5. **Domestic direct suppliers**: Co-design and co-development activities are realized in the framework of the joint projects, concerning developing critical sub-components and/or new products, with the local- or foreign-capital direct suppliers operating at home.
6. **Companies from whom expertise support is obtained on certain topics**: This includes obtaining consultancy services, training and various assistance from the companies specializing in specific engineering topics (such as problems encountered during product development and production, new technologies, software, design, and R&D) and also bringing in the specialists of those companies to Turkey for a certain period, when required. It has been discovered that such activities play a very important role in KTTs.
7. **Test centers located abroad**: There are certain tests that are employed intensively by the AMMs in their activities (such as production, production processes, and R&D) and that involve advanced technologies like performance tests, durability tests, and tests evaluating the conformance with the regulations. The engineers of the AMMs make visits to the special test centers and laboratories located abroad to test the parts and work together with the foreign specialists-engineers at those centers. As a consequence of such mutual relationships, know-how exchange occurs and knowledge of the local engineers is enhanced.
8. **Universities**: The AMMs realize various (joint) projects with the universities in Turkey,[13] in the fields such as new product development, improving existing products, developing production process, and test methods under the scope of TUBITAK, TEYDEB, SANTEZ, and R&D incentives. In addition, some other collaborations in the academic sense, involving recruitment of the students from the engineering departments of those universities for a certain period at the plants as project students, consultancy provided by the academic personnel of the university for the AMM concerned on the projects requiring specialty, and

[12] There exist agreements on engines, concluded both currently and in the past among global AMMs (such as ISUZU, GM, TOYOTA, NISSAN, MAZDA, RENAULT, AUDI, VW, BMW, and OPEL). The latest of these is the mutual collaboration agreement on electrical vehicle systems and diesel engines, concluded between TOYOTA and BMW at the end of the year 2011. That's why the same engines, same safety systems, etc. are seen in different global brands. Similarly, many car manufacturers could use the brake, air-conditioning, airbag, etc. systems produced by the same global-mega suppliers.

[13] The universities from which such collaborations are provided are basically ITU, METU, Uludag and Marmara.

the like are seen. Although such collaboration activities between universities and industry are not so intensive, they nevertheless ensure mutual KTTs.

9. **Overseas authorized dealers and services**: Especially in the importing countries, the dealers and services play a very important role in terms of proper commercialization, sale, marketing, publicity and promotion activities of the vehicles manufactured, and they constitute the contact points of the AMMs who are in touch with that country. At the same time, the knowledge and experience of such dealers and services prove beneficial for the AMMs in terms of providing guidance in certain issues such as vehicle design, marketing, sales, and post-sales services.

In sum, it could be said that the AMMs obtain the KTT from partners discussed above, first of all the multinationals or global mega suppliers (Bosch, Autoliv, Magna, Delphi, etc.) via developing joint projects, and enhance their know-how at home and thus develop high value-added new products involving advanced technology. As we pointed out before, the most important partner in terms of the transfers is the multinationals as foreign partner. KTTs from the partner multinationals to domestic AMMs take place due to the strategic collaborations in many subject fields such as products, production processes, supervision, investment, quality, testing, and training. The transfers are realized via certain channels such as collaboration activities and joint projects with the multinationals; implementation of the multinationals' production, management and quality systems at the domestic plants; auditing, supervision and assessment activities of the multinational; training activities; and various assistances. Those channels are reviewed in detail below.

6.4 Channels of the KTTs from MNCs to AMMs Operating in Turkey at Intra-firm Level

In this section, the channels of KTTs from multinationals to the AMMs in Turkey at intra-firm level, and the factors playing an important role in these transfers are discussed.

6.4.1 Via Collaboration Activities with MNCs

It is observed that since the know-how and technologies of the partner multinationals are at a higher level, issues such as major investments, designs, parts manufacturing, and R&D activities in Turkey are realized in collaboration with the multinational via joint projects. Although the systems, strategies, principles and planning are the same under the roof of the multinational as the foreign partner, every plant exhibits conditions and circumstances specific to itself, and a system being implemented at a plant of the foreign partner multinational cannot be implemented successfully on an identical basis at another plant of the multinational located in somewhere else in the world. Therefore, despite the fact that the AMMs with foreign ownership structure on a joint-venture basis obtain intensive support from the multinationals as their foreign partners during their joint projects,[14] it is their own local engineers who are supposed, and required, to ensure adaptation of those projects into the domestic plants, their operation and their continuity, provided that the principles of the multinational systems are adhered to. Such projects that are executed in collaboration with the multinationals ensure the transfers in terms of R&D, production and production processes, and management structure, playing a role in enhancing the skills of the local engineers and at the same time providing accumulation of important know-how and knowledge at home. One example of high significance among others which could be given in this respect is the collaboration made with the multinational as the foreign partner during the foundation process of one of the AMM's plant in Turkey. Since the plant to be established in Turkey will be the first and sole bus-manufacturing plant of the multinational in the world, the multinational has no cut-and-dried system which is ready to be used specifically for the buses. Therefore, during the foundation process of the plant, the local engineers have worked very closely with the multinational's engineers/specialists when establishing the production processes of the plant and developing solutions to the problems arisen, thus playing a highly important role. This enabled the local engineers to obtain knowledge and know-how, enhancing their technological skills via KTTs from the multinational. In fact, it is expected that, during the planning phase to establish a bus manufacturing plant in another country, the multinational will benefit from the knowledge of the local engineers participating in this project in Turkey.

In sum, it is noticed that the AMMs are in close collaboration with the partner multinationals in almost all kinds of activities, the said collaboration being more intensive in the joint project activities. Within the framework of the collaboration, various assistance such as required raw materials, software, hardware, know-how, tools and equipment, engineering, financial support, test facilities, and consultancy services could be provided by the multinationals in formal or informal ways. In other words, since they are a part of the multinationals concerned, the AMMs benefit from the advantages of being included in the global production network of the multinationals to the utmost degree.

[14] Activities such as establishing the production lines and production processes of the plant, developing new products, designing new models, R&D activities, and improving the existing products.

6.4.2 Via Production, Management, and Quality Systems of MNCs

Based on the interviews, it has been determined that the AMMs that are fully owned by foreigners and are in the form of joint-venture capital partnership take the various production, quality and supervision systems called *World Class Manufacturing* (WCM) developed by the multinational as basis, and use them for their business management, production, production processes, management and quality systems.[15] These are self-proven systems involving advanced know-how and technology that have been established based on the multinational's business culture and systems, created in the light of experience, knowledge and developments gained and obtained over long years, and they are implemented at all plants of the multinational worldwide. It is targeted via these systems that the companies will develop themselves continuously on the topics such as high quality, efficiency, supply and safety, that they will reduce costs and increase the performance, and that they will achieve the highest quality in the production.[16] To be brief, such systems aim at performing the job faster, more smoothly, and with higher quality, at a less cost, and utilizing fewer workforces. These systems are implemented across all departments of the company, being subjected to a continuous supervision and improvement. Continuous and successful implementation of the systems involves all the employees, from the top level of management to the workers working at the assembly lines. It has been determined that a highly intensive and significant support is provided by the multinational for the implementation of these systems successfully at the domestic plants. The most important and effective support in this respect is to assign the AMM's personnel at the multinational's headquarters for a certain period (3 months, 6 months or 1 year) so that they observe the operation of the system in situ and receive training, and/or to assign the specialists from the multinational's headquarters a duty at the plant in Turkey for a certain period. Furthermore, training and seminars are delivered to the AMM's personnel by the multinational's specialists at certain intervals, and continuous benchmarking is applied between the multinational and the AMM, and the system is continuously supervised, improved and updated via the knowledge exchange between the local and foreign personnel. For instance, whenever a "best practice" which improves a system takes place at any plant of the multinational, this method is immediately shared with other plants around the world and the system is updated accordingly.

[15] Such as Toyota Production System (TPS) of TOYOTA also known as KAIZEN, Manufacturing Management System (IMM) of ISUZU, Total Quality Management, and Lean Management Systems.

[16] All vehicles will be at the required quality standards; efficiency will be at the highest level; work that does not generate any added value will be removed; it will be ensured that no loss occurs in the production; the parts supply methods (JIT) and the safety of the workers and work will be at the highest level achievable; and problems and occupational accidents will be at minimal levels. The main philosophy could be summarized as "manufacturing the best vehicle in the shortest time, at minimum cost, and in compliance with the quality standards set forth".

It has been discovered that in the framework of the above-discussed activities, and intensive know-how, KTTs from the multinationals to the local engineers of the AMMs take place, enhancing the knowledge of the local engineers.

6.4.3 Via Auditing, Assessment, and Awarding Activities of MNCs

As we pointed out in the previous section, the AMMs with foreign-capital implement the global production, management, quality and auditing systems of the partner multinationals at their own plants. It is observed that each and every kind of activity of the AMMs is monitored, audited, evaluated and approved by the partner multinational under the scope of such systems. In other words, all processes of the AMMs (such as the quality, production, monitoring, and auditing) are intertwined with the partner multinational. For example, depending on the multinational's production and quality systems, the AMMs have certain targets on a daily, weekly and monthly basis for aspects (such as defect rate, quality, efficiency, delivery on time, and number of accidents) and the entire personnel of the company work together from head to toe to achieve those targets that are continuously monitored by the multinational.[17] In addition, the organizations of the AMMs (such as production, assembly, and purchasing) are regularly monitored and inspected by the multinational auditors to ensure that such targets are achieved. During the inspections, the plant in question is required to successfully pass the checklist based on the multinational's production and quality systems and achieve highest levels thereof. If, upon inspections, it is discovered that there are some incomplete aspects or problems, or the company fails to achieve the targets, improvement and development supports are provided by the multinational for solution of the problems.

Moreover, there are manufacturing assessment activities conducted by multinationals across their own plants worldwide, within the framework of the *WCM* system. The multinational evaluates their world-wide plants based on the activities and the most successful plant is financially or morally rewarded (plaques, certificates, letter of commendation, etc.) by the multinational. It has been determined, according to the information obtained from the interviews, that some of the AMMs in Turkey have surpassed the other plants of the multinational all around the world and have been awarded silver and bronze medals that are highly valued. In addition, it has been discovered that three of the AMMs in Turkey have won the silver award since they are the multinational plants where the most modern and quality production is executed in the world, other than those multinational headquarters winning the gold award.

[17] Monitored by means of indicators such as DPM (delivery per million) and PPM (parts per million).

Due to the reasons explained above, the plants of the multinationals all around the world compete among themselves every year to achieve the predetermined targets. All these monitoring, auditing, assessment and awarding processes of the multinationals ensure that AMMs in Turkey (the plants of MNCs at home) develop themselves continuously by competing against the other plants of the multinational on the global scale. In conclusion, it has been determined that KTTs are provided to the AMMs and local engineers by means of mutual collaboration with the specialists of multinational in implementing, improving and developing all those systems.

6.4.4 Via Training and Education Activities of MNCs

It is observed that the AMMs have fairly considerable opportunities to provide training, and that they attach great importance to continuous training and improvement of the technical capabilities of all their employees, first of all their R&D personnel and engineers.[18] In this context, they subject their personnel continuously to in-house training and follow up their status in this respect by using various indicators to measure their individual proficiency. For instance, every engineer in the AMMs with foreign-capital has a scale which is established according to the multinational systems and composed of specific topics, and his/her proficiency is monitored continuously based on the scale. With the evaluations made together with the foreign partner multinational, the achievement of certain targets are checked by reviewing the scales, and personal development plans are devised accordingly.

Another significant finding is that the proficiency of the R&D personnel in the area of design is considered an important learning process. It has been determined that for this reason, the more projects a member of the R&D personnel is assigned at, the more his/her proficiency in the area of "design" is considered to have enhanced and increased and the more he/she gains knowledge and experience in such area. In parallel, as the R&D personnel are assigned duties under the scope of the new projects, they obtain more knowledge by communicating with the multinational; they develop themselves more, and they gain more experience. It has also been determined that the AMMs apply various reward systems based on material and moral incentives to encourage the training of their personnel.

The findings concerning main training activities that are performed by the AMMs and that play a role in making the personnel specialize in a certain field and in enhancing their know-how and technical capability are as follows:

[18] It is determined that in the AMMs that have an R&D center within the framework of the R&D Law No. 5746, the cost of such training activities are met from the R&D center's own budget, or included in the projects realized at such centers. It is also expressed by the AMMs that one of the requirements to be met by the R&D centers is continuous training and self-development.

- Under the scope of a new vehicle or a new product development project, local engineers' visiting the multinational headquarters and assuming duties in such projects, and staying abroad for a certain period (1 month to 1.5 years) during the design and product development phase,
- Making visits periodically to the headquarters of the multinationals and/or global suppliers from whom parts are procured (such as Germany, Japan, France, Italy, and USA) and R&D personnel's working together with the foreign engineers for a certain period during the testing phase of the parts,
- R&D personnel's visiting the test centers located abroad, for the testing of some major parts manufactured locally, and their working together with foreign engineers for a certain period in these centers,[19]
- Teams' visiting the multinational headquarters for the training on new systems, developments, and applications, periodically every few months, and their staying at the headquarters for a certain period (2–3 weeks or a couple of months), depending on the nature of the training,
- Regular visiting of the foreign engineers/specialists from the multinational headquarters to Turkey, for training, supervision, consultancy and discussion purposes and to follow up the processes,
- Delivering online training and testing via the multinational to the personnel of the AMM, at certain intervals, and by means of advanced communication tools. The points scored as a result of these training activities are entered in the score cards of the personnel and monitored. If the required criteria cannot be met when the results of the training are evaluated, new training development and advancement activities are drawn up accordingly. In other words, it is observed that there exists an ongoing communication with the multinationals,
- It is understood that close relationships with the suppliers enable also the AMM's personnel to obtain knowledge from the suppliers, which in turn contributes to their self-development. For example, during the design phase of a product to be manufactured by the suppliers, the AMM's engineers are assigned duties at the production facilities of the suppliers and gain an on-the-job insight into how the product is manufactured, and therefore make better design feasible for production,
- There are various training centers available for the blue- and white-collar personnel, and miscellaneous courses for the production and management skills, seminars and software training, etc. are delivered within the structure of these centers. In addition, the newly recruited personnel are provided with on-the-job training, being subjected to the orientation training for a certain period at these centers,
- Receiving education on several topics as required from domestic universities,
- Participation of the managers, technicians, engineers and R&D personnel in various training, conferences, expositions and courses at home and abroad, within their R&D budgets,

[19] For example wind tunnel, durability, crash, exhaust-emission, aerodynamics and road tests.

- Providing incentives for self-development of the personnel, encouraging them to take post-graduate and doctoral degrees in the relevant fields of engineering. It is observed that scholarships are granted under this scope, and that the personnel are allowed to utilize the R&D facilities of the company for the projects they will prepare for their post-graduate or doctoral degrees. Also, it has been determined that the personnel is rewarded financially when they make an academic presentation, participate in conferences, and publish a paper.

In consequence, all these training and education activities enhance further the transfers from the multinationals and other resources; and develop the local personnel's experience, practice, skills, know-how and knowledge. Moreover, it has been determined that the increased skill level of the personnel raises the absorptive capacities of the AMMs as well, facilitating absorption of such transfers.

6.5 General Characteristics of the Direct Suppliers of the AMMs Interviewed

This section analyzes the relationship of the AMMs with their direct suppliers in Turkey, and their general characteristics. In addition, it evaluates the channels/determinants of the KTTs at inter-firm level realized by the AMMs to their direct suppliers through backward linkages.

6.5.1 General Information About the Direct Suppliers

Based on the analyses of the interviews, it has been determined that, in general, the AMMs set and establish the design, basic systems, quality criteria and design of the production lines of the vehicles they manufacture, and procure all other systems and sub-components composing the vehicle from the domestic or overseas direct suppliers and then assemble them. In this context, the AMMs inform their suppliers of the blank size[20] of the parts requested according to the design, their current regulations and the *sine qua non* requirements, and tell them *you will set the design of the parts and deliver your suggestions, illustrations and drawings to us by the predetermined date*. In turn, the suppliers make their own design related to parts in accordance with the requirements such as size, quality, and test values that are notified to and requested from them, and present their designs to the AMMs. All the processes within this stage can be realized in the computer environment by employing design software.[21] Once the design of the requested product or part is

[20] The limit dimensions of the part which it should have with respect the whole vehicle design.

[21] It has been determined that making the *design, assembly and crash tests* of a vehicle completely has become possible in a computer environment, thanks to the software such as CATIA, ROBCAD, CAD, CAM, and CAE.

approved by the AMM in Turkey after certain processes, serial production is launched. In this context, detailed design, design verification, development, prototype, quality tests, durability tests, etc. concerning the parts as well as their production are carried out by the suppliers and the final product is delivered to the AMM. It has been determined that, due to such close collaboration between the AMMs and their suppliers, the direct suppliers are considered a part of the AMMs; in other words, AMMs envisage a life-time collaboration with these suppliers. Consequently, there exists cooperation between the suppliers and the AMMs based on mutual trust, and the direct suppliers are considered to be strategic partners and a vital part of the production chain. Naturally, the duration of collaboration/working together with some suppliers could be quite long, even going back to 20–30 years.

6.5.2 Number and Origin of the Direct Suppliers of AMMs

Table 6.1 shows the number of the domestic and overseas direct suppliers of the AMMs based on their foreign-ownership structure according to the information obtained from the top-executives interviewed.

According to the table:

- The total number of the direct suppliers of the AMMs ranges between minimum 150 and maximum 700; on average 338.
- Among the total direct suppliers, the number of the domestic direct suppliers ranges between 83 and 345 while the number of the overseas direct suppliers ranges between 25 and 400; and the average number of domestic and overseas direct suppliers is 212 and 126, respectively.
- The rate of the overseas suppliers in the total number of direct suppliers of the AMMs ranges between 10 and 57 %; on average, 34 % being overseas suppliers. Therefore, it is observed that the AMMs work with the domestic suppliers by two-thirds and overseas suppliers by one-third.
- The number of the direct suppliers with whom the AMMs work changes depending on the types of the vehicles manufactured, increasing gradually from personal car production towards light commercial vehicle production and then bus production. In other words, since the production of the commercial vehicles requires more parts than that of the personal cars, the AMMs manufacturing the commercial vehicles work with more suppliers.
- It is observed that the rate of overseas suppliers in the total number of the suppliers gradually increases from the domestic-capital AMMs (33.91 %), to joint-venture (34.03 %) and fully foreign-capital AMMs (45.03 %). In other words, there exists a positive correlation between the rate of foreign capital in AMMs and the ratio of overseas direct suppliers they have. In conclusion, affiliates of MNCs (FS = 100 %) work with more overseas suppliers (45.03 %).

Table 6.1 Number of direct suppliers of AMMs by foreign share of AMMs and origin of direct suppliers

			Domestic		Overseas		Total
Foreign share (FS)	AMMs[a]		N	%	N	%	N
FS > 50 %		A	300	42.86	400	57.14	700
		B	180	64.29	100	35.71	280
		C	200	86.96	30	13.04	230
		D	83	46.63	95	53.37	178
	Ave.		**191**	**54.97**	**156**	**45.03**	**347**
29 % < FS < 42 %		E	236	54.63	196	45.37	432
		F	161	61.69	100	38.31	261
		G	225	90.00	25	10.00	250
		H	134	66.01	69	33.99	203
	Ave.		**189**	**65.97**	**98**	**34.03**	**287**
FS = 0 %		I	345	63.89	195	36.11	540
		J	350	70.00	150	30.00	500
		K	120	80.00	30	20.00	150
	Ave.		**272**	**68.49**	**125**	**31.51**	**397**
General ave.			**212**	**66.09**	**126**	**33.91**	**338**

Source: Based on interviews conducted with AMMs
[a]Letters are used to keep the identity of the AMMs confidential

According to the information obtained, the most important overseas direct suppliers are global-mega suppliers who fall within category A, the highest level in the manufacturing classes. The parts procured from these class-A suppliers could directly be sent to the production line and mounted onto the vehicle, without performing any quality control. The parts procured from these suppliers and other overseas suppliers are in general the most important parts of the vehicle, called *system parts*, such as engines, gearboxes, brake systems, electronic assemblies, and safety equipment.[22] These are mostly the electronic parts which use know-how and technology extensively. Therefore, the rate of the foreign direct inputs in the value of the total inputs used in manufacturing of a vehicle elevates up to 40–50 % (see Sect. 6.2.4). However, the rate of the imported inputs used in manufacturing of a vehicle rises given the fact that the domestic direct suppliers procure some part of their inputs from abroad. Moreover, it has been determined that on average 80 % of the total input costs of the AMMs is paid to the domestic and overseas direct suppliers who are 100 or fewer in number.

Lastly, it has been determined that 80 % of the domestic direct suppliers given in Table 6.1 are situated in provinces Istanbul, Kocaeli, Bursa and Izmir. The findings in this section explain why the sample for the supply industry firms surveyed as discussed in Chap. 3 has been chosen in such four provinces, and indicate once

[22] Such parts are called *system parts* while the companies who manufacture and provide them to the AMMs are called *system suppliers*. System suppliers also assume the whole responsibility (warranty, etc.) for the product, following the sale of that product.

more the success of the data on the 132 ea. direct supplier firms obtained from the survey.

6.5.3 Capabilities of the Direct Suppliers

Turkish automotive supply industry now witnesses the third generation, and there are enormous differences in various aspects between the supply industry of 40 years ago and the current one. It has been determined that, today, the Turkish supply industry has become relatively institutionalized with its current structure of developed technology, knowledge and experience, having been globalized and thus become capable of exporting 50 % of its products, having established its R&D and know-how systems, and possessing advanced engineering services.[23] Moreover, it has become an industry which possess skilled and trained workforce and which is capable of manufacturing high-quality products and delivering them on time, which has a competitive power in the international arena and a capability to participate in the international expositions and fairs continuously. For these reasons, AMMs consider the existence of such a robust supply industry to be one of the most important advantages of the automotive industry in Turkey. The following statement expressed by a top-level manager during the interview summarizes the case best:

> The supply industry and the main industry advance further hand in hand; if there is no main industry, there will not be a supply industry either, and in a country whose supply industry is weak, the automotive industry would not and could not develop, or even exist at all.

When we look at the domestic direct suppliers from the viewpoint of the AMMs, we see that they are highly skillful, carrying out quality and high-technology manufacturing with requested characteristics and at competitive prices and delivering the products just in time (JIT). In addition, they conduct R&D activities and collaborate with the AMMs in co-design and co-development activities. The abovementioned situation coincides with the findings stated in Sects. 4.4 and 4.7. Moreover, it has been determined that the domestic direct suppliers who are successful in the industry establish joint-venture partnerships with the global mega-suppliers and foreign companies, and participate in various joint projects with other suppliers and AMMs both at home and abroad, for production, etc. purposes.

Another significant finding obtained from the interviews, which verifies the findings in Chap. 4 is that the domestic direct suppliers of a foreign-capital structure are more capable than the ones with local capital, that they have higher absorptive capacity and higher-quality workforce, that they engage more in R&D activities, that they have higher potential to export, and that they perform knowledge and

---

[23] It has been pointed out that the Turkish supply industry is a robust industry in terms of parts production, especially in sheet metal parts manufacturing and plastic injection activities.

technology-intensive production. Furthermore, the suppliers falling into this category at home are able to produce complete systems instead of just parts, to conduct type approvals and type tests of the parts, and to sell them to the multinationals under the scope of turn-key projects.[24] It has been determined that some direct suppliers in Turkey actually have reached a structure greater than those of some AMMs operating in Turkey in terms of their turnover.

6.5.4 Expectations from the Direct Suppliers

As pointed out before, the direct suppliers are considered a major link in the production chain, and therefore establishing a long-term strategic relationship with the suppliers based on mutual trust becomes essential. The most important anticipation and target of the AMMs in this respect is to develop long-term collaborations with their suppliers, and their expectations from the suppliers could be summarized in general as *delivery of the requested product with the highest quality, at the lowest or target price and just in time*. Given the fact that a vehicle manufactured is composed of thousands of parts, it could be assumed that the main philosophy adopted by the AMMs—especially on the suppliers of the critical parts—is that each supplier should specialize in the part he produces, that he should produce the part at the highest quality as per the international criteria, and at the lowest cost and deliver it just in time. The suppliers are expected to have specific and general certifications in their product range concerned, to have competitive power, and to have design, R&D, technical, engineering and manufacturing capabilities. That being said, and for the reasons stated above, the suppliers are required to conduct activities aiming at continuous quality improvement, to ensure flexibility in manufacturing, to develop their technological and engineering capabilities continuously, to watch their rivals and new technologies closely, to enhance their design skills, to carry out R&D activities, and to have a strong financial structure.

6.5.5 Supplier Selection Criteria of the AMMs

It has been discovered that it is extremely difficult for a new supplier to be included in the supply chain of the AMMs as direct supplier and/or to come and take place in the market entirely from scratch. This is due to the fact that the AMMs have been working with their existing direct suppliers included in their supply chains for long years, that those suppliers, being highly capable, quality and trustworthy, have

[24] Some domestic direct supplier firms with foreign capital could concurrently produce parts for the multinationals abroad (global AMMs) such as BMW, MERCEDES, RENAULT, JAGUAR and FORD.

proven themselves in terms of production consistency, and that they could manufacture continuously at the competitive prices in terms of costs. It is observed that remaining in the supply chain is as difficult as being included in the first place. Therefore, it is imperative that suppliers continuously develop themselves, diminish their costs, compete, and produce quality products.

Managing to become a direct supplier of the AMMs depend on the applications and efforts made by the suppliers, rather than the AMMs' searching for a supplier (this case coincides with the findings stated in Sect. 4.7.1). The AMMs always prefer to work with the supplier firms primarily in their existing supplier pools. Although seldom observed, searching for a new direct supplier occurs only when it becomes necessary to do so for the development of a new vehicle/part when it becomes unavoidable (bankruptcy of the existing supplier, his parting ways with the company of his own accord, discontinuing the supplied part concerned, etc.), and/or when an existing supplier is excluded from the pool of the direct suppliers by the AMM.[25] The AMMs have stated that, in such cases, they search for the domestic and overseas supply industry companies in the product range concerned, via sourcing, and thus are able to find a supplier.

It is also determined that the multinationals who have made investments and taken production decisions in Turkey choose the suitable domestic suppliers among the direct suppliers of the other AMMs that are currently operating in Turkey. As it is pointed out in detail in Sect. 6.5.6, some of the overseas direct suppliers of MNCs have also come and invested in Turkey together with MNCs in order to supply goods to these companies.

Becoming a direct supplier of an AMM is an extremely challenging process. The criteria to become a direct supplier of the AMMs change from one AMM to the other in general, and every AMM has their own selection criteria and policy in this respect. The direct supplier selection processes of the AMMs could be summarized as follows:

- Candidate firms make a direct application via the standard forms, based on the preliminary evaluation criteria and procedures of the AMMs,
- After the application process, the candidate firm is, on paper, subjected to the preliminary evaluation called *Initial Assessment* by the relevant departments.[26] During the initial assessment, the characteristics of the candidate firm such as organizational structure, financial structure, technological capabilities, and

[25] It is observed from the interviews that in cases where a supplier constantly falls behind in terms of quality, or fails to deliver the parts in a timely manner, or fails to achieve the defined targets in terms of costs, or fails to reach the levels requested and/or scores extremely low points during the inspections, or experiences commercial and financial problems, or fails to meet the requirements concerning the product and production process, the AMM warns the supplier severely and grants him a certain period of time to correct his shortcomings and finally parts ways with the direct supplier firm if the problems will not be resolved within the given period of time despite all the warning. All these findings overlap with the findings in Sect. 4.7.3.

[26] Departments such as purchasing, supplier relations, quality, engineering, and supplier development.

specific certificates (e.g. Ford Q1), as well as the AMMs for whom he is the supplier[27] and his other references are reviewed in detail and marked by the relevant departments of the AMM.[28]

- The candidate firm successfully passing the initial assessment is subjected to detailed practical inspection and auditing in terms of technical and technological capability (such as manufacturing ability, production process, quality, cost, delivery, occupational safety, and productivity systems) by the AMM's specialists paying visits at certain periods. During these inspections at the candidate firm, the way the production processes work, the quality and the work site is reviewed, and the availability of the firm's own inspection and control mechanism is sought. The AMM should be assured of the quality of the product the candidate firm will produce and the stability and continuity of this quality, and convinced that the product will be delivered in time. The processes could take some time, from 1 to 5 years, depending on the product range concerned and the candidate firm.

Another finding is that, in the MNC affiliates (FS = 100 %) in Turkey, the inspections of the candidate firms are carried out by the multinational's own specialists located at home or abroad, and that the final decision on accepting of the candidate firm into the pool of suppliers lies with the multinational headquarters. The *central purchasing* departments at these headquarters engage in such activities. If the candidate firm is approved and included in the pool of suppliers, then all the global plants of the multinational could utilize from this firm.

As to the AMMs in the form of a joint-venture capital partnership, it has been determined that the specialists of the AMMs located at home and/or multinational headquarters are assigned duty in the selection processes, depending on the type of the part concerned, and that the decision on accepting the candidate firm into the pool may be taken by the AMM located at home, as well as by the multinational headquarters. While in the AMMs with minority foreign ownership capital structure, the multinational headquarters in general do not take part in the selection process for the domestic direct suppliers, in the AMMs with majority foreign ownership structure, the multinational headquarters play a role in the decision making process concerning the supplier if the part is to be used globally, or otherwise, the center located in Turkey, if the part is to be used only for the Turkish market. In addition, the overseas direct suppliers of the AMMs with the joint-venture capital structure are determined by the multinational headquarters abroad, and the parts procured from these suppliers are delivered officially over the foreign partner multinational.

[27] It appears highly difficult for a domestic supplier currently not working with any other AMM in Turkey to become a new direct supplier of the AMM concerned.

[28] The partnership structure, capital, turnover, R&D and design capabilities, quality and work systems, the quality certificates the firm possesses (ISO9000, ISO14000, ISO/TS16949, etc.), human resources, the projects performed before, etc.

Once purchasing specifications, legal contracts and confidentiality agreements are concluded and signed with the candidate firm that has been approved to be included in the pool of suppliers after all these inspection processes, the serial production process begins. It has been observed that if the candidate firm concerned is discovered to be poor and incompetent, either the AMM parts ways with that candidate or—if the candidate is a critical firm—the AMM's specialists determine the topics which need correction and improvement and the development plans on those topics are devised together with the specialists of the candidate firm.

The main criteria sought for in becoming a direct Supplier could be summarized as follows, according to the findings obtained:

- Ability to take part in every stage of the parts' development process and interoperate with the AMM's production systems, since it will produce the parts developed by the AMM,
- Having quality, specialist, trained and qualified human capital,
- Bearing high-technology capability in manufacturing,
- Possessing co-design, co-development, R&D and engineering capabilities, depending on the product ranges,
- Delivering just in time and supplying defect-free products,
- Being successful in the areas of quality, quality control, cost and management,
- Having sophisticated and well-developed functions and proficiency in terms of production processes, equipment pool, sub-suppliers, safety philosophy, production and production support groups,
- Having a financial structure that enables investments for future and self-development,
- Adopting and implementing the working criteria of the AMM well,
- Being open-minded, innovative and open to collaboration,
- Allowing transparency and effective communication.

6.5.6 Location of the Direct Suppliers

Based on the findings under this topic, it is observed that the AMMs prefer an intensive and close communication with their suppliers at every stage of the collaboration between them, and that they consider the *go and see* (genchi genbutsu)[29] principle involving going and observing any problem or issue on site as one of the most important working principles. Furthermore, when a supplier is located close to the AMM, it brings significant advantages especially in terms of communication, delivery on time, storage, reacting quickly, detection of manufacturing problems on site, problem-solving, inspection and logistics costs. These advantages are among the major reasons why the domestic suppliers are

[29] It is an important working principle of the *Toyota Production System* and means *go and see*.

concentrated in the Marmara region (Bursa, Istanbul, and Kocaeli) where in general the AMMs operate. Due to these advantages, some AMMs established considerably large "supplier parks" next to their own plants and have assisted and allowed some of the suppliers to establish their own production facilities in these parks.[30] In addition, since it is more advantageous both in terms of cost and time, the products are procured by the AMMs by means of the "milk-run system"[31] from the suppliers who are located close to them. Furthermore, some of the suppliers and global-mega suppliers located abroad have come and invested in Turkey to supply goods to the AMMs operating in Turkey again due to these advantages. Multinationals' preferring Turkey as a regional production base has also played a major role in this process. In other words, the investments made by multinationals in Turkey have encouraged/led the overseas direct suppliers of these multinationals to also invest in Turkey.[32]

Although geographical proximity of the suppliers to the AMM bears the advantages discussed above, it is not considered the most important criterion by the AMMs in selecting the suppliers. The product quality of the supplier, its production systems, cost structure, R&D activities, co-design skills, ability to design, know-how structure, technological capabilities, workforce quality, and organizational and financial structure play a more significant and important role in the selection criteria than the location. In sum, it appears that under the competition conditions of the global economy, the AMMs do not actually care so much about the geographical location a supplier performs the production, as long as the supplier produces more quality, more comfortable, technological and less expensive parts, and delivers these just in time.

6.6 Channels of the KTTs from AMMs to Direct Suppliers in Turkey at Inter-firm Level

In this section, the channels of the KTTs from the AMMs to the domestic direct suppliers, and the factors playing a role in these transfers are discussed.

[30] When selecting the supplier firms who will manufacture in these parks, what is considered and taken into account is that the parts to be supplied are of a fairly large size and that the logistics cost increases in case of an emergency.

[31] In this system, the transport of the parts to be supplied is assumed by the AMMs due to the advantages in terms of cost, time and procurement. An 18-wheeler belonging to the AMM visits the suppliers in turn every day and carries 1- or 2-day stocks to the AMM.

[32] Some of the AMMs have stated that they themselves had brought their overseas suppliers in Turkey.

6.6.1 *Role of Being the Sole Supplier of the AMMs*

It has been determined that being the sole supplier of the AMM in a certain product range plays a highly significant role in the transfers. It has been observed that the AMMs work mainly with a single supplier in the procurement of the critical parts and systems, while the number of the suppliers per item purchased increases going down the sub-systems and more simple parts. The AMMs conduct many activities (such as product development, co-design, R&D, quality, and tests) together with the firms who are their sole supplier, and provide them with every support on such issues.

Among the most important reasons for working with only one supplier per item purchased are the lack of alternative suppliers in some products, buyer-supplier trust, scale economies, long-term strategic relationships with the suppliers, considering the supplier as the strategic partner of the AMM, and high investment and technological capability required by some parts. That is to say, the AMMs prefer to work with only one supplier for the majority of the parts they procure while they prefer to work with more than one supplier for only some of the parts (some off-the-shelf products such as tires, and ceiling lights). Risk distribution in terms of procurement, production and cost, poor capacity, weak financial structure, raw material procurement problems and technological inabilities of the suppliers have been indicated by the AMMs among the most important reasons for working with more than one supplier per item purchased. On the other hand, it has been discovered that the direct supplier firms may be the supplier of more than one AMM, and that they may be the supplier of only one AMM for a certain product range while they may work with more than one AMM for some product ranges. In other words, the direct supplier firms could be the sole supplier of more than one AMM. Also, the findings obtained in this respect coincide with the findings in Sect. 4.7.3.

6.6.2 *Via Joint Projects*

Intensive collaboration takes place between the AMMs and their domestic direct suppliers through the joint projects in the activities concerning developing a new product or improving an existing product, increasing the quality, and developing the production and occupational safety processes. Within the framework of such collaboration, meetings and discussions are continuously held among the personnel of both companies, and the suppliers are visited by the AMM's specialists. Likewise, the personnel of the supplier are allowed to visit the AMM and observe the line of production and use the laboratories and test benches of the AMM. The specialists of the companies perform the trial runs and tests of the product together, and determine the shortcomings and positive aspects. In consequence of these activities, mutual knowledge exchanges occur between the parties and the suppliers enhance their own know-how and R&D levels in line with the feedback they get, and they attain a better position and apply the lessons they have learned in every project to other AMMs during the next project, thereby developing themselves continuously. In summary, as the supplier firm specialists conduct joint projects with the AMMs, they receive the opportunity to develop themselves and gain knowledge. These findings verify the transfers related to product via the joint projects on the product, as discussed with reference to the surveyed firms in Sect. 4.2.2.

6.6.3 Via Co-design Activities

It has been observed that other than engines and powertrains, the AMMs produce many systems and/or sub-parts that are critical and that require know-how together with the co-designer suppliers (this coincides also with our findings in Sect. 4.4.6). The AMMs work together with the supplier firms in co-design activities (such as parts design, design verification, development, and tests) and the cost, profit margin and sales price of the parts to be manufactured are determined together with the supplier firm.[33] Such activities also enable a new vehicle to be produced in a shorter period. Due to the reasons stated above, the co-designer suppliers are considered strategic partners by the AMMs, and long-term business relationships are established with those suppliers.

It has been determined that the rate of the domestic direct suppliers with whom the AMMs conduct co-design activities in the total number of suppliers is 20–30 %. Among the most noticeable common characteristics of these supplier firms are their possession of advanced engineering skills, their R&D department, and their performance of high-technology production, their quality human capital, and their having a foreign-capital structure for the most part. In addition, the suppliers with whom co-design activities are carried out are in general the sole supplier of the AMMs for the product range concerned. Since ultimately it is the AMM who will be the final user of those parts, the AMMs share all kinds of knowledge and technology with their co-designer suppliers in areas such as production of the parts, their design, development, establishment of the production lines, and production processes. In this respect, it is observed that the AMMs;

- Could bring together the multinational headquarters, global-mega suppliers, other foreign companies and co-designer suppliers they collaborate with during the development process of a new vehicle,
- Could deliver on-the-job and/or off-the-job training/seminars/courses to the personnel of the suppliers by means of their own specialists and could assign their own engineers in a certain period at the supplier's plants for support purposes,

[33] It has been determined that once the AMMs have an R&D department, their co-design activities with suppliers increase.

- Could obtain technical support and know-how from the multinational headquarters for the supplier, and could provide assistance on various issues such as procurement of raw materials, equipment and tools, and obtaining the required technologies from abroad,
- Could provide their suppliers with advance payments for the projects and financial assistance for high investment costs.[34]

The AMMs do not specialize in the production of the low-tech parts in sub-system levels (such as seats, door parts, signal arms, mirrors, air-conditioners, and internal trim materials), and specializing in such parts is not deemed a must, nor a necessity, either. As to the manufacturing of such parts, the AMMs inform their suppliers of the quantity, blank size, and quality standards required as well as their project and cost targets, and are not further involved in the details such as the design, test, and development of those parts. They leave all those issues to the supplier's discretion and request timely delivery and assembly of those parts in the framework of turnkey contracts so that the parts are delivered to production line whenever required. The AMMs' collaboration with and assistance to their suppliers in such activities are relatively little when compared to the co-design activities. Both the AMM and the suppliers assume mutual responsibility for those parts.

In conclusion, the efforts made for the co-design activities constitute the most important leg of the collaboration between the main and supply industry companies in Turkey. Such activities, via KTTs from the AMMs, enhance the suppliers' capabilities in the fields of design, quality, product development, R&D, engineering and technology, and increase their know-how and absorptive capacity levels, making a highly positive contribution to their performance.

6.6.4 Via Auditing, Assessment, and Awarding Activities

One of the important channels through which the transfers are realized between the AMMs and the suppliers is auditing, assessment and awarding activities. Every AMM utilizes various monitoring and auditing systems to closely follow their direct suppliers. It is observed that these systems are basically the same, although different for each AMM. The AMM's own specialists[35] visit all the suppliers at certain intervals (every month, every 6 months or once a year) and audit them during the production and collaboration activities.[36] Audits are in generally divided into two sections:

[34] The AMMs could provide the suppliers with various financial assistance in cases where the equipment pool of the part to be manufactured requires high amounts of investment or in cases where the die cast of the part to be manufactured is costly.

[35] These specialists are called "supplier quality engineer" or "supplier technical assistant".

[36] The specialists are assigned duties at the departments such as purchasing, quality, and supplier development.

Process audits: It covers head-to-toe checking of how the parts are manufactured by the supplier firm.

System audits: It is the inspection of the quality, production and management processes of the suppliers. Here, a system similar to the *WCM* production systems (production, quality, management and supervision systems) applied by the MNC headquarters to the AMMs in Turkey is applied this time by the AMMs to the suppliers (see Sect. 6.4.2). The success of such systems, which have been generated by the multinationals with the experience and knowledge accumulated over long years and which involve high level of know-how in terms of their application on the AMMs, depends on whether they are also applied successfully on the direct suppliers included in the supply chain. The most important reason for this is the fact that the AMMs are organically tied to their direct suppliers and that even the smallest problem encountered by suppliers would naturally be reflected on the AMM's production lines. Therefore, the suppliers are also monitored constantly by the AMMs in accordance with those systems, and their performances in terms of production, quality, delivery, cost, productivity and competitiveness are inspected very closely.[37] The findings obtained in this respect also coincide with the findings in Sect. 4.7.2.

Based on the findings from the interviews, the auditing and assessment stages conducted by the AMMs could be summarized as follows:

- Once approved as the direct supplier at the very beginning, the suppliers are given a score between 0 and 100, depending on the results of the test and evaluation criteria of the AMM, and are classified into categories as A, B, C or D, A being the highest grade.[38]
- The suppliers are subjected to the assessment systems called *Manufacturing Side Assessment* (MSA). According to this, the suppliers are constantly monitored and subjected to assessments by the auditors[39] of the AMM at certain intervals (every month or every 6 months), based on the procedures determined by the audit programs of the AMMs. The audits are performed under certain topics such as logistics, quality, prices, production processes, and delivery on time. Following each audit, the suppliers are graded once more over a scale of 100 points, based on the checklists prepared in accordance with the topics.

---

[37] Since it is a matter of serial production, the quality of the parts to be mounted onto the vehicles is not inspected generally on the production line. Quality control exists at every stage of the production process but the inspections and audits are performed from scratch. During such audits, the suppliers are monitored on weekly, monthly and annual basis via the indicators such as timely delivery under the scope of DPM (delivery performance management) and defect-free product procurement in PPM (parts per million). Less than 100 ppm for the personal cars and less than 500 PPM for the buses is accepted as standard.

[38] Such categorical classifications vary among the AMMs.

[39] Auditors from the departments such as purchasing, quality, production, supplier relations, and supplier development.

- The level of the supplier firm is re-assessed on the basis of the reports and recommendations are drawn up as a result of the evaluations. Afterwards, the firms may upgrade to the next higher level, may be warned or excluded from the pool of the suppliers, depending on their success in achieving the predetermined targets.

The target and mission of the supplier firms is to proceed to the next higher category (to category C if it is currently in category D, or to B if it is currently in category C) in the short run and to achieve and settle down in grade A in the long run. It is observed that the detailed and constant audits of the AMMs continue until the suppliers reach grade A. The main goal from the AMMs' point of view is the advancement of their suppliers in category C and D essentially to category B in the short run and finally their becoming a grade-A supplier in the long run. However, since the criteria to become a grade-A supplier are set quite high, it seems rather improbable to realize that goal. The firms achieving the top level category, namely grade A, are in general global-mega suppliers, foreign capital firms, large-scale and institutionalized firms that perform manufacturing by utilizing superior technological capabilities. The suppliers of grade A are also subjected to the high-quality assessments constantly by the AMMs at certain periods, but their audits are realized at longer intervals. The firms falling in this category A are considered as the trusted firms and their products are transferred directly to the production line without conducting input quality control by the AMM. In addition, it has been determined that, along with the practical audit and assessment activities, the AMMs continuously test the suppliers' parts that are included in the production line and ready for assembly at certain intervals by means of random selection method.

The suppliers are recommended to increase the quality of their production and/or to develop their processes, based on the reports drawn up as a result of the audits. Some suppliers are requested to improve their scores and the suppliers who display poor performance and who have been discovered to have certain problems and/or deficiencies are warned and they are given a certain period of time for correction. Also, some suppliers are included in the scope of improvement work by the special teams of the AMMs. The AMMs have stated they prefer to be interested in and support more the suppliers who remain below a certain level, who encounter problems, rather than the suppliers who are already in good status, due to limited resources. The main reason for this preference is that the means and capabilities allocated to such suppliers would ensure higher quality and increase in the productivity on average, in terms of the return obtained. Based on the findings, the improvement activities performed with the direct suppliers could be summarized as follows:

- Generation of work plans and schedules by the teams established within the AMM, together with the specialists of the suppliers, determining the topics which need to be improved in the suppliers, and their assisting the suppliers during the application of such plans,
- Inviting the engineers of the suppliers to the AMM and delivering seminars, training and support to them in line with the audit outputs,

- The AMMs' acting as an intermediate agent between the multinational headquarters and suppliers to provide support from such headquarters,
- Performing KTTs concerning the quality, product, production processes and systems of the AMM with a view to improving and developing the suppliers.

The AMMs also employ various rewarding systems also for the purpose of making their suppliers compete among themselves and raising their technology and productivity levels. To this end, relevant departments[40] of the AMMs review the data on certain indicators of the suppliers (such as delivery on time, quality, defect rate, cost, product development, and project management) every year, and then grade them.[41] Suppliers who display an outstanding performance according to these scoring criteria are awarded financially and morally as the *supplier of the year* due to their success, through a ceremony.[42] In addition, the suppliers determined by all AMMs (15 ea.) are assessed again according to certain criteria by OSD, and the top three are awarded *the best supplier of the year* by OSD. It is observed that winning such rewards play a vital role in developing new business relationships especially with the foreign firms abroad, becoming the direct supplier of other companies, being known across the industry, and being a reference, and is considered highly important for the suppliers.

In sum, it has been determined that the AMMs monitor, assess and reward their direct suppliers, the most important element in the supply chain, by means of continuous audits and evaluations so that they are assured of the product quality, timely delivery of the product and its costs. According to the information obtained from the AMMs, the output from such activities motivates the suppliers, raises their production and technological capabilities and their productivity, quality and production levels; promotes their R&D and innovation activities; encourages them to increase their know-how and technological skills continuously; creates competition among the supplier firms and ensures the transfers from the AMMs to the suppliers.[43] In conclusion, such activities lead to increases in the performance levels of the suppliers. The findings obtained in this section coincide also with the findings in Sects. 4.7.2 and 4.9.2.

[40] Departments such as purchasing, supplier relations, and supply industry development.

[41] These criteria may be summarized as follows: No problem encountered concerning supplier's delivery on time during the course of that year; no return of any product due to quality problems; supplier's scoring high or acceptable points as a result of the audits; supplier's success in drawing their costs back to the predetermined price levels; and supplier's success in increasing the productivity.

[42] Suppliers are awarded for three ranks: gold, silver and bronze awards. Certificates, plaques, letter of commendation, etc. are given as awards.

[43] It is necessary to question the suppliers in order to ensure how much these situations reflect the reality.

6.6.5 Via Global Certificates of MNCs

It has been discovered that the direct suppliers' obtaining special quality certificates of the multinationals (Q1 of Ford, Toyota, etc.) affects positively the transfers (see also Sect. 4.4.2). Especially the multinationals abroad set a precondition that the suppliers they will work with should have the special quality certificates issued by them. Becoming entitled to the special quality certificates and being able to maintain them are only possible by achieving extremely high success and assessment criteria at the end of a long-term and planned process. Domestic suppliers could be entitled to such special certificates only upon approval of the multinationals abroad, as a result of the joint assessments of both AMMs in Turkey and multinationals as foreign partner. Therefore, these certificates are valid globally and whether the firms obtaining these certificates are originally Italian, German, American or Turkish firms makes no difference in terms of their quality and production capability.

The suppliers possessing these certificates are included in the direct suppliers pool of the multinational and are also able to sell their products directly to the multinational plants located in other countries. However, assessment and audit processes for the firms who possess these certificates are based on more strict conditions and requirements, and it is possible to take back these certificates in case of any problem. Since these certificates are of global nature, they are regarded as an extremely important reference and reward by the suppliers who possess them; at the same time, they constitute a benchmark for other firms and play a major role in procurement of products to the firms at home or abroad or executing joint projects with them. Consequently, it has been observed that possessing the special quality certificates issued by the multinational is highly important for the suppliers to develop themselves and to ensure KTTs.

6.6.6 Via Training and Education Activities

One of the most important channels through which KTTs to the domestic direct suppliers are realized is certainly education and training-related activities delivered and performed by the AMMs.[44] It has been determined that the AMMs continuously deliver on-the-job and off-the-job training to the personnel of their direct suppliers, on topics such as production, production process, R&D, quality, cost reduction, productivity, technology, and production systems. Such training activities are mostly in the form of seminars, workshops, courses and consultancy services, and are executed at the training departments or supply industry

[44] Although the terms "training" and "education" are generally used interchangeably, they are not the same at all. Companies may use both terms depending on their activities. Training simply refers to the practical activities in order to gain skills in the best use of new equipment, while education refers to teaching activities to help someone solve a problem.

development departments of the AMMs. It has been determined that such training could also be delivered by the specialists of the AMMs at the plants of the suppliers when necessary. It has also been determined that some AMMs with foreign-capital assign their own specialist personnel[45] at the plants of the suppliers for a certain period (between 1 week and 1 month, etc.) in order to assist in various subjects (especially in quality problems encountered during serial production), under the scope of the activities aiming at improvement of the suppliers' quality. These assigned personnel are mainly focused on resolving the problem; however, they also work on the issues such as management of the sub-suppliers[46] who are considered to be in need of improvement, maintenance of the equipment and machinery, production and quality systems applications of the AMM, field of delivery, etc. The above-mentioned findings obtained from the AMMs also verify the information obtained in Sect. 4.2.4.

Activities included in the training delivered by the AMMs to their direct suppliers' engineers, production and management personnel could be summarized as follows:

- Making training arrangements by the AMM's specialists in the form of meetings, seminars and workshops for the purpose of ensuring adaptation to the AMM's production, quality and audit systems,
- On-the-job demonstration of the work principles of the AMM, concerning the issues such as tests, quality, audit, and delivery just in time,
- Delivering training on 3D design and the use of certain software such as CATIA, ROBCAD, CAD, CAE, and CAM,
- Delivering training on the occupational safety and safety of the workers,
- Executing intensive and extensive training activities and holding seminars and meetings during the processes of co-design, joint projects, and production of new vehicles and parts,
- The engineers and R&D personnel of the suppliers' working on the job together with the engineers of the AMM for a certain period concerning a new part to be produced or developed,
- Conducting extensive training activities to ensure speaking the same language and enhance communication between the AMM and suppliers,
- Delivering on-the-job training to the production personnel of the supplier on the production lines of the AMM for a certain period,
- Holding and performing workshops, meetings and training on the problem-solving techniques for resolution of the common problems of more than one supplier and for prevention of "operational blindness", and on finalization of the projects and works,
- Introducing and allowing visits to the production lines, R&D departments, and test laboratories of the AMMs.

[45] Quality engineer, quality group leader, supplier production management or quality engineer, etc.

[46] Second and third-tier suppliers.

Such training activities should improve the quality and production processes of the domestic direct suppliers, raise their technological levels and increase their accumulation of know-how and knowledge by means of transfers to them via their personnel, and therefore affect their performance positively. However, to what degree such training activities delivered by the AMMs provide KTT to the supplier firms and its effect on the suppliers' performance would essentially depend on the technological capabilities and absorptive capacities of the suppliers as well as on to what extent they could internalize and digest such training.

6.6.7 Via Other Assistances by AMMs

During the interviews made with the top-executives of the AMMs, it was also determined that, apart from the above-mentioned supports, there were various forms of assistance provided to the domestic direct suppliers. Such activities indirectly ensure self-development of the suppliers and lead to raises in their know-how, technology, productivity and performance levels. For this reason, these could also be regarded as some kind of KTT from the AMMs. Such assistances are mainly as follows:

- Acting as an intermediary agent to help the supplier firm to find raw materials from abroad, if the raw material necessary for the relevant part is not available at home,[47]
- The AMM's intervening in and concluding some wholesale contracts themselves, concerning some raw material resources to be utilized by more than one suppliers due to high cost advantages,
- Supporting financially and providing advance payments in the new projects requiring high level of investment and in production of costly parts,
- Providing guidance in purchasing various software programs,
- Allowing the suppliers to use their own test laboratories and benches during tests of the parts requested from the suppliers,
- Providing guidance to the supplier in purchasing new machinery-equipment from abroad, in selection of the production equipment and providing some materials and tools in this respect,
- Assisting the supplier in finding some sub-parts if the supplier has a difficulty in finding those sub-parts that are required for manufacturing a part,
- Acting as an intermediary agent to help the supplier in accessing overseas companies with which the supplier could establish technological cooperation, with respect to a part to be manufactured,
- Ensuring that the new projects and future plans are discussed mutually by means of holding meetings once or twice a year to bring together the top-level

[47] Since the AMMs have more extensive contacts and communication channels abroad compared to the domestic supplier firms because of the fact that they are included in the production networks of the multinational, they could develop business relations with the companies abroad more easily.

managers of the suppliers, and trying to resolve the problems concerning production and buyer-supplier relations within the framework of mutual cooperation, by means of creating opportunities to express such problems and relevant expectations,

- Providing various assistance on logistics, documentation, quality control and know-how,
- Carrying out workshops constantly on issues such as productivity enhancement, quality control, new production methods, and cost reduction,
- Providing assistance and guidance for the problems encountered in the R&D activities,
- Bringing together the suppliers who have high technology capabilities under joint projects and ensuring that they support one another and establish cooperation among themselves. In these activities mutual trust is very important and a prerequisite among companies.
- Promoting the foreign direct suppliers in Turkey, who are included in the direct supplier pool of the AMM, to lead the local suppliers operating in the same field and share their knowledge and experience with them. The aim of such supports is to ensure that the local suppliers establish contacts with the foreign suppliers, who have an advanced level of production & technology capabilities in the same field and enhance their own capabilities concerning production, quality, delivery, design, defect rate, etc. by means of know-how they will obtain from those foreign suppliers. It is thought that supports of this kind are far more important than other assistance in that they ensure KTTs from advanced-level foreign suppliers to local suppliers, and these could be considered a kind of horizontal technology spillovers.

6.7 SWOT Analysis

This section will try to point out the strengths, weaknesses, opportunities and threats concerning the Turkish automotive main and supply industry that will affect the transfers from multinationals to home based on the findings obtained from the interviews made with the top-executives of the AMMs.[48] The findings obtained contain the generalizations drawn upon the personal opinions, views and contributions expressed and made by the managers concerning the questions remaining under the title E of the interview guide (see Appendix I) within the framework of the study herein. When the findings are put forward, the opinions and views of the managers have been revised and reviewed by the author and tried to be presented in

[48] With regard to this subject, see also SPO (2007), MITI (2010, 2011), and various TAYSAD (http://www.taysad.org.tr) and OSD (www.osd.org.tr) publications on the industry, as well as Wasti, Kozan, and Çınar (2009) and Erdil, Pamukçu, Erden, Göksidan, and Kepenek (2011).

a compilation. The questions formulated with regard to the industry are composed of the following topics in general:

- Current status and future of the Turkish automotive industry,
- Opportunities and challenges concerning the industry and how the Turkish automotive industry will be affected by them,
- Capabilities, strong and weak aspects of the supply industry firms, in comparison to the rival countries,
- Development of the strategic relationships between the AMMs and their suppliers,
- Government policies towards the industry,
- The status of the cooperation activities among the Universities, Industry and Public,
- Opportunities and challenges lying before becoming a base of R&D, design and excellency, rather than being a regional production base in the automotive industry,
- The role of the multinationals in the industry, and how the direct investments that are knowledge-intensive and that involve advanced technology could be drawn from multinationals into home,
- Obstacles lying before the transfers from the multinationals to the domestic firms, and how such transfers could be realized in a more effective manner.

In the SWOT analysis, the findings put forward in each category are put in order according to the level of significance commonly expressed by more than one manager.

6.7.1 *Strengths (Internal-Positive)*

The "strong aspects" of the industry as expressed by the participants concerning the transfers are as follows:

- Geo-political location and demographic power (young population) of Turkey,
- Young, well-educated, quality, enthusiastic, entrepreneurial, and relatively inexpensive human capital: It is observed that, although the blue-collar human resource is inexpensive in Turkey, well-educated and quality white-collar (engineer) human resource does not contain inexpensive work force anymore; on the contrary, it is more expensive than those of the Central and Eastern Europe countries, newly developed China, India, etc.; which, in turn, creates a disadvantage in competition. For the time being, Turkey offsets this disadvantage with her quality human capital; however, if well-tailored policies will not be developed in the short and medium run, the said countries' increasing quality level in human capital could become a serious threat in the near future.
- Almost 50 years of experience and culture in the automotive industry,

- The automotive industry in Turkey having become a highly developed and strategic industry,
- Existence of 15 AMMs operating in Turkey and their utilizing Turkey as a regional production base,
- The majority of the exports being to the EU countries where extremely high production and quality standards are applicable,
- Existence of the provinces which specialize in the automotive industry and which have advanced logistics infrastructure (such as Bursa, Istanbul, and Kocaeli), and the existence of clustering of organized industrial zones (GOSB, DOSAB, NOSAB, etc.),
- Existence of a strong supply industry and chain of suppliers,
- Accumulation of experience, practice and knowledge in the areas of production, R&D and design, which, in turn ensures transfers; as well as existence of R&D centers to be able to execute joint projects with the multinationals, and also high R&D expenses and innovation activities made by the AMMs,
- Existence of qualified and quality engineers who will be able to work together with foreign engineers,
- The interest in the advanced technology and technologies generating value added, as well as in R&D, which has increased recently in the industry; in addition, establishment of the R&D departments and increase in the R&D expenses and investments, and establishing collaborations on these issues with foreign firms,
- As to the direct supplier firms in Turkish automotive supply industry,
  - Their being creative and entrepreneurial,
  - Their having strong accumulation of experience, practice, know-how and knowledge in the areas of production such as sheet metal parts, plastic injections, glass, tires, rubber, lighting, seats, and cast & die-cast products,
  - The development in their commercial, technical and manufacturing capabilities enabling them to manufacture products at international standards,
  - The quality of the products they manufacture having developed to a great extent while the costs having diminished,
  - Their having been globalized and having their potential to export and having begun to consider their target market as the world,
  - Their having begun to attach importance to the innovation and technology activities and establishing cooperation and concluding technology agreements with foreign firms at home or abroad to this end.
- Investment and export incentives of the government; project incentives of TUBITAK, TEYDEB and SANTEZ; and the fact that the R&D incentives have been begun to be granted on a legal basis upon enactment of the law no. 5746,
- The importance of drawing more multinational investment into home and thereby obtain KTT, also the importance of conducting innovation activities and thereby manufacturing mid-tech/high-tech products that are knowledge-intensive and that involve high value added, has been well understood both by the government and the actors in the industry,

- Relatively low investment costs,
- The fact that Turkey has a developing and growing economic structure with increasing exports.

6.7.2 Weaknesses (Internal-Negative)

The "weak aspects" of the industry as expressed by the participants concerning the transfers are as follows:

- The competence in the fields of powertrain, engines, electrical and electronic equipment is poor and inadequate; also, inability to master such fields,
- Insufficient number of qualified, well-educated and specialized intermediate staff/engineer/R&D personnel are required by the industry,
- Inability to utilize and benefit from the scale economies due to the lack of a robust and adequate-size domestic market,
- The cost of the energy inputs (fixed costs) are high, which, in turn creates a disadvantage in competition with the European, Far East and Asian countries,
- The taxes imposed on the industry are extremely high and that inappropriate taxing policies towards the vehicles restrict the production as well as innovation and R&D activities,
- Insufficient regulatory role of the government to move and push forward the industry, to increase the international competitive power, to draw FDI flows from the multinationals and thereby ensure transfers to the domestic firms,
- Lack of basic test facilities required to generate a vehicle from scratch (wind tunnels, crash tests, special drive courses, exhaust-emission tests), and thus remaining dependent on abroad in this respect. It has been determined that such test facilities affect directly the transfers, R&D and design activities and MNC investments. Once such test facilities become available, it will be possible to design and manufacture a vehicle from scratch at home. Furthermore, it is expected that availability of those test facilities will ensure that overseas firms come and invest in at home, that the AMMs will be able to draw more projects from the multinationals, and that the co-design/co-development activities performed with the multinationals will increase in quality and quantity; also, availability of such facilities would contribute to Turkey's becoming a base of R&D and excellency in the near future, rapidly increasing the technology and know-how. For these reasons, it is regarded extremely important and vital for the future of the automotive industry that such test facilities, which require high level of investments that cannot be assumed solely by the industry, should be developed at home immediately.
- Being dependent on abroad in some vital raw material resources (such as plastic, sheet metal, batteries, and sponge),
- Despite the increase in recent years, the collaboration among the public, industry, universities is still rather weak; the universities' research projects still remain at the theoretical and literature research to a great extent, rather than

practical projects aiming at development; the academic personnel fail to follow the innovations realized in the industry and they remain distant from and are unfamiliar with the industry; the universities' academic personnel participate very rarely in the projects executed in collaboration with the industry; there exist discouraging bureaucratic obstacles in front of the academic personnel to prevent them from serving as consultants in the industry; as well as failure of delivering the university students at the engineering departments on-the-job training via internship,

- Although currently more favorable (especially with R&D Law no. 5746), the R&D incentives still remain inadequate compared to those in developed countries. There is a lack of a strategy and incentive system especially for the small-scale automotive supply industry, or there is an inadequacy of such a system. The R&D incentives are granted on the basis of product and there exist no basic, long-term scientific R&D supports for developing advanced technologies,
- The government's policies concerning the industry are inconsistent and not supplementary, although there have been some developments,
- The engineers' lack of good level of knowledge of several foreign languages,
- The fact that the majority of the supply industry;
  - Is small scale,
  - Still seems to have a structure that is far from professionalism, that is non-institutionalized and that is poor and weak in financial structure,
  - Has limited competence in terms of production, design, engineering and R&D,
  - Has extremely poor resources reserved for the R&D and innovation activities,
  - Has high level of turnover among the firms, in terms of workforce such as engineering, and has failed to obtain and acquire know-how, with continuing brain drain,
  - Has inadequate and weak competitive power in the international arena,
  - Is heavily dependent on the domestic AMMs, having inadequate contacts with and links abroad,
  - Has quite inadequate and poor organizational activities,
  - Has excess capacity,
  - Is composed of too many firms manufacturing similar parts; the firms that do not care about merging and/or clustering policies in order to strengthen their financial structure and to increase their technological levels as well as competitive power,
  - Fails to follow closely the technology and lacks vision.
- The *simultaneous engineering*[49] (SE) activities have not developed at all or are inadequate,

---

[49] It is a process widely used in the automotive industry and refers to complex simulation and development projects for the manufacturing of a new vehicle. It is defined in business dictionary as "concurrent new product development through employing cross-functional teams to reduce cycle time" (see www.businessdictionary.com).

- The industry has no macro plans and road maps containing the mid- and long-term future strategies,
- Worthy investments and R&D activities in the industry are poor and inadequate; the fact that the financial resources reserved for R&D are extremely low compared to the developed countries (USA, Germany, Japan, France, Italy, etc.),
- Lack of or inadequate pre-competition collaboration strategies among the firms,
- Turkey is still perceived by the developed countries as a third-world country where poor-quality and low-cost production is realized with inexpensive workforce; the facilities and capabilities Turkey possesses in terms of production, investment, technology, R&D, workforce and infrastructure have not been promoted and explained satisfactorily in the developed countries,
- The strategic relationship between the main and supply industry is poor and inadequate,
- Transportation and shipment costs are extremely high due to inadequacy of the logistics and transport infrastructure (such as railways and seaways) compared to rival countries,
- The domestic or foreign investment incentives are not at the requested levels,
- Works on patents and patent applications are not satisfactory.

6.7.3 Opportunities (External-Positive)

The "opportunities" for the industry as expressed by the participants concerning the transfers are as follows:

- The demand for environmentally-friendly high-tech and mid-tech products has increased to a great extent all over the world and governments provide substantial supports and incentives for such technologies,
- Due to the global financial crisis and increased prices for engineering in the developed countries, the multinationals have shifted their R&D and direct investments to the developing countries where incentives and inexpensive workforce are available,
- The multinationals tend to develop their R&D and design activities not only in their headquarters but also other centers in terms of developing new and differentiated products, and making changes—adaptive and technical—to products aiming at target markets,
- Quite sound and long-term strategic cooperation activities with the multinationals have been achieved as a result of almost a 50-year experience in the Turkish automotive industry; which in turn enables new business relationships with the multinationals at further levels,
- The strategic approaches and enterprises aiming at becoming an R&D, design, test and excellency base in the automotive, rather than remaining a production base, have been expressed both by the government and the industry,

- The focus is now on developing and producing knowledge-intensive products at home that generate high value added and that involve R&D, design, test and advanced technologies, rather than being on the labor-intensive production,
- The transportation facilities have become easier and less expensive,
- Ability to follow the technologies easily thanks to the advanced communication facilities,
- The government has a target to create local brand and trademarks,
- Existence of economic and political stability.

6.7.4 Threats (External-Negative)

Lastly, the "threats" against the industry as expressed by the participants concerning the transfers are as follows:

- Existence of rival countries[50] with low-cost energy inputs and advantages in terms of workforce and investment costs,
- The fact that newly developed countries such as China and India have enormous domestic markets; which, in turn, enables attraction of more MNC and R&D investments into those countries. Knowledge and technology could be attained only via the multinationals possessing advanced technologies and R&D investments. Such investments are extremely costly; therefore, existence of a fairly large and robust domestic market to support those investments is a must. This situation poses a serious threat to Turkey in terms of attraction of multinational investments to enable transfers, and carrying out sophisticated R&D and innovation activities and hence production of knowledge-intensive, high value-added products because of small domestic market,
- The fact that the rival countries have been implementing, strictly and on the basis of a long-term planning, extensive investment and export incentives and strategic industry policies to attract the multinationals with advanced technologies and R&D investments to themselves. The fact that such countries also support the industry's R&D activities and activities aiming at developing advanced technologies (such as electrical cars) by means of extensive incentives, public procurement[51] and tax applications,

[50] China, India, South Korea, Thailand, Malaysia and East Europe countries (such as Bulgaria, Slovakia, Hungary, Romania, Czech Republic, and Poland).

[51] In the developed and some developing countries, for instance, municipalities and various governmental organizations create a demand for the new products via billion-dollar tenders and guarantees purchase, for the purpose of encouraging and promoting new technologies involving high investment costs; and international institutions offer highly favourable loans and financing means and facilities for production of such technologies, too.

- The rival countries have a resource of well-educated, well-equipped, qualified and less expensive human capital composed of individuals who speak several foreign languages and who specialize in computer programs (such as India),
- Raw material resources are abundant in the rival countries,
- In the rival countries, the public-university-industry collaboration has been implemented for a long time successfully as required by the industry and at higher levels,
- In the rival countries, the government plays a regulatory role in the automotive industry and the government's policies towards the industry are more strategic, more extensive, longer-term and more supplementary in those countries.

6.8 Summary

This section summarizes the findings of the analyses of the interviews which were conducted with the top-executives of 11 AMMs and which represent nearly 90 % of the Turkish automotive main industry successfully.

Turkey has become the 16th largest automotive manufacturer in the world in 2010, with 15 AMMs achieving a vehicle production capacity of more than 1.6M vehicles a year (total production of more than 1.1M ea.), attracting significant amount of FDI into the automotive industry within the last 40 years. She has witnessed a highly rapid development especially in the bus and commercial vehicle production and achieved the first-rank as the production base in these two classes across the EU and third-rank in the truck class, becoming therefore a strategic industry.[52] The information obtained from the interviews summarizes this situation best: While in the past the AMMs with domestic-capital that manufacture bus and commercial vehicles could produce only 3–5 vehicles per week on solely assembly basis at each plant, by mounting driver's cabin onto the truck chassis, today it has become possible to develop and produce vehicles that are of local design and in compliance with the EU regulation standards[53] for export to the EU countries, up to an amount of 12 vehicles per day at the domestic manufacturing plants. In addition, it has become possible to conduct some major tests[54] concerning the vehicles by the AMMs' own personnel at home, instead of having those tests conducted by the

[52] It has been determined that approximately 80 % of the vehicles manufactured in Turkey by M.A. N, Mercedes-Benz Türk, TEMSA, OTOKAR, B.M.C and ISUZU is exported to the EU countries. This could be regarded as a highly significant criterion in that it clearly shows the development level of the bus and commercial vehicle production especially and the achievement of the Turkish automotive industry across Europe.

[53] Ensuring compliance with the regulations of the EU is considered vitally important since the most important exportation market is composed of the EU countries. Such regulations affect each and every stage; from planning of the vehicle to its design and production. Therefore, there exist AMMs that have separate departments focused on the EU regulations.

[54] Industrial designs, 3D analyses, finite element analyses, durability tests, etc.

specialist firms and laboratories abroad. Especially as a result of the developments witnessed during the last decade, the automotive main industry today offers 20–30 % of the vehicles manufactured to the domestic market, while the rest is exported to first of all the EU countries and to many countries across continents, such as Eastern Europe, Middle East, Asia, North Africa, and America. Today, the Turkish automotive main industry has become a strategically important industry of Turkey with its current performance making it the industry achieving the highest export figures in Turkey. Nevertheless, in gaining this importance it should be taken into account that the level of the foreign inputs utilized in the production within the vehicle costs as we discussed before (see Sect. 6.2.4) is 40–50 % for the AMMs and reaches 50–60 % when considered together with the supply industry.

The AMMs interviewed are divided into three groups depending on their foreign capital structure: (1) the companies without foreign capital but manufacturing under foreign license (three companies), (2) the companies with partial foreign ownership (or joint ventures with the MNCs) (six companies), and (3) affiliates of the MNCs (or companies with full foreign ownership) (two companies). The headquarters of the AMMs with foreign-capital is located in the partner multinational's country and basic R&D and design activities are performed and all kinds of strategic decisions are taken by such headquarters. Since AMMs with foreign-capital operating in Turkey are in a way the multinationals' own companies, they are open to all types of knowledge exchange, and there is no discrimination against personnel, be they local or foreign citizen. Furthermore, the multinationals utilize Turkey as a production base and make exports to the surrounding countries. Therefore, the AMMs with foreign-capital in Turkey are in fact a part of the multinationals and included in their global production networks. Another important finding is that the Turkish automotive main industry is under the control and domination of such multinationals.

It has been determined that the industry highly specializes in the production of commercial vehicles rather than personal cars, and that it could export 80 % of its products. For this reason, 90 % of the R&D, innovation and technological activities of the AMMs aim at the commercial vehicles. Yet another significant finding is that while both the AMMs with domestic capital and the AMMs with minority foreign capital conduct R&D and design activities at home, AMMs with majority foreign capital conduct such activities mainly in their headquarters abroad (see Appendix H).

Production structures of the AMMs interviewed may be summarized in three main categories: (1) adaptive or minor technical changes on complete imported vehicles, (2) manufacturing the MNC's vehicles at home by localization, and (3) designing and manufacturing the vehicle completely at home. It could be stated that the AMMs manufacturing personal cars operate in production structures no. 1 and 2, while the AMMs manufacturing the commercial vehicles operate in production structures no. 2 and 3. As to the production structure no. 3, if the AMM is of a foreign-capital structure, the final design and production approvals of the vehicle are given by the multinational headquarters. The most critical finding here is that, irrespective of the production structure, knowledge-intensive and quite costly basic system parts/components are procured from the multinational headquarters and/or global-mega suppliers, which, in turn, renders the industry dependent on abroad in terms of such parts.

It is observed that all of the AMMs conduct R&D activities; furthermore, except one (TOYOTA), all of them has an R&D center in their structure, within the framework of the R&D Law no. 5746. However, the most significant finding is that the AMMs are engaging in short- and mid-term R&D activities aiming at product improvement and development, instead of long-term basic scientific projects. When conducting such activities, the AMMs with foreign-capital benefit from the advantage of being included in the global production networks of the multinationals to the utmost extent, and establish collaboration with the multinationals and obtain all kinds of support, in this respect. This may be interpreted as substitution of local R&D by international linkages.

KTTs to the AMMs with domestic-capital and the AMMs with joint-venture capital partnership are realized via the partners with whom they establish strategic collaboration. When put in order of significance, the most important ones are the partner multinational, licensor multinational and global-mega suppliers from whom transfers in many critical areas (mainly such as product, production processes, design, R&D, engineering, and training) are realized. It appears that the most critical knowledge and technology resource in this respect is the partner multinationals. Transfers from the multinationals are realized through certain channels such as joint projects, co-design and co-development activities, implementation of the multinationals' quality-production-management systems at the domestic plants, monitoring and auditing by the multinationals, and training delivered by the multinationals to the personnel of the AMMs. The training activities performed by the multinationals ensure that the local engineers enhance their skills and capabilities, thereby facilitating internalization and absorption of such transfers.

The AMMs procure the system parts and sub-components from the direct suppliers located at home and abroad and assemble them, and these suppliers constitute a vitally important part of the production and supply chain of the AMMs and are considered as a strategic partner. The average number of domestic and overseas direct suppliers of the AMMs is 338; the number of suppliers with whom the AMMs work is higher for the AMMs engaging in producing commercial vehicles, and they work with overseas suppliers by one-third. In addition, it is observed that there is a positive correlation between the foreign capital share of the AMMs and the ratio of the overseas direct suppliers they work with.

From the viewpoint of the AMMs, the domestic direct supplier firms appear highly developed, capable, competitive and strong; they conduct R&D activities and have a high level of absorptive capacity, establishing collaboration with the AMMs in the co-design/co-development activities. The most important expectations from these suppliers are summarized as delivering the requested products just in time, at the highest quality and at the lowest price. Due to the reasons discussed above, being selected as the direct supplier of the AMMs and included in the supply chain becomes possible only after a challenging process. It is observed that once these suppliers are included in the supply chain, they perform various collaboration activities with the AMMs.

Another significant finding in this section is that the findings obtained from AMMs concerning their direct suppliers coincide with the findings obtained from the suppliers surveyed in Chap. 4. Especially, the topics such as the transfers concerning product and production processes, financial transfers, training, monitoring and auditing, co-design, co-development collaboration activities, capabilities of the direct suppliers and their absorptive capacities, the importance of being the sole supplier in the product range concerned, selection criteria for the direct suppliers, the advantages of having the special quality certificates of the AMMs, and being the direct supplier coincide with each other.

It has been determined that the transfers from the AMMs to the domestic direct suppliers are realized through certain channels such as joint projects, co-design/co-development activities, monitoring and auditing systems, the process of obtaining the global certificates of the AMMs, delivering training by the AMMs, and that being a sole direct supplier of an AMM in certain product range affects such transfers positively. The most important channels in this respect are co-design and training activities, from the perspective of the content of such transfers. However, to what extent such transfers realized are absorbed by the supplier firms and what effects such transfers would create on their performance essentially depend on the technological and absorptive capacities of the suppliers.

It has been determined that the AMMs establish more intensive and more quality collaboration with their co-designer suppliers, that they share necessary knowledge with them and that such suppliers are in general the sole supplier of the AMM in the product range concerned. The most important factor here is that the level of the technological capabilities of the co-designer suppliers is fairly high. As a matter of fact, to have qualified, skilled and specialized human capital with high production capabilities is demanded by the AMMs and they are considered the precondition for the suppliers for realization of such collaboration with the AMMs.

The strengths of the industry which would affect the transfers from the multinationals to home could be stated as the geo-political location of Turkey and the fact that she possesses an automotive culture of nearly half a century long, while being dependent on abroad in basic system parts and not having an economically-strong and large domestic market could be regarded as its weaknesses. The major opportunities for the industry are increased interest in the new technologies on the global scale as well as the fact that multinationals have shifted their investments to the developing countries due to increased investment and workforce costs in the developed countries. On the other hand, existence of rival countries who have a large domestic market, who hold fairly high advantages in terms of production costs, and who have extensive incentive policies being applied for the purpose of attracting significant R&D and technological investments of the multinationals within the framework of the strategic collaboration of university-industry-public could be considered as a major threats.

Chapter 7
Conclusion

7.1 Introduction

The main purpose of this study is to evaluate whether FDI occurring in the Turkish economy has any effect on domestic firms, especially whether and to what extent it leads to the KTTs by paying attention to the role of multinationals. Findings of the studies[1] on Turkish economy indicate that foreign firms have played a major role especially in the automotive industry through horizontal and backward linkages; hence, this also justifies the aim of this study in terms of choosing the automotive industry as a case study to analyze the transfers at both inter-and intra firm level. In this context, the study mainly aims to reveal the complex nature of these transfers, in which ways these occur, what these are, what the factors are that give rise to the transfers between firms or lack of it, their impacts on the domestic firms, what the role of the firm characteristics and strategic collaborations between firms in these transfers is. For these aims, a very detailed and comprehensive case study research was conducted at the firm-level in the industry.

The study has two main purposes in terms of the transfers at inter-firm level. The first one is to examine whether FDI in Turkish automotive industry give rise to transfers from multinationals (customers) to their direct suppliers through backward linkages at inter-firm level. The second one is to analyze the relative importance of the transfers that took place and the effect on the performance level of the suppliers. Therefore, the main purpose at inter-firm level is to analyze what kinds of transfers related to *product* (design, co-design, joint activities, various documentations, etc.), *production process* (various know-how, R&D, logistics, etc.) and *training*

[1] For instance, the study of Sönmez and Pamukçu pointed out the importance of foreign presence (especially multinationals) especially in automotive industry which is in the first place among all manufacturing industries in Turkey in terms of number of foreign firms operating (17 %), and their shares in employment (55 %), gross output (80 %) and value added (80 %) values. In addition, authors revealed that the industry has the highest horizontal (50 %) and backward spillovers value (45 %), and also it has the third highest value in forward spillover value (21 %) across all manufacturing industries (for details see Sönmez & Pamukçu, 2012).

A. Sönmez, *Multinational Companies, Knowledge and Technology Transfer*, Contributions to Management Science, DOI 10.1007/978-3-319-02033-4_7,

(on-the-job, off-the-job) issues have been provided by multinationals to suppliers operating in Turkish automotive supply industry and the effects of these transfers on suppliers' performance level. In addition, the following issues were also included in the analysis:

- Determining the firm characteristics that play an important role in these transfers,
- Revealing and understanding channels and determinants of such transfers,
- Obtaining clues about the place of the suppliers in the global value chain of the MNCs,
- Analyzing the effects of such transfers on suppliers performance level by using a series of performance indicators,
- Evaluating technology, R&D and innovation activities of the suppliers,

In addition, the aim of the study in terms of the transfers at intra-firm level is to analyze the channels of the transfers realized by the multinationals to the AMMs operating in Turkey, the characteristics of the AMMs, their cooperation activities with the MNCs as their foreign partners (global AMMs), and R&D & technology-related activities.

To achieve these purposes, two different research methods were used to collect detailed data and information at the firm-level from both suppliers and AMMs in Turkey:

- Firstly, a face-to-face questionnaire survey was successfully conducted with the top executives of 166 automotive supplier firms (132 out of them are direct suppliers) in order to collect quantitative data—that corresponds to a 55.7 % response rate (given that the main sample consists of 298 firms). With this collected survey data, we tried to analyze our objectives in terms of the suppliers in automotive supply industry.
- Secondly, in-depth interviews were successfully carried out with the 19 top-executives of 11 AMMs in Turkey based on semi-structured interviewed guide. With the collected qualitative data, we tried to analyze our objectives in terms of AMMs. In other words, we examined and looked at our objectives from two perspectives: one from the suppliers' perspective as a recipient of these transfers and one from the customers' perspective as a source of the transfers at inter-firm level.

When it is considered that this kind of studies needs greater budget, time, research team, and include very detailed, comprehensive and confidential information on the firms, and therefore needs very effort to be successful, it might be claimed that the study was carried out with success. To the best of our knowledge, in this context, this is one of the first, and rare empirical studies carried out successfully on FDI-related transfers at both inter- and intra-firm level.

7.2 Main Findings

If the main findings in terms of the transfers at inter-firm level in Chap. 4 are summarized, it seems that the transfers occur from the AMMs to their domestic suppliers mainly on *providing documentations, assistance for logistic management—quality control—know-how—and R&D, co-design and co-development activities, designing,* and *cost reduction.* Compared to foreign suppliers, local suppliers tend to be involved in those production-product-training related transfers which are less knowledge-intensive and of a lesser quality. On the other hand, being a direct supplier of AMMs in Turkey and therefore being more close to customers in the supply value chain exerts a positive effect on the number, quality and intensity of the transfers. Moreover, various types and modes of training activities have also been provided to personnel of suppliers by customers. These trainings are mostly given to *production/operation staff* via *off-the-job training activities.* However, foreign suppliers are more involved in these training activities than local firms. This situation does not mean that foreign firms need more training because of lower capability of their human capital; on the contrary, it indicates that customers prefer to work mostly with foreign suppliers because of both their advanced technological capabilities and absorptive capacity level. These kinds of the transfers are generally provided by customers on the basis of new product development process. Therefore, this confirms that strategic relationships between foreign supplier firms and customers are very strong and it reflects that it is necessary to have very qualified personnel to benefit from these activities. Finally, it seems that customers do not generally prefer providing financial assistances to supplier firms, but if they do, these assistances are generally given in the form of pre-financing of machinery and prepayments for orders before delivery.

When the findings in Chap. 5 are reviewed, five factors come to the fore:

- It is observed that *being a direct supplier* (first-tier supplier) of at least one AMM in the automotive main industry will not trigger the transfers in terms of both the product and production process, but triggers the ones realized via training (see Tables 5.33 and 5.34). This may indicate that the technological capability levels and absorptive capacities of the direct supplier firms are not in good shape. The other side of the coin is that *being a direct supplier* will decrease the probability of establishing collaboration with the lower-tier suppliers and that the second or third-tier suppliers could not benefit from such transfer channels (see Table 5.34). In addition, an interesting point is that we could not find any positive effect of *being a direct supplier* on the performance of these firms.[2]
- Especially *R&D capabilities of the foreign firms* in the surveyed sample are more advanced and higher compared to the local firms. This has been reflected in

[2] It should be taken into account that a majority of the firms participating in our survey are direct suppliers (first-tier suppliers) and may have affected these results.

some way or other on the performance (average costs) of the foreign firms, as well (see Sect. 4.8.3, and Tables 4.36 and 5.34). Another finding is that the probability for the foreign firms to collaborate with other companies is lower compared to the local firms[3] and that they obtain the technologies they utilize from the headquarters of their parent companies (MNCs) (another indicator to depict the technology dependence on abroad). The regression analyses in Chap. 5 put forward that the foreign supplier firms establish collaboration for production and/or R&D purposes with neither the firms in their own industry (automotive supply industry) nor the AMMs in Turkey; on the contrary, a negative effect is in question (see Table 5.33). This results in the fact that the foreign firms in supply industry will not share their technological knowledge with the local firms.

- More than 90 % of the supplier firms surveyed *export* to other countries. This rate reflects the trend that has become widespread after the second half of the 1990s. The findings obtained from Chap. 5 show that the increase in the share of production aiming at the world market in the total sales triggers the innovation and co-design activities of the supplier firms for that product. On the other hand, it is observed that it does not affect the transfers realized by the customers concerning the production process and does not affect the supplier firms' performance positively, either. It could even be said that it has a negative effect on being a direct supplier (see Tables 5.33 and 5.34). These findings point out the fact that exportation and increase in the share of production aiming at the world markets in the turnovers alone will not suffice to enhance the production, design and R&D capabilities of the firms. In our opinion, these findings require that the status of the automotive supply industry firms in the global production networks of the multinationals should be questioned.
- We see that the *firm size* affects many factors positively (see Tables 5.33 and 5. 34). It affects the transfer processes in terms of product and finance positively but does not affect the performance of the firm (capabilities and developing production capability). In addition, it has positive effects also on the probability of being a direct supplier, performing R&D activities, and establishing co-design and technology collaboration with other firms for the product.
- Lastly, the findings of the econometric analyses show that *performing R&D* and *innovation activities in terms of product* is of vital importance for many factors, first of all to *obtain KTTs from the AMMs* and to *develop business relationships with other companies* (see Tables 5.33 and 5.34). Moreover, the findings in Chap. 6 obtained from the interviews verify that those two factors are highly important especially to be able to become a direct supplier and to perform co-design activities with the AMMs. These are also assumed as important indicators of absorptive capacity of the suppliers and point out to the importance of technology competences for those suppliers which desire to be a co-designer,

[3] In addition, the probability of the customers to realize KTT in terms of both the production process and products to the foreign firms is lower compared to the local ones.

and to obtain transfers. In these respects, these two factors are found more important than any other factor.

The findings in Chap. 6, obtained from the AMMs as a result of the interviews, coincide in general with the findings in Chaps. 4 and 5, obtained from the suppliers via questionnaire survey. We could summarize those findings as follows:

- It is understood from the findings we have obtained that the AMMs have an intensive and close relationship with most of their suppliers, first of all their direct suppliers with whom they carry out co-design activities. It is observed that various collaboration activities between the AMMs and their suppliers have been conducted on subjects such as product co-design and product development activities—agreements—discussions, design of the production tools, development of the quality & control methods, cost reduction, and material designs. Since ultimately it is the AMMs who will be the final users of the supplied parts, they want to be assured of the quality and timely delivery of those parts, because even the most minor problem encountered in the production and supply chain could bring the production lines to a halt. In other words, any problem, even the most minor ones, encountered on the side of the suppliers reflects directly on the AMMs due to the production chain. Therefore, the AMMs monitor their suppliers closely, audit and assist them in resolving their problems.
- The AMMs share their knowledge and technologies with their suppliers in general via certain channels such as joint projects, co-design/co-development, and training activities. The AMMs perform co-design activities with their suppliers during the development phase of a new vehicle and part, and provide them required support concerning product and production processes. Such buyer-supplier relationships ensure KTTs to the suppliers through backward linkages and may enhance and develop their technological capabilities, quality levels, productivities and performance levels.
- When problems arise in terms of production process, product, projects implemented, program follow-up and applicability, the AMMs deliver the training they deem necessary to their supplier firms by means of their engineers, through various methods such as conferences, informative meetings, seminars, and on-the-job training held at the AMM's premises. In addition, the AMMs assign their engineers at the relevant supply industry firm for a short period under highly specific circumstances (for problem-solving purposes or when requested). Visits paid by the supply industry engineers to the AMM's plants are generally performed in the case of production of a new part, in order to ensure that no problem occurs at the production line of the product, to detect problems if any, and to support the material flow. The engineers or employees of the supply industry firms also apply the information they obtained during those activities afterwards to the firms they work for.
- At the same time, the monitoring and auditing systems, requested high quality standards, price target, and cost reduction requests of the AMMs in turn raise the levels of the quality, technology and R&D capabilities of the suppliers and force them to work up to more advanced standards. For example, one of the AMMs

who deals with bus production stated that since it employs 3-D design in production, it also requests from its suppliers to use the same technology and to provide AMM itself with not only the illustration of the part but also its 3-D data as well. With every new project developed by the AMMs together with their suppliers, the AMMs' standards and demands are notched up another degree and this, in turn, ensures KTTs from customers to the supplier firms, contributing greatly to accumulation of knowledge.

- Furthermore, the competition among the supply industry firms both at home and abroad causes them to develop themselves constantly. In sum, the supply industry firms have to develop their technological abilities continuously so that they can compete in the fields of quality, R&D, costs, prices, design and productivity, and if they fail to do so, they will be faced with being wiped out from the market. The supply industry firms obtain know-how and knowledge by enhancing their technological capabilities within the framework of the constant improvement and development system and by getting KTTs.
- The AMMs have also stated that they support the supply industry firms when required, on issues such as selection of production-equipment, selection of machinery, purchasing press-moulds, and purchasing new software. However, it has been stated that there is no such application in terms of financial supports, but that advance payments are made if production of a new part is in question or financial supports are provided in case of purchasing the parts (software, benches, press-moulds, etc.) that are vital for the production.

The AMMs have underlined that the above-discussed collaboration activities provide an important information exchange between the suppliers and the AMMs, contributing greatly to the development of the supply industry firms and ensuring accumulation of knowledge. In this context, it could be argued that the transfers from the AMMs to the supply industry firms are realized via joint-projects, training and agreements. Ultimately, the suppliers' exhibiting a better performance also increases the performance levels of the AMMs indirectly, and makes it possible for them to obtain better-quality products on-time. All these processes constitute a gain from the supplier's perspective.

The findings in terms of KTTs at intra-firm level show that transfers in many critical areas are provided to AMMs by their partner multinationals. These transfers are realized through joint projects; co-design and co-development activities; implementation of the MNCs' quality, production and management systems at the domestic plants; monitoring and auditing by the MNCs; and training delivered by the MNCs to the personnel of the AMMs. The training activities performed by the MNCs ensure that the local engineers enhance their skills and capabilities, thereby facilitating internalization and absorption of such transfers.

In the light of the findings we have obtained, we see that the Turkish automotive industry has made a great deal of development in the last 40 years and has been transformed from production in small quantities on complete assembly basis under licenses and not quite open to competition to being able to sell its own vehicles, especially commercial vehicles it designs and produces with its own R&D, to many countries in the world. Moreover, it has been observed that the main and supply industry firms establish strategic partnerships with foreign companies abroad, that their technological capabilities enhance and that they follow the technologies and innovations closely, being integrated into the world. However, despite all these positive developments, Turkish automotive main industry still appears to be unable to create its own global trademark in the personal car, continuing to produce under foreign licenses, and it seems that the MNCs dominate the domestic market completely in this area. As a result of the interviews made with the top-executives, it has been determined that, irrespective of the foreign capital ratio, all the AMMs with foreign capital are dependent on the headquarters of MNCs as foreign partner,[4] and that all kinds of basic product development, design and R&D activities are carried out at these headquarters abroad. On the other hand, it has been observed that even if the MNCs have R&D departments in their local affiliates in Turkey (AMMs), these departments generally work on the issues such as adaptive changes, partial modifications and improvements of the products, making them appropriate for the domestic market and production line.

In conclusion, the multinationals make investments in the developing countries for reasons such as inexpensive labor force, raw materials, large size of the domestic market (and high growth rate), and incentives concerning investment, export, and R&D. The knowledge and technology possessed by the multinationals are generally transferred to the local firms in host country through backward linkages, increasing the accumulation of knowledge and technological capabilities of those firms. Furthermore, the presence, the culture and experience of the multinationals create a competitive effect on the industry they operate and force the local firms to develop themselves, affecting their productivity and technological capabilities positively. However, the multinationals maintain their most critical product development, technology, design and basic R&D departments at their headquarters abroad and remain unwilling to share the knowledge they have in such fields with the firms other than those they naturally establish strategic partnership and collaboration with. The local firms' obtaining the multinationals' knowledge and technologies becomes possible via joint projects and technology or collaboration agreements. Yet, internalizing and absorbing the knowledge and technologies obtained in this way largely depend on the absorptive capacities and capabilities of the local firms in the areas of technology and engineering. For this reason, attracting the multinationals' investments involving high level of knowledge and technology to host country does not suffice alone, but it is also required that those multinationals develop intensive business relationships with the local firms and that the local firms have a high level of technological capabilities.

[4] Naturally, the AMMs dealing with production under foreign license in Turkey become dependent on the headquarters of the MNCs as foreign partner (the licensor MNC located abroad) in almost all strategic decisions.

7.3 Policy Implications and Recommendations

This section, as another object of the study, suggests some policy implications for the Turkish automotive industry, as well as practical recommendations for the public institutions, top-executives in emerging country, and multinationals in order to conceive and implement science and technology policies in supporting the transfers. These should be read along with the findings obtained from long-lasting research conducted under the scope of the study.

It appears that the Turkish automotive main and supply industries with their history of 40 years have succeeded in attaining significant amount of knowledge, technology and capability especially in production area. The production proficiency established in a period during which import-substitution industrialization strategy was implemented and made only a limited development has started to develop in the 1980s and reached an advanced level especially after the second half of the 1990s. Now, the priority issue for the firms in the supply industry is to launch R&D activities, to closely follow the technology developing rapidly in the industry or in their product ranges, to achieve proficiency in product design, and to establish collaboration with the main industry companies at home or abroad in order to achieve all these priority issues. Since the buyer-supplier collaboration has the potential to trigger the transfers to the firms in the supply industry, it paves the way for notching up the R&D and innovation activities of these firms and their technological capabilities concerning production process and product design to the next level. Given the fact that such transfers benefit both sides, the sooner such collaboration activities established the better. Moreover, with a view to increasing the transfers to the domestic firms in Turkish supply industry in quantity and quality, the ways and opportunities to develop further the technological collaboration with the multinationals and foreign firms abroad should be sought, and the existing collaboration should be moved to the next level. In this context, the main and supply industry firms in Turkey should take part in the joint projects to enhance their technology and know-how with foreign companies abroad by using the worldwide business channels and contacts of the multinationals in Turkey.

It has been determined that the Turkish automotive industry has specialized in, focused on and accumulated knowledge about the production of commercial vehicles rather than personal cars. In this area, the main and supply industry firms seem powerful as a result of the production, technology, R&D and design capabilities and able to compete in the international markets. Therefore, it appears that Turkey maintains to be the production base of the commercial vehicles in the medium term, even able to create her own global trademark in the field. To this end, it would be highly beneficial in terms of existence in the global market in the long run and to utilize scale economies if strategic collaboration is established among the AMMs producing commercial vehicles, although it seems good for them to compete among themselves in the domestic and overseas markets in the short term.

Despite the fact that university-industry-public collaboration has developed in Turkey in recent years, the information gathered depicts that the relationships of both the main industry and the supply industry firms with the universities are still rather poor. For this aim, especially within the framework of the university-industry-public collaboration, it is essential that the dedicated, extensive and planned joint projects and road maps for the industry be devised, by closely following and analysing the strategic and technological advancements in world.

The foundation of the global competitive power is quality, qualified and specialized workforce. The recent establishment of automobile engineering departments in some universities in Turkey is considered a positive development by the industry in this respect. However, the industry chooses the staff it needs among the successful university graduates and the requested experience and practice expected from the staff realized via training and participating in various projects is attained only in the medium and long run. Schooling and raising the quality, specialist and qualified engineers and R&D staff required by the industry in a short time could be possible by not only delivering theoretical education as has long been practiced by the universities, but also through delivering practical and on-the-job training within university-industry collaboration. For these reasons, it is mandatory that the government attach importance to raise qualified workforce specialized in the subject area, both at the universities and at the vocational high schools and devise long-term strategic planning, for raising and schooling the personnel such as engineers, R&D staff, and technicians. Therefore, realizing the required arrangements at once, based on the examples of the countries who display superior success in this respect (such as Germany, Japan, France, and South Korea) would also enhance the competitive power of the industry.

It is undeniable that the government's incentives for R&D, investment and export (subsidies such as TUBITAK, TEYDEB, KOSGEB, and SANTEZ) are of critical importance in terms of increasing the technological capabilities and competitive powers of the main and supply industry firms. Especially the R&D Law no. 5746, enacted in 2008, is a highly critical step in this respect, and it has been determined that the companies who did not or could not perform R&D activities before now have launched their R&D activities and that most of the AMMs have established their separate R&D centers, all thanks to the supports provided by the said law. Under this law, the two basic rules to benefit from the tax incentives granted for R&D are (1) existence of an individual R&D building separate from the business establishment and (2) recruitment of minimum 50 equivalent R&D personnel to work at that building. While it seems difficult for large-scale companies to benefit from the law due to these two basic rules, it seems impossible for the small and medium-scale (SME) supply industry firms for the time being. This constitutes a considerable obstacle in front of the firms to benefit from the law, especially given the fact that the majority of the firms in the automotive supply industry falls within SME category. Yet, the fact that only a few firms in the supply industry have adequate R&D capabilities is expressed by the AMMs interviewed. The findings of the econometric analyses in Chap. 5 point out that *gaining R&D capability* by the supplier firms is of vital importance to establish collaborations with other companies and realize KTTs from the customers to the firms in the supply industry (see Table 5.34). Since the public incentives granted for R&D (RDSUB) have a positive effect on *launching R&D activities* (see Table 5.33), the existing R&D incentive system may have to be changed, taking the necessities of the automotive supply industry firms into account. Also, the level of the incentives and project subsidies remain too low compared to the developed countries. In this context, it is required that the amount of these incentives and subsidies be increased; the SME firms be made to benefit from the R&D, investment and export incentives more effectively; and that the incentives be widespread via various associations or organizations in the industry. In addition, it is important for these firms to cluster and act jointly, adjust their production capacities accordingly and be in constant interaction with each other, in order to increase their global competitive power.

Another significant finding obtained under the scope of the study is that since personal cars are considered luxury goods in Turkey, more tax is levied on the new and high-engine power cars, which, in turn, decreases the total demand for the new, environmentally-friendly and advanced technologies[5] and therefore restricts the innovation and R&D activities with respect to these products at home. In contrast to the tax policies applied in Turkey, the policies in the developed and rival countries (such as levying less tax on electrical and hybrid vehicles, levying higher taxes on aged vehicles older than a certain limit, and levying taxes on vehicles not based on their engine power but on their exhaust emission values) increase the total demand for the new, environmentally-friendly and technological products, and therefore keep the R&D and innovation activities alive in these areas by promoting them. Therefore, it would be critical to apply also in Turkey similar tax policies aiming at increasing the total demand that would promote the said activities and ensure equivalent competition conditions with the rival countries, on condition that at least total tax income is kept constant. This kind of policies may also play an important role in attracting the multinationals with advanced technology to the country and to enhance the transfers to domestic firms.

The labour cost especially wages paid to the white-collar in the Turkish automotive industry is not so inexpensive although it remains behind what is applicable in Europe and USA. It has been determined that especially when compared to the Eastern Europe countries and newly developing markets (such as China, India, Malaysia, and Thailand), the labour cost is considerably high and that Turkey has no wage advantage any more over those countries. For the time being, Turkey offsets this disadvantage with her know-how, R&D, more quality products and workforce. That is, the expression *quality and inexpensive workforce* is not valid anymore for Turkey. Even if this case still seems as an advantage for Turkey in the short run, it is inevitable that continuously increasing quality and technology levels of those countries combined with their inexpensive workforce structures would pose a major threat against Turkey in the medium and long run.

[5] Such as electrical and hybrid cars, products less harmful to the environment, and active safety assistance systems for driver.

Production of a vehicle in the automotive main industry is possible by bringing together a great number of parts, systems and processes. Today, it is impossible for just one company to produce a vehicle from scratch. As a result of the information gathered from the interviews, it is determined that the AMMs stand at the top of the pyramid as planners and organizers, manufacturing basic powertrain parts, determining the basic designs and performing developments for the product, and identifying specifications of the parts to be supplied. In this context, other systems and parts required by the main industry are manufactured by the direct suppliers (first tier suppliers) according to specifications determined by the main companies. For these reasons, we are of the opinion that specializing in just one of the thousands of the parts composing a vehicle; developing new products; acquiring high level of knowledge and technology in any part; developing test equipment concerning the products; and focusing on the design, R&D and engineering activities or rendering service in these fields are more strategic ways to gain superiority in the industry, instead of trying to produce a complete vehicle from scratch at home. In this respect, assuming a role and specializing in the development of, for instance, knowledge-intensive new technologies concerning the robots and lasers utilized in the vehicle production lines, and concerning the batteries that are the most critical component of the electrical vehicles, or concerning active and passive safety systems, etc., creates, from our point of view, considerably much more value added and bear essentially vital importance.

It has been understood from the interviews that only a few multinationals we call as global-mega suppliers (such as BOSCH, DELPHI, DENSO, AUTOLIV, AVL, SIEMENS, and MAGNA) are dominant globally in the production of the specific system parts like steering systems, braking, safety, electrical and electronic parts and that they supply these major systems to all the global AMMs (OAMM). At the same time, it is observed that certain multinationals in global automotive supply industry (such as BOSCH and SIEMENS producing brake systems, electrical and electronic parts like electronic control units and processors) have become globally the single and common supplier of the global AMMs in these fields, and that reached a greater structure than some of the global AMMs. In addition, the market composed by these multinationals (global mega-suppliers) has oligopolistic market characteristics. They conduct all kinds of R&D, design, product development, etc. activities at their headquarters abroad and work with a large number of R&D personnel and a quite high R&D budget. For instance, the number of the R&D personnel of the BOSCH only is greater than the total R&D personnel of the Turkish automotive industry, and the company's R&D budget is also higher than the total budgets of the firms operating in main and supply industry in Turkey. For these reasons, it seems rather improbable to develop new products and to become competitive in these fields in the short and medium terms. In our opinion, certain fields such as plastic injection, sheet metal parts, lighting system parts, rubber parts, and interior equipment seem more open to develop for the Turkish automotive supply industry.

As discussed before, the Turkish automotive main industry has no dominant power over the major system parts (such as powertrain, engine, steering, safety, and electronic processor) that constitute the most important input costs of a vehicle, and that the industry has to import these parts from the multinationals as foreign partners and/or global mega suppliers abroad. These are highly expensive parts that have been manufactured by global AMMs and/or global-mega suppliers (multinationals) as a result of long-term R&D activities, involving extensive knowledge and know-how. In order to increase the Turkish automotive industry's competitive and technological level, to attain the desired costs and finally to create its own trademark, it is necessary to develop and manufacture such knowledge-intensive parts domestically. This, in turn, seems probable only in the long run (10–15 years) by ensuring KTTs as a result of the strategic collaborations established by the multinationals abroad and increasing the investments in the R&D and innovation activities, as well as know-how accumulation. Unless such activities are performed, it seems rather unlikely to produce such parts that require superior production, know-how and technological capabilities as well as higher investment costs domestically in the short and medium term. This leads to the fact that the automotive industry would remain dependent on abroad in the medium term and that it may lose its advantages to the newly-developing markets.

It has been determined that the production of the labour-intensive commercial vehicles (such as buses, trucks, and light commercial vehicles) has shifted from the west to the east, to countries such as Eastern Europe, Turkey, India, and China, due to the increase in the operating costs, economic recessions and production shrinkage in the developed countries resulting from the recent global financial crisis. It has been expressed during the interviews with the AMMs that the multinationals regard Turkey as a production base due to her proximity to the Middle East and EU countries; her quality and relatively inexpensive workforce; her inexpensive raw materials; quality and strong supply industry; and that they invest in Turkey as a result. For these reasons, it is observed that nowadays Turkey has attained a commercial vehicle production base status in the automotive industry. From the multinationals' point of view, investing in Turkey means that they expect Turkey to display significant progress and contributions in the companies' battle they are waging against their competitors in the Far East and Asia. Today, it is clearly foreseen that, as well as the developed countries leading the automotive industry, the manufacturers located in the developing Far East and Asia countries will direct the world automotive industry in the near future. We are of the opinion that the large domestic markets of such developing countries with their more dynamic economies, their domestic market demands growing day by day (with the increase in income levels), and their inexpensive workforce and raw material will strengthen their increased export potential further, and that in this way, by increasing their quality levels and technological skills, they would become serious competitors against Turkey in the near future.

In the light of the findings obtained from the interviews, it has been determined that especially the countries in the Far East and Asia provide the multinationals with significant incentives for FDI investments, but on condition that the multinationals realize the conditions to introduce new technologies in their countries and establish various technological collaborations with the local companies. In this sense, it would be important for the Turkish government to play a similar regulatory role in the industry and to develop strategic planning in the areas such as increasing the skills and capabilities of the local firms, ensuring accumulation of knowledge, providing transfers from the multinationals, and to develop strategic investment decisions. This would also facilitate Turkey's ability to compete under equal conditions against the rival countries.

It is observed that the newly developing markets (such as India, Malaysia, Taiwan, and Thailand), and especially China, would become a serious threat against the Turkish automotive industry in the near future due to their competitive capacities. It is highly probable for the reasons we discussed before that the multinationals shift their plants and certain departments (such as R&D and design) from headquarters to such countries in the near future. It has been determined that such countries have not been able to realize the necessary regulations and quality criteria in the vehicle production yet, but that they improve and develop themselves continuously in those areas, raising their technological capabilities. In addition, it is observed that they are able to produce a high number of vehicles and sell them at considerably lower prices thanks to their significantly low costs and large domestic markets. Especially China is predicted to become the world's automotive production base in the medium and long term (10–15 years) because of her significant advantages such as low production and raw material costs, and large domestic market. Moreover, China seems that she will be the most serious rival of Turkey in terms of becoming the R&D and excellence centre target by realizing intensive transfers from the multinationals and developing her technological capability. The fact that these newly developing markets, for the time being, are still following Turkey in the areas of technology, quality, know-how and qualified workforce, appears as an advantage for Turkey. Nevertheless, the ability of the Turkish automotive industry to maintain its advantages and status will certainly depend on its continuous self-development, becoming a design, test, R&D and excellence base, creating its own trademark, enhancing its technological capability level through transfers from the multinationals and producing technological products that create high value added. Otherwise, it appears that it is highly probable for the Turkish automotive industry to experience a serious bottleneck in the near future (10–15 years).

With the globalization process, now the market is the world from the companies' point of view, and competition is realized on a global scale. What is important from the economic standpoint is to produce the best at the least cost in the world. In this context, benefiting from the scale economies is vital for both the main industry and the supply industry. The long-term costs of a company producing several millions of vehicles (parts) in a year (such as Toyota) and a company producing 10,000 or a 100,000 of vehicles (parts) in a year are not the same, and the scale economies provide the former with significant superiorities over the latter in terms of competition. For these reasons, it is required that the newly and rapidly developing countries of the Far East and Asia should not be considered as a threat against Turkish automotive industry; instead, the AMMs and suppliers in Turkish automotive industry should establish production bases in those countries, by investing in them with strategic collaborations, and thereby they should turn the threat into an opportunity. Especially, the capabilities and facilities the countries possess, such as inexpensive labour force, raw material resources, large domestic market, and possibility to export to the surrounding countries, should be put to good use.

Today, in the automotive industry production, Turkey has achieved the rank of 16 in the world and 7 in Europe. According to the information obtained from the interviews, maintenance and sustainability of the current success of the industry would, in the medium and long term, depend on adopting a target of becoming a design and R&D base rather than being a production base, and attaining the target in the medium term. In other words, just being a production or assembly base would mean nothing for the Turkish automotive industry in the near future.[6] The situation is such that nowadays China has become a global production base for many products and multinationals, and it is possible to see the mark *Made in China* almost on all products worldwide. However, design, basic R&D, and development activities of all these products that are manufactured in China are conducted in the headquarters of multinationals concerned in developed countries, and therefore, such products also bear the marks *Designed in XXX* and *Developed in XXX*.[7] This clearly reveals the fact that the knowledge-intensive products with high value added are designed and developed in the developed countries and the production process which means the labour-intensive part is realized in the newly developing markets where inexpensive labour force is available. The highest value added part within the products being produced in such developing countries is the design and development part of the products that are knowledge-intensive, and this value turns back to the developed countries at the end via transfer of the profits. Therefore, the design, development, quality, technology, know-how and production cost of the product, rather than the manufacturing, have become more important. In the present situation, the most vital product is the knowledge from the viewpoint of both the companies and the governments; generating and possessing knowledge becomes the most important source of power. Thus, it has now become mandatory for Turkey that the new and advanced technologies be followed and observed, that those technologies be obtained from the multinationals and be developed further, and that high value-added, knowledge-intensive products be developed through her own R&D and innovation activities in the collaboration of university-industry-public. In other words, the case could be summarized clearly by this motto, *in*

---

[6] For instance, the most important part (hardware) of the electrical vehicles that have been started to be produced by some AMMs with foreign capital in Turkey is battery. It involves intensive knowledge and technology and its cost is very high in the total inputs of the vehicles. This battery is imported from abroad, just as the powertrain parts of other vehicles being produced in Turkey, and assembled into the vehicles. These parts have been developed by a few MNCs as a result of long years of advanced R&D studies that cost billions of dollars; such parts involve extremely intensive knowledge and technology. Therefore, the assembly of these parts into the vehicles by importing at home means actually nothing to be production base.

[7] Here, XXX (USA, Germany, France, England, etc.) expresses the developed countries other than China, where the product design and development is realized.

today's global economy, if you know how to produce the best product demanded by the market at the lowest cost, then where you produce the product is not so important any more.

In sum, Turkey should become a country where accumulation of knowledge is realized, by attaining a status of being a design and R&D base that creates high value added across Europe, along with being a competitive, developed automotive production base in the world. To this end, it is necessary to develop long-term strategic plans which, within the collaboration of public-industry-university, will ensure developing the technological capabilities and absorptive capacities of the domestic firms; realizing future foresights and projections; providing accumulation of knowledge and therefore realizing more technological innovations and developments at home. For this aim, global technologies should be followed and observed; FDI investments to provide advanced technology and knowledge accumulation should be attracted to home; state-of-the-art and latest knowledge and technologies should be transferred to home; and they should be well absorbed. Otherwise, the foreign capital and technology would go to the developing rival countries where it can find inexpensive workforce, engineering and raw materials, high investment incentives (such as China and India), and thus the current advantage would be lost.

In conclusion, today, the Turkish automotive industry is a developing industry that is one step ahead in the global sense; however, her possibility to maintain, and push forward that advantageous position seems to depend on developing new, knowledge- and technology-intensive products that have high value added at home by means of R&D and innovation activities. This, in turn, could be possible only by closely following the global technologies, by learning them via transferring the know-how involved, and by ensuring accumulation of knowledge at home. In other words, it is not necessary to reinvent the wheel; instead, to be informed of and to have heard of the fact that the wheel has been invented, to know how it is produced, to develop and produce a better one or its alternative should be the targeted goal. Turkey appears that she is capable of achieving that with her geo-political location; her human capital, infrastructure, logistics facilities; her industrial structure integrated into Europe via customs union; her 50 years of automotive production culture; knowledge accumulation; strong main and supply industry, and her vision.

7.4 Implications for Further Research

This study contributes to the literature as one of the first, rare empirical evaluations to understand the main channels and determinants of the transfers between AMMs and suppliers in the Turkish automotive industry. The interviews with the AMMs and survey conducted amongst suppliers in the industry, and the econometric analysis conducted here using data collected via the survey is the first study of its kind for the industry. The data used in the analysis is only cross-section; however, it may take a long time for the transfers to produce their final impact on the productivity level of firms. For further research, the same study can be repeated in the near future in order to test accuracy of such study, and this could shed more light as whether the transfers actually exist or not. Moreover, similar industry specific case studies using different target industries or different research methodologies should be undertaken to address the issues industries will have to tackle with in the following decades. Furthermore, the findings of this study in terms of determinants of the transfers, R&D and innovation activities, collaborations among firms, etc. can be developed as separate hypotheses, and they can then be tested empirically as further research.

Appendix A: Cover Letter for Face-to-Face Questionnaire Survey with Suppliers in Turkey Automotive Industry

MIDDLE EAST TECHNICAL UNIVERSITY
Department of Economics & Science and Technology Policies Research Center
Knowledge and Technology Transfer Research in Turkey Automotive Industry

Dear/Company Name_ _ _ _ _ _ _ _ _ _ _ _ DD.MM.YYYY

First of all your firm is a part of the sample chosen specially and carefully in the automotive industry.

This study titled *Knowledge and Technology Transfer Research in Turkey Automotive Industry* is supported as a research project by TUBITAK, OSD, TAYSAD and its application has been conducted by XXX Co. Inc.[1]

Preface

In the 21st century, the most important factors that determine firms' power and position in international competition area are firms' capabilities on improving/developing competitive new technologies – products - production processes, and on adaptation of and follow up closely new technologies. Therefore firms have undertaken increasingly innovation and R&D activities, and they have increasingly rely on collaborations with external partners (such as MNCs, customers, suppliers, universities, and private institutions) in order to obtain technological innovations. At the same time, governments, firms, institutions and universities have come

[1] The name of the company conducted cooperation for the survey is a research enterprise engaged in many national and international projects, and a specialist and pioneer in its sector. The name of the company is kept confidential by us here due to data privacy, however surveyed firms are informed in detail about this company.

A. Sönmez, *Multinational Companies, Knowledge and Technology Transfer*, Contributions to Management Science, DOI 10.1007/978-3-319-02033-4,

together to support these activities and to develop national science and technology policies for these aims in many developed and developing countries.

The Purpose of the Study

This research has four main aims; (i) to reveal what kind of knowledge and technology transfers have been provided to the suppliers by their customers, and their determinants, (ii) to reveal characteristics and technology capabilities of the suppliers in Turkey automotive industry, (iii) to examine innovation and R&D activities of the suppliers, and (iv) to analyze the qualitative of collaboration between AMMs and their suppliers with respect to transfers and design activities.

This scientific research is of great significance for the firms in the sector - **especially yours** - in terms of continuing the current position and competitive environment in the future. At the same time, we think that it may be an important source for the public and various private institutions (such as SPO, KOSGEB, TUBITAK, OSD, and TAYSAD) to develop knowledge and technology policies in support of R&D, innovation and technology transfer activities relating to sector.

In the countries in which there are developed automotive industry and highly concentrated cooperation among public, university and industry (such as USA, Germany, France, South Korea, Thailand, Malaysia, and China), similar researches have continuously been implemented in a successful way, and most important results have also obtained in terms of development of the sector. With this respect, one important aim of the study is to make a contribution by closing this deficiency in our country, and the other one is to prevent losing its leader role to other countries by providing to the sector to capture sustainable competitive environment that will lead to the sector to be regional production and excellence base in the near future.

Research Methodology

Considering its scope and characteristic, the research needs to collect detailed information and data at the firm level. For this purpose, a questionnaire form was prepared under cover of international scientific literature and it consists of 9 sections and 43 detailed questions. It has been careful to keep the questionnaire short; and apart from a few questions the majority of it is formed with closed-ended multiple-choice questions. Average response time of the questionnaire is about **60 minutes** and it will be applied to members of TAYSAD and leading firms in the sector like as your firm.

Your participation to this research study as an important leader and executive in the industry is very critical for the success of the study. We hope that you will agree to participate and you will want to have a voice in this study because of contribution of the research and also future success of Turkey automotive industry.

We will happily send you -free of charge- a copy of the final report.

It must be emphasized that all information and data obtained from this questionnaire will be kept in strictly confidential, and it will not be published in a form like report or article which would identify you or your firm without your consent.

The success of the study, and obtaining accurate results and policies for the sector depend on your responses' accuracy and completeness, otherwise all efforts related to this study will fail.

Please do not hesitate to contact with me from the following communication channels if you have any question or problem about the survey.

Alper Sönmez
Middle East Technical University,
Department of Economics
Room No: A202, 06531 Ankara
E-Mail: salper@metu.edu.tr
Office Phone: (0312) 210 30 7.
Mobil: 0 533 xxx xx xx
Fax: (0312) 210 79 6.

Yours Sincerely,

Appendix 1: Letter of support from OSD
Appendix 2: Letter of support from TAYSAD

*****Thank you for your kindly cooperation and concern about the research*****

Appendix B: Cover Letter for Interview Request from the AMMs in Turkey

MIDDLE EAST TECHNICAL UNIVERSITY
Department of Economics & Science and Technology Policies Research Center
Knowledge and Technology Transfer Research in Turkey Automotive Industry

Subject: Request for an Interview

Dear/Company Name_ _ _ _ _ _ _ _ _ _ _ _ DD.MM.YYYY

This study titled *Knowledge and Technology Transfer Research in Turkey Automotive Industry* is supported as a research project by TUBITAK, OSD and TAYSAD.

Preface

In the 21st century, the most important factors that determine power and position of the firms in international competition area are firms' capabilities on improving/ developing competitive new technologies – products - production processes, and on adaptation of and follow up closely new technologies. Therefore firms have undertaken increasingly innovation and R&D activities, and they have increasingly rely on collaborations with external partners (such as MNCs, customers, suppliers, universities, and private institutions) in order to obtain technological innovations. At the same time, governments, firms, institutions and universities have come together to support these activities and to develop national science and technology policies for these aims in many developed and developing countries.

The Purpose of the Study

This research has four main aims; (i) to reveal what kind of knowledge and technology transfers have been provided to the suppliers by their customers, and

A. Sönmez, *Multinational Companies, Knowledge and Technology Transfer*, Contributions to Management Science, DOI 10.1007/978-3-319-02033-4,

their determinants, (ii) to reveal characteristics and technology capabilities of the suppliers in Turkey automotive industry, (iii) to examine innovation and R&D activities of the suppliers, and (iv) to analyze the qualitative of collaboration between AMMs and their suppliers with respect to transfers and design activities.

This scientific research is of great significance for the firms in the sector - **especially yours** - in terms of continuing the current position and competitive environment in the future. At the same time, we think that it may be an important source for the public and various private institutions (such as SPO, KOSGEB, TUBITAK, OSD, and TAYSAD) to develop knowledge and technology policies in support of R&D, innovation and technology transfer activities relating to sector.

In the countries in which there are highly developed automotive industry and highly concentrated cooperation among public, university and industry (such as USA, Germany, France, South Korea, Thailand, Malaysia, and China), similar researches have continuously been implemented in a successful way, and most important results have also obtained in terms of development of the sector. With this respect, one important aim of the study is to make a contribution by closing this deficiency in our country, and the other one is to prevent losing its leader role to other countries by providing to the sector to capture sustainable competitive environment that will lead to the sector to be regional production and excellence base in the near future.

Research Methodology

Considering its scope and characteristic, the research needs to collect detailed information and data at the firm level. For this purpose, it is planned to make an interview with the top executives (managers of R&D, production, purchasing etc.) of the AMMs in the Turkey automotive industry, and it will take place from February to June. The interview will be based on **interview guide with open-ended questions** (see Appendix 1), and it will be conducted in a semi-structured way to give you ample opportunity to express your opinions and it should take **about 120 minutes**.

Your participation to this research study as an important leader and executive in the industry is very critical for the success of the study. We hope that you will agree to participate and you will want to have a voice in this study because of contribution of the research and also future success of Turkey automotive industry. In anticipation of your co-operation and participation to this research, we are writing to request **a factory visit and interview with you**, or one of your colleagues recommended by you. In the arrangement of interview date, you can contact with me as interviewer from the following communication channels.

We will happily send you -free of charge- a copy of the final report.

It must be emphasized that all information and responses obtained from this interview will be kept in strictly confidential, and it will not be published in a form like report or article which would identify you or your firm without your consent.

Please do not hesitate to contact with me from the following communication channels if you have any question or problem about the study.

Alper Sönmez
Middle East Technical University,
Department of Economics
Room No: A202, 06531 Ankara
E-Mail: salper@metu.edu.tr
Office Phone: (0312) 210 30 7.
Mobil: 0 533 xxx xx xx
Fax: (0312) 210 79 64.

Yours Sincerely,

Appendix 1: Interview Guide
Appendix 2: Letter of support from OSD
Appendix 3: Letter of support from TAYSAD

*****Thank you for your kindly cooperation and concern about the research*****

Appendix C: Letter of Support from OSD (in Turkish)

OSD 35.yıl

OTOMOTİV SANAYİİ DERNEĞİ
AUTOMOTIVE MANUFACTURERS ASSOCIATION

Atilla Sokak, No:10, Altunizade 34676 İSTANBUL/TÜRKİYE Tel.: (0216) 3182994/6Hat Fax: (0216) 3219497
E-Mail: osd@osd.org.tr Internet: www.osd.org.tr

Sayın Doç.Dr.M.Teoman PAMUKÇU
ODTÜ TEKPOL
ANKARA

İstanbul, 23.09.2009
Ref: 2009/0648

Konu:. "Türkiye Otomotiv Sektörü'nde ArGe, Yenilik, Bilgi ve Teknoloji Transferi" Projesi Hakkında.

Orta Doğu Teknik Üniversitesi tarafından yönetiminizde "Türkiye Otomotiv Sektörü'nde ArGe, Yenilik, Bilgi ve Teknoloji Transferi" konulu bilimsel bir araştırma projesi gerçekleştirildiği memnuniyetle öğrendik.

Söz konusu araştırma, Türkiye otomotiv sektöründe faaliyet gösteren firmaların, yenilik ve ArGe faaliyetleri ile yeni teknolojileri öğrenme-uyarlama konusunda yeteneklerinin ortaya çıkarılmasını, ana sanayi ve çokuluslu yabancı firmalar ile yan sanayi arasındaki işbirliğinin niteliği ve yoğunluğunun saptanması ve incelenmesini hedeflemektedir. Bu açıdan sanayimiz için yararlı sonuçlar verecektir.

Araştırma, kapsamı ve özelliği itibariyle ana ve yan sanayi firmalarından ayrıntılı veri ve bilgi toplamaya ihtiyaç duymaktadır. Araştırmanın ülkemize sağlayacağı katkı ve Türkiye otomotiv sektörünün gelecekteki başarısı açısından önemine inanıyoruz ve Otomotiv Sanayii Derneği olarak destekliyoruz.

Söz konusu araştırmanın başarısı açısından Firmalarımızın araştırmayı gerçekleştiren proje ekibine gerekli yardımı ve desteği göstereceğinize inanıyoruz.

Şirketlere sunulacak Anket Formu'nun "Veri Gizliliği" ilkesi kapsamında olması kaydı ile Projeyi desteklediğimizi bilgilerinize sunar, çalışmalarınızda başarılar dileriz.

Saygılarımızla,
OTOMOTİV SANAYİİ DERNEĞİ

Prof.Dr. Ercan TEZER
Genel Sekreter

Yönetim Kurulu Başkanı a.

A. Sönmez, *Multinational Companies, Knowledge and Technology Transfer*, Contributions to Management Science, DOI 10.1007/978-3-319-02033-4,

Appendix D: Letter of Support from TAYSAD (in Turkish)

taysad

TAŞIT ARAÇLARI YAN SANAYICILERI DERNEĞİ ASSOCIATION OF AUTOMOTIVE PARTS & COMPONENTS MANUFACTURERS
TOSB-TAYSAD Organize Sanayi Bölgesi 41490 Şekerpınar, Gebze, Kocaeli / Turkiye Tel: +90.262. 658 98 18 Fax: +90.262. 658 98 39
www.taysad.org.tr - info@taysad.org.tr

Sayın Doç.Dr.M.Teoman PAMUKÇU 16.04.2010
ODTÜ TEKPOL
ANKARA

Konu: "Türkiye Otomotiv Sektörü'nde ArGe, Yenilik, Bilgi ve Teknoloji" Projesi Hakkında.

Orta Doğu Teknik Üniversitesi tarafından yönetiminizde "Türkiye Otomotiv Sektörü'nde ArGe, Yenilik, Bilgi ve Teknoloji Transferi" konulu bilimsel bir araştırma projesi gerçekleştirildiği memnuniyetiyle öğrendik.

Söz konusu araştırma, Türkiye otomotiv faaliyet gösteren firmaların, yenilik ve ArGe faaliyetleri ile yeni teknolojileri öğrenme-uyarlama konusunda yeteneklerinin ortaya çıkarılmasını, ana sanayi ve çokuluslu yabancı firmalar ile yan sanayi arasındaki işbirliğinin niteliği ve yoğunluğunun saptanması ve incelenmesini hedeflemektedir. Bu açıdan sanayimiz için yararlı sonuçlar verecektir

Araştırma, kapsamı ve özelliği itibariyle ana ve yan sanayi firmalarından ayrıntılı veri ve bilgi toplamaya ihtiyaç duymaktadır. Araştırmanın ülkemize sağlayacağı katkı ve Türkiye otomotiv sektörünün gelecekteki başarısı açısından önemine inanıyoruz ve Taşıt Araçları Yan Sanayicileri Derneği olarak destekliyoruz.

Söz konusu araştırmanın başarısı açısından Firmalarımızın araştırmayı gerçekleştiren proje ekibine gerekli yardımı ve desteği göstereceğinize inanıyoruz.

Şirketlere sunulacak Anket Formu'nun "Veri Gizliliğe" ilkesi kapsamında olması kaydı ile Projeyi desteklediğimizi bilgilerinize sunar, çalışmalarınızda başarılar dileriz.

Saygılarımızla,
Taşıt Araçları Yan Sanayicileri Derneği

Özlem Gülşen
Genel Sekreter

CLEPA
European Association of Automotive Suppliers
Avrupa Otomotiv Yan Sanayicileri Derneği

A. Sönmez, *Multinational Companies, Knowledge and Technology Transfer*, Contributions to Management Science, DOI 10.1007/978-3-319-02033-4,

Appendix E: Fieldwork Instructions for the Questionnaire Survey

1. Questionnaire survey will be applied to 298 automotive supply firms which are primarily the members of TAYSAD and operate in various industrial zones (BTSO, UIB, NOSAB, DOSAB and GOSB) in Istanbul, Izmir, Bursa and Kocaeli.
2. Addresses and contact information of the firms were acquired primarily from TAYSAD and various communication channels in terms of sector such as web pages of industrial zones, web pages of firms, and web pages of Istanbul Chamber of Industry.
3. The questionnaire will be applied by means of face-to-face interview with the top-executives of the firms.
4. Top-executives (respondents of the questionnaire) may be senior managers such as general director, production manager, factory manager, R&D manager, sales or marketing manager, deputy managers, product development manager, and owner of the factory.
5. Firms should be accessed by means of required communication devices in order to obtain the name/e-mail addresses of the top-executives specified in article 4, provide information to the person about the importance and details of the survey and request appointment.
6. In the event that the lists of the firms obtained is not up-to-date, in phone calls,

   (a) It should be verified that the main operational field of the firm is automotive industry. If the main operational field of the firm is another one (service industry, food industry, electrical and electronic industry, white appliances industry, defense etc.), the questionnaire should not be applied.
   (b) If the firm has withdrawn from the manufacturing or it has ceased the manufacturing for a period of time, the questionnaire should not be applied.
   (c) Name, address and contact information of the firms to be surveyed should be verified in order to avoid problems.

7. Top-executives should be informed regarding the objective, scope and method of the survey within the framework of the information included in the cover letter (see Appendix A) and appointment should be requested by contacting

A. Sönmez, *Multinational Companies, Knowledge and Technology Transfer*, Contributions to Management Science, DOI 10.1007/978-3-319-02033-4,

them via phone. Following phone calls, cover letter organized in her/his own name should be sent to the e-mail address together with the support letters of OSD (see Appendix C) and TAYSAD (see Appendix D) regarding the survey.

8. Appointments taken from the firms must be joined in time. Making appointment for more than one firm in the same day should not prevent the pollster arriving in time for the appointments. If it's not possible to the place of an appointment in time due to various reasons, respondent must be informed about the situation beforehand and if necessary new appointment should be taken.
9. It should be noted that the questionnaire survey is a scientific research; any information regarding the participant firms will not be explained or released.
10. Respondents may give their business cards if they want.
11. Full name of the pollster and interview date should be specified on each questionnaire form.
12. Success of the questionnaire survey is dependent on the accurate and complete response for each question. Therefore, pollsters should attach utmost importance to the survey. If there is any missed point or a situation that needs explanation or if there are any problems during the interview and so on, **Alper Sönmez** must be contacted on the telephone via the phone number **0533 XXX XX XX**.

Appendix F: Structure of the Questionnaire Form

A. **Questions for Supplier Firm Characteristics: Numerical Indicators**

- Establishment date (age of the firm)
- Size of the firm (total sales, total number of workers)
- Distribution of total sales according to main customers/buyers (used also in supply chain)
- Share of exports in total sales (export or domestic oriented firm, export intensity)
- Major export markets
- Related to products:
  - Major three products manufactured/assembled and their share in total sales (technological complexity level of the products and firm)
  - Major sources of raw materials (suppliers of the firm)
- Ownership status (share of foreign capital)
- Origin of foreign partner or firm

B. **Knowledge and Technology Transfer Questions (explicit and tacit)**

- Types and degrees of KTTs related to product
- Types and degrees of KTTs related to production process
- Types and degrees of KTTs related to training (type of training activities provided, on the job or off the job training, visual training)
- Types and degrees of KTTs related to financial assistances
- The reasons for not making sufficient amount of KTTs

C. **Questions for Supplier Firm Capabilities: Technological Capabilities related to Design, R&D, Innovation and Production**

- Rate of skilled workers to total workers (share of engineers/white collar workers in total employment)
- Share of exports in total sales (used also in B)

A. Sönmez, *Multinational Companies, Knowledge and Technology Transfer*, Contributions to Management Science, DOI 10.1007/978-3-319-02033-4,

- Technology level of the products (used also in B)
- R&D and innovation activities (R&D department, total R&D expenditures)
- Quality certifications
- Number of patents
- Technology agreements
- Sources of technologies used or acquired
- Design capability of the firm
- Various indicators (such as design capability, quality control, maintenance, and JIT)
- Co-product design activities and partners in these activities

D. **Strategic Cooperation Questions**

- Intensity of relationship among main buyers, suppliers, various firms and institutions
- Main reasons for collaborations with external parties
- Length of relationship with most important customer (year)
- Subcontracting agreements (share of subcontracting agreements in total contracts)

E. **Questions for Direct Suppliers**

- How to be a direct supplier of AMMs
- Benefits of being a direct supplier of AMMs
- Strategic relationships with AMMs (governance meetings, design-production process- production development activities, send its expert staff, KTTs, etc.)

F. **Questions in terms of R&D and Innovation Activities**

- R&D and innovation activities and partners in these activities
- Have a separated R&D department in firm
- Support from government or various institutions related to R&D
- Number of patents obtained and applied
- R&D Expenditures (TL) and/or share of R&D expenditures in total sales

G. **Questions related to Performance of Supplier Firm (for last three years)**

- Related to production capability (defect rate, average cost, delivery and cycle time)
- Related to supports of customers (design, productivity, export, quality control etc.)

H. **Questions Related to Supply Chain**

- Main customers (buyers) and suppliers of the firm (used also in B)
- Major export markets (used also in B)
- A subsidiary (or affiliate) firm dependent on any domestic/overseas firm/group/holding?

- A part of a MNC
- MNCs among customers
- Purchasing strategy of the most important customer (buyer)

I. **Questions for Future Projections (Related to Technology Acquisition and Firm Activities)**

Appendix G: Questionnaire Form Applied to Suppliers in Turkey Automotive Industry

Section 1: General Information

Questionnaire Number:	
Form Number:	

1. **Please indicate your firm's establishment date.**

Year of Establishment	______

2. **Please indicate your firm's average number of employees in 2007 and 2008 according to the following table.**

	2007	2008
1- Engineers	People	People
2- White-Collar	________People	________People
3- Blue-Collar	________People	________People
Total Number of Employees (1+2+3)	________People	________People

3. **Please indicate approximately your firm's total sales (turnover) in 2008.**

2008	**T.L________________Million**

A. Sönmez, *Multinational Companies, Knowledge and Technology Transfer*, Contributions to Management Science, DOI 10.1007/978-3-319-02033-4,

4. **Please indicate average distribution (%) of your firm's total sales according to the following categories (main customers of your firm).**

1. Domestic automotive main manufacturers (DAMM)	%
2. Overseas automotive main manufacturers (OAMM)	%
3. Domestic suppliers (DS)	%
4. Overseas suppliers (OS)	%
5. Parent company	%
6. Other affiliates of parent company	%
7. Retailers	%
Others, (please specify): ------------------	%
Total	100 %

5. **Please indicate the share of exports (%) in your firm's total sales for 2008.**

Share of Exports in Total Sales in 2008	______%

6. **If there is a foreign capital in your firm, please indicate the share (%) and origin of this capital (for example; Germany, Italy, France, Spain, Japan, England, USA, South Korea etc.).**

		Origin*
Share of Foreign Capital	______%	______
Completely Foreign Ownership	[ ]	______

*In case of more than one foreign partner, please indicate the origin of the foreign partner that has the highest capital share in the firm

7. **Please specify the names of three major products manufactured/assembled by your firm and the approximate share (%) of these in your firms' total sales.**

Major 3 Products	Share (%) in Total Sales
1.	______%
2.	______%
3.	______%

Section 2: Knowledge and Technology Transfer (KTTs)

1. **Please indicate how often the following types of transfers related to product are provided by your customers.**
   Attention: The implied meaning of the word "**Assistance**" used in items related to product and production process transfers (following sentences) is every help such as providing technical support-training- advice-knowledge and technology transfers.

Cronbach's α = 0.785; Inter-Item Correlations: mean = 0.422, min: 0.273, max: 0.572

Types of knowledge and technology transfer related to product	Neve	Rarely	Frequently
1. Assistance related to product designs	[ ]	[ ]	[ ]
2. Joint operation related to product	[ ]	[ ]	[ ]
3. Joint design activity related to product	[ ]	[ ]	[ ]
4. Product specifications	[ ]	[ ]	[ ]
5. Technical specifications, original design or technical drawings for products	[ ]	[ ]	[ ]
Others, (please specify): ________________	[ ]	[ ]	[ ]

2. **Please indicate how often the following types of transfers related to production process are provided by your customers.**

Cronbach's α = 0.853

Types of knowledge and technology transfer related to production process	Never	Rarely	Frequently
1. Assistance for R&D activities	[ ]	[ ]	[ ]
2. Assistance for logistic management	[ ]	[ ]	[ ]
3. Provide documentations	[ ]	[ ]	[ ]
4. Know-how	[ ]	[ ]	[ ]
5. Supply of machinery, tools and equipment	[ ]	[ ]	[ ]
6. Supply of raw material	[ ]	[ ]	[ ]
7. Customer sent its expert staff to stay at your plant for a certain period of time for assistance in solving problems in the production process	[ ]	[ ]	[ ]
8. Assistance for business management	[ ]	[ ]	[ ]
9. Assistance for quality control methods	[ ]	[ ]	[ ]
10. Patent and/or license rights granted	[ ]	[ ]	[ ]
11. Assistance for design	[ ]	[ ]	[ ]
12. Assistance for productivity-related problems	[ ]	[ ]	[ ]
13. Consumer assigned its expert staff in the establishment of production process of the plant	[ ]	[ ]	[ ]
Others, (Please specify): ________________	[ ]	[ ]	[ ]

3. **Please indicate how often the following types of financial transfers are provided by your customers.**

Cronbach's α = 0.712; Inter-Item Correlations: mean = 0.389, min: 0.241, max: 0.578

Types of financial transfer	Never	Rarely	Frequently
1. Unilateral financial aid	[ ]	[ ]	[ ]
2. Loans with low interest rates	[ ]	[ ]	[ ]
3. Contribution to risk capital	[ ]	[ ]	[ ]
4. Pre-financing of machinery, equipment and tools	[ ]	[ ]	[ ]
5. Prepayment for orders before delivery	[ ]	[ ]	[ ]
Others, (Please specify): ------------------	[ ]	[ ]	[ ]

4. **If you think that enough knowledge and technology transfers have not been provided by your customers, please specify the reasons for this.**

(More than one choice can be selected)

1. Not working with automotive main manufacturers (AMMs)	[ ]
2. Reluctance of automotive main manufacturers (AMMs) in knowledge and technology transfer	[ ]
3. Reluctance of foreign firms in knowledge and technology transfer	[ ]
4. Realizing co-designer and/or our own design activities	[ ]
5. Needed technologies can be obtained by using reverse engineering methods	[ ]
6. Realizing our own R&D activities	[ ]
7. Technologies are strictly confidential	[ ]
Others, (please specify):-----------------	

Section 3: Training

1. **Please indicate how often the following types of trainings are provided by your customers to your fims' staff.**

Cronbach's α = 0.835; Inter-Item Correlations: mean = 0.627, min: 0.600, max: 0.661

Types of training	Never	Rarely	Frequently
1. Training on technologies used in production	[ ]	[ ]	[ ]
2. Training of production/operation staff (engineers, technicians etc.)	[ ]	[ ]	[ ]
3. Training of management staff	[ ]	[ ]	[ ]

2. **Please indicate <u>how often</u> the following <u>training modes</u> are provided by your customers in the training of your firms' staff.**

Cronbach's α = 0.721; Inter-Item Correlations: mean = 0.234, min: 0.069, max: 0.574		**Never**	**Rarely**	**Frequently**
A) Visits to customers' plants (training is provided as visual)		[ ]	[ ]	[ ]
B) <u>Off-the-job training</u> Seminars and courses	**1.** at customers' plants	[ ]	[ ]	[ ]
	2. at our plant	[ ]	[ ]	[ ]
	3. at other private specialized institutes	[ ]	[ ]	[ ]
<u>On-the-job training</u> Working at **<u>the costumers' plant</u>** for a period of time for taking theoretical and/or practical training related to product and production processes	In Turkey	[ ]	[ ]	[ ]
	Abroad	[ ]	[ ]	[ ]
<u>On-the-job training</u> Taking theoretical and/or practical training **<u>in-house</u>** related to product and production processes through expert staff assigned by customers for a certain period of time		[ ]	[ ]	[ ]

Section 4: Market and Competition Structure

1. **Please rank <u>your firm's major markets</u> in order of importance,** *(1>2>3>4>5).*

Domestic Market	[ ]
European Union:_________	[ ]
USA	[ ]
Middle East	[ ]
Asia	[ ]
Africa	[ ]
Others, (please specify):____________________	

2. **Please indicate the location of your firm's <u>main competitors</u>.**

	(Please tick as appropriate)
In Turkey	[ ]
Abroad	[ ]
Both	[ ]

3. **Please indicate how much you either agree or disagree with each of the following statements related to your firm's business environment.**

***0**-No response, **1**–Strongly disagree, **2**-Disagree*
***3**-Neither agree nor disagree, **4**-Agree, **5**-Strongly agree*

1. Domestic competition is intense	0	1	2	3	4	5
2. Global competition is intense	0	1	2	3	4	5
3. Relationships between customers and suppliers are strong	0	1	2	3	4	5

Section 5: Technology Capabilities related to Production and Design

1. **Please indicate your firm's capability level on a five-point scale ranging from 1(low) to 5(high) for each category when compared to other rival or leader companies in the sector.**

Cronbach's α = 0.751 *0- No response, 1-Low <------------------> 5- High*

1. R&D capability	0	1	2	3	4	5
2. Reach to lower prices	0	1	2	3	4	5
3. Expertness on CAD-CAM-CAE*	0	1	2	3	4	5
4. Co-designer capability	0	1	2	3	4	5
5. Quality control capability	0	1	2	3	4	5
6. Testing and analytical capability	0	1	2	3	4	5
7. Design capability	0	1	2	3	4	5
8. Product improvement capability	0	1	2	3	4	5
9. Automation level in production process	0	1	2	3	4	5
10. On-time delivery	0	1	2	3	4	5
Others, (please specify): ------------------	0	1	2	3	4	5

***CAD:** Computer Aided Design, **CAM:** Computer Aided Manufacturing, **CAE:** Computer Aided Engineering

2. **Please indicate international quality certifications obtained by your firm.**

(More than one choice can be selected)

ISO/TS 16949	[ ]
QS 9000	[ ]
ISO 9001	[ ]
ISO 9002	[ ]
ISO 14000	[ ]
ISO 14001	[ ]
ISO 18000	[ ]
Specific Certificates of Automotive Main Manufacturers (AMMs)	[ ]
Others, (please specify): --------------	

3. **Please indicate what kinds of <u>technology agreements</u> your firm undertakes.**

Cronbach's α = 0.726 *(More than one choice can be selected)*

1. Turn-key projects	[ ]
2. Management contracts	[ ]
3. Licensing agreement	[ ]
4. Know-how agreement	[ ]
5. Purchasing of engineering services	[ ]
6. Joint venture agreement	[ ]
7. Purchasing of license and/or patent	[ ]
8. Agreements on personnel exchange	[ ]
9. Technical assistance agreement	[ ]
10. International subcontracting agreements	[ ]
11. Agreements on the determination of product design and/or technical specifications	[ ]
Others, (please specify):_________________	

4. **Please indicate <u>the sources of technologies</u> acquired and/or used by your firm.**

Cronbach's α = 0.779 *(More than one choice can be selected)*

1. Domestic automotive main manufacturers (DAMM)	[ ]
2. Overseas automotive main manufacturers (OAMM)	[ ]
3. Domestic suppliers (DS)	[ ]
4. Overseas suppliers (OS)	[ ]
5. Parent company	[ ]
6. Universities	[ ]
7. Private engineering and consultancy firms	[ ]
Others, (please specify):_________________	

5. **Please indicate which of the following statements <u>describes</u> your firm's <u>design capability.</u>**

(More than one choice can be selected)

1. All technical specifications, design and quality standards of products produced are determined by customers	[ ]
2. Although basic designs are determined by customers, we can add details and/or make joint designing with customer (co-designer capability)	[ ]
3. Although our firm makes **<u>all</u>** or **<u>most</u>** of the designing, customer approval is necessary for final designs	[ ]
4. Our firm is **<u>entirely responsible</u>** for all stages of product design	[ ]

6A. **Does your firm carry out co- design activities related to products with customers?**

Yes	[ ]	No	[ ]	→ Go to section 6

6B. **If your firm carries out co-design activities, please indicate your partners in these activities.**

Cronbach's α = 0.730 *(More than one choice can be selected)*

1. Domestic automotive main manufacturers (DAMM)	[ ]
2. Overseas automotive main manufacturers (OAMM)	[ ]
3. Domestic suppliers (DS)	[ ]
4. Overseas suppliers (OS)	[ ]
5. Parent company	[ ]
6. Universities	[ ]
7. Private engineering and consultancy firms	[ ]
Others, (please specify):_______________	

7. **If your firm carries out co-product design activities, are you included to this process from the beginning?**

Yes	[ ]	No	[ ]

Section 6: Supply Chain

1. **Please answer the following four questions as Yes or No**

	YES	NO
1. Is your firm a subsidiary (or affiliate) firm dependent on any domestic firm/group/holding?	[ ]	[ ]
2. Is your firm a subsidiary (or affiliate) firm dependent on any overseas firm/group/holding?	[ ]	[ ]
3. Is your firm part of a multinational company?	[ ]	[ ]
4. Are there MNCs among your customers?	[ ]	[ ]

2. **Please rank primary source(s) of inputs (raw materials and/or intermediate goods etc.) used in the production of major products in order of importance,** ***(1>2>3>4>5)***

Domestic automotive main manufacturers (DAMM)	[ ]
Overseas automotive main manufacturers (OAMM)	[ ]
Domestic suppliers (DS)	[ ]
Overseas suppliers (OS)	[ ]
Parent company	[ ]
Other affiliates of parent company	[ ]
Others, (please specify):_________________	

3. **Please indicate the purchasing strategy of your most important customer.**

(Please choose only one option)

Customer prefers to study with **only one supplier** per item purchased	[ ]
Customer prefers to study with **many suppliers** per item purchased	[ ]

4. **Please indicate approximately how many years you have been working together with your customer which you have most-intense cooperation with.**

Duration of working	______ **Years**

5. **Please indicate the share (%) of subcontractor contracts in your total contracts.**

The Share of Subcontracting Contracts in Total Contracts	______ **%**

Section 7: Cooperation

1. **Please indicate the degree of importance of the following factors for carrying out cooperation activities with other firms.**

0- No response, 1– Very unimportant, 2- Unimportant
3- Neither important nor unimportant, 4- Important, 5- Very important

Cronbach's α = 0.925; Inter-Item Correlations: mean = 0.580, min: 0.453, max: 0.819

1. Carrying out R&D activities	0	1	2	3	4	5
2. Know-how transfer	0	1	2	3	4	5
3. Reducing/sharing production costs/risks	0	1	2	3	4	5
4. Replacing technologically phased out products with the new ones	0	1	2	3	4	5
5. Establishing long-term strategic partnership	0	1	2	3	4	5
6. Improving product quality	0	1	2	3	4	5
7. Opening up global markets	0	1	2	3	4	5
8. Entering new technology fields	0	1	2	3	4	5
9. Learning about new technologies	0	1	2	3	4	5
Others, (please specify):_________________	0	1	2	3	4	5

2. **Please indicate the degree of improvement and/or development in the following areas as a result of customer assistances (support, advice, training, knowledge or technology transfers etc.) during the last 3 years**

0- No response, 1–Very decreased, 2- Decreased
3- Neither increased nor decreased, 4- Increased, 5- Very increased

Cronbach's α = 0.929	During the last 3 years					
1. Productivity	0	1	2	3	4	5
2. Profitability	0	1	2	3	4	5
3. Design capability (design new products and processes)	0	1	2	3	4	5
4. Engineering capability related to product	0	1	2	3	4	5
5. Improve production process and/or capability to find solutions for production problems	0	1	2	3	4	5
6. Improvement in quality control methods	0	1	2	3	4	5
7. Improvement in business management	0	1	2	3	4	5
8. Performance on export	0	1	2	3	4	5
9. Reach to lower output prices	0	1	2	3	4	5
10. Learning about new technologies and production processes	0	1	2	3	4	5
Others, (please specify):_________________	0	1	2	3	4	5

3. **Is your firm a direct supplier of one or more automotive main manufacturer (AMM) in Turkey?**

Yes	[ ]	
No	[ ]	→ Go to section 8

If you answered **Yes** to previous question, please carry on;
If you answered **No**, please **go to section 8**

4. **Please indicate how your firm became a direct supplier of automotive main manufacturer(s).**

(More than one choice can be selected)

1. Our firm attempted and asked for order	[ ]
2. Through advices and introduces of our other customers	[ ]
3. Automotive main manufacturer (AMM) attempted and offered business plan	[ ]
4. Our firm applied exhaustive and long-term program to work with automotive main manufacturers (AMMs)	[ ]
Others, (please specify):_________________	

5. **Please indicate how important you evaluate the following benefits of being a direct supplier of automotive main manufacturer(s) based on your experiences in this area.**

0- No response, 1– Very unimportant, 2- Unimportant
Cronbach's α = 0.951 *3- Neither important nor unimportant, 4- Important, 5- Very important*

1. Developing new business relationships with other firms	0	1	2	3	4	5
2. Good reputation and familiarness in sector	0	1	2	3	4	5
3. Stable trade	0	1	2	3	4	5
4. Learning/improving new production processes	0	1	2	3	4	5
5. Learning/improving new quality control methods	0	1	2	3	4	5
6. Better understand the customers' blueprint and technical specifications	0	1	2	3	4	5
7. Improvement in testing and analyzing capabilities/techniques	0	1	2	3	4	5
8. Attendance to early stage in the design activities and developing prototype	0	1	2	3	4	5
9. Increase in design capability	0	1	2	3	4	5
10. Access to international developed markets	0	1	2	3	4	5
11. Improving in ability and knowledge about identifying and finding solutions to problems encountered in the production	0	1	2	3	4	5
12. Increase in productivity	0	1	2	3	4	5
13. Reducing risks involved with making decision to invest in new technology and/or machines	0	1	2	3	4	5
14. Learning about new technologies	0	1	2	3	4	5
Others, (please specify): _________________	0	1	2	3	4	5

6. **Please answer the following seven questions as Yes or No**

AMM=Automotive Main Manufacturer	YES	NO
1- Does AMM sent its expert staff to your firm for giving support or assistance about various topics?	[ ]	[ ]
2- Does AMM arrange regular management and/or governance meetings?	[ ]	[ ]
3. Did AMM give any support in the establishment process of your firm? (Cooperation in capacity and production planning/ in purchasing machines/ in technical organization etc.)	[ ]	[ ]
4- Does AMM make regular visits to evaluate your firm about quality, cost and on-time delivery etc.?	[ ]	[ ]
5- Does AMM provide regular training activities for your staff?	[ ]	[ ]
6- Does AMM transfer any knowledge or technology related to product/production process/design?	[ ]	[ ]
7- Does your firm have to provide detailed information to AMM(s) about production and/or quality control processes?	[ ]	[ ]

Section 8: R&D and Innovation Activities

1. **Please answer the following six questions as Yes or No**

	Yes	No
1. Does your firm carry out R&D activities?	[ ]	[ ]
2. Is there a separated R&D department in your firm?	[ ]	[ ]
3. If your firm is a direct supplier of automotive main manufacturer, are joint R&D activities being carried out with these manufacturer(s)?	[ ]	[ ]
4. Are joint R&D activities being carried out with other customers?	[ ]	[ ]
5. Do your customers help or support your firm related to R&D and design?	[ ]	[ ]
6. Does your firm get any support from government or various institutions related to R&D (KOSGEB – TÜBİTAK – TEYDEB – TTGV - EU etc.)?	[ ]	[ ]

2. **If your firm carries out R&D activities, please indicate R&D expenditures (TL) and/or share of R&D expenditures in total sales (%) for 2008**

	R&D Expenditures (TL)	and/Or	Share of R&D expenditures in total sales
2008	TL______________		________%

3. **If your firm carries out R&D activities, please indicate who started these activities.**

(More than one choice can be selected)

1. Domestic automotive main manufacturers that we are the supplier of	[ ]
2. Overseas automotive main manufacturers that we are the supplier of	[ ]
3. Foreign partner	[ ]
4. Parent company	[ ]
5. Own choice	[ ]
Others, (please specify): ________________	

4A. **Does your firm undertake innovation activities?**

Yes	[ ]	**No**	[ ]	→	**Go to question 5**

4B. **Please indicate how intensely you cooperate with the following external parties in your innovation activities.**

Degree of cooperation

0- No response, 1-No cooperation, 2- Little, 3– Average, 4-Intense, 5- Very intense

Cronbach's $\alpha = 0.718$; Inter-Item Correlations: mean = 0.336, min: 0.110, max: 0.572

1. Domestic automotive main manufacturers (DAMM)	0	1	2	3	4	5
2. Overseas automotive main manufacturers (OAMM)	0	1	2	3	4	5
3. Domestic suppliers (DS)	0	1	2	3	4	5
4. Overseas suppliers (OS)	0	1	2	3	4	5
5. Parent company	0	1	2	3	4	5
6. R&D institutions	0	1	2	3	4	5
7. Universities	0	1	2	3	4	5
8. Private engineering and consultancy firms	0	1	2	3	4	5
Others, (please specify): ________________	0	1	2	3	4	5

5. **Please indicate the number of patents applied and obtained by your firm.**

1. Number of patents applied	______
2. Number of patents obtained	______

6. **Has your firm developed and/or introduced following innovation activities (TPP innovation activities*) into the market during the last 3-years?**

	Yes	No
1. New or significantly improved methods for production process(es)	[ ]	[ ]
2. New or significantly improved product(s)	[ ]	[ ]
3. New or significantly improved service(s)	[ ]	[ ]
4. New or significantly improved logistics, delivery, and distribution methods	[ ]	[ ]
5. New or significantly improved supporting activities related to processes (maintenance systems, and operations for purchasing/accounting/computing)	[ ]	[ ]

*

"**Technological product and process (TPP) innovation activities** are all those scientific, technological, organizational, financial and commercial steps which actually, or are intended to, lead to the implementation of technologically new or improved products or processes.
A technologically new product is a product whose technological characteristics or intended uses differ significantly from those of previously produced products.
A technologically improved product is an existing product whose performance has been significantly enhanced or upgraded.
Technological process innovation is the adoption of technologically new or significantly improved production methods, including methods of product delivery" (see OECD, 1997; Oslo Manual: 32-39).

Section 9: Performance and Future Projections

1. **Please indicate how your firm's production capability has been improved in the following areas during the last 3-years.**

0- No response, 1– Very increased, 2- Increased
3- Neither increased nor decreased, 4- Decreased, 5– Very decreased
Cronbach's α = 0.745; Inter-Item Correlations: mean = 0.379, min: 0.319, max: 0.508

	For the last 3 years					
1. Cycle time*	0	1	2	3	4	5
2. Defect rate	0	1	2	3	4	5
3. Average costs	0	1	2	3	4	5
4. Duration of on-time delivery	0	1	2	3	4	5

* Cycle time: Period required to complete one cycle of an operation; or to complete a function, job, or task from start to finish. (from businessdictionary.com)

2. **Please indicate <u>the effectiveness degree</u> of the following factors in the improvement of your firm's <u>production capability</u> and/or in developing <u>new technologies</u> related to product/process in recent years.**

***0**- No response, **1**– Very ineffective, **2**- Ineffective*
Cronbach's α = 0.795 *3- Neither effective nor ineffective, **4**- Effective, **5**- Very effective*

1. Being a direct supplier of automotive main manufacturer (AMM)	0	1	2	3	4	5
2. Knowledge and technology transfers from domestic automotive main manufacturers (DAMM)	0	1	2	3	4	5
3. Knowledge and technology transfers from foreign manufacturers	0	1	2	3	4	5
4. Undertaking R&D activities	0	1	2	3	4	5
5. Hiring skilled-specialist employees from other firms	0	1	2	3	4	5
6. Skill and capability improvement of employees through in-house training	0	1	2	3	4	5
7. Improvements in our suppliers' quality level	0	1	2	3	4	5
8. Cooperation with universities and other institutions* in Turkey	0	1	2	3	4	5
9. Using of new production methods	0	1	2	3	4	5
10. Acquisition of new software, hardware, equipment and tools	0	1	2	3	4	5
Others, (Please specify): _______________	0	1	2	3	4	5

***Other institutions**: Various research and consultancy institutions/firms, private or public research laboratories etc

3. **Please indicate your firm's <u>future projections</u> related to <u>acquisition of new technologies.</u>**

(More than one choice can be selected)

1. Undertake R&D activities and/or Improve existing R&D activities	[ ]
2. Make joint R&D activities with automotive main manufacturers (AMMs)	[ ]
3. Hiring expert staff from automotive manufacturer(s)	[ ]
4. Sign agreements to provide skill and knowledge transfer	[ ]
5. Analyze rival products and processes	[ ]
6. Get training	[ ]
7. Get consultancy from Universities	[ ]
8. Get consultancy from domestic	[ ]
9. Get consultancy from abroad	[ ]
10. Bring experts from abroad	[ ]
Others, (please specify): _______________	[ ]

Note: If you want to share your other comments and proposals related to research through face-to-face interview with research team in the appropriate date, please put a mark ✓ in the following box

We will gladly send you a copy of the final report
*****Thank you for invaluable information, your kindly cooperation and concern about the research *****

Information on Firm and Respondent,
*****If you prefer, you can give your business card**

Name of Firm:	
Name of Respondent(s) and their position(s):	
Telephone/Fax Number:	
E-mail-WEB Adress	

Name of Interviewer (obligatory)	
Date and duration of interview	__/__/___ ______Min

Appendix H: List of the Domestic Automotive Main Manufacturers (DAMM)

	DAMM	FS (%)	Establishment Date	Location	Licence	Types of Vehicles Manufactured	Capital (TL)	Total Capacity (2011)	Total Production (2010)	Sales (TL) (2010)
FS > 50%	Toyota	**100**	1994	Sakarya	Toyota	P.CAR	150M	150,000	83,286	2,186M
	Honda Türkiye	**100**	1997	Kocaeli	HONDA MOTOR EUROPE	P.CAR	180M	50,000	20,305	473M
	M.A.N Türkiye	**99.9**	1966	Ankara	MAN Truck & BUS AG	BUS	65M	1,700	1,132	393M
	Mercedes-Benz Türk	**85**	1968/1985	İstanbul/Aksaray	Mercedes-Benz	Road Tractor / Truck / Bus	275M	18,500	14,480	2,069M
	Hyundai Assan	**70**	1997	Kocaeli	HYUNDAI Motor	P. Car / Pick Up / Minibus	206M	125,000	77,000	1,500M
	Oyak Renault	**51**	1971	Bursa	Renault	P.CAR	323M	360,000	307,083	6,450M
29% < FS < 42%	Ford Otosan	**41**	1983/2001	Eskişehir/Kocaeli	FORD	Truck / Pick Up / Minibus	351M	330,000	242,070	5,753M
	Tofaş	**37.8**	1971	Bursa	FIAT	P. Car / Pick-Up	500M	400,000	312,245	5,409M
	Türk Traktör	**37.5**	1954	Ankara	-	Farm Tractor	53M	35,000	28,277	984M
	Anadolu Isuzu	**29.74**	1966	Kocaeli	ISUZU	Truck / Pick Up / Midibus / Minibus	25M	13,155	3,292	226M
FS = 0%	Otokar	**0**	1963	Sakarya	Land Rover / Fruehauf	Pick Up / Bus / Minibus / Midibus	24M	6,700	2,236	516M
	B.M.C	**0**	1966	İzmir	-	Truck / Pick Up / Midibus / Minibus / Bus	500M	20,000	3,342	681M
	Karsan	**0**	1966	Bursa	Karsan / Peugeot / Hyundai Motor / Renault Trucks / Breda Menarini Bus	Truck / Pick Up / Minibus / Midibus / Bus	196M	95,050	24,719	514M
	Temsa Global	**0**	1987/2008	Adana/Sakarya	Temsa / Mitsubishi / Fuso Truck & Bus Corp.	Pick Up / Bus / Midibus	210M	10,750	3,367	377M
	Hattat	**0**	2002	Tekirdağ	Valtra, Universal, Hattat	Farm Tractor	40M	10,500	2,148	107M
						Total	**3,098M**	**1,626,355**	**1,124,982**	**27,638M**

Source: Constructed by Author depend on OSD data; Legend: FS (Foreign Share); P.CAR (Personal Car)

: Interviewed DAMM

Contributions to Management Science, DOI 10.1007/978-3-319-02033-4,

Appendix I: Semi-Structured Interview Guide used in Interviews with the AMMs in Turkey

MIDDLE EAST TECHNICAL UNIVERSITY

Department of Economics & Science and Technology Policies Research Center

Knowledge and Technology Transfer Research in Turkey Automotive Industry

A. **General Information**

- Please indicate your company's ownership structure.
  - Is your company a part of a MNC or network?

    If yes, please state headquarters and other locations of these?
    What is the role and importance of your company within this MNC or network?
- What are your company's main products?
- What are your major markets? (domestic, country, region)
- What are the major sources of inputs (raw material, components etc.) used in production?

B. **Technology Activities**

- What are the most critical technologies used in production?
- What are the sources of these technologies (explicit or tacit) used in production?
- What kind of innovation and technology activities does your company carry out in order to improve or change your products- production process – organization - business management or to develop new ones? (R&D, patent, design, design verification, test and innovation activities etc.). Please specify.

C. **Strategic Collaboration**

- Who are your company's strategic partners? Where are they located?

A. Sönmez, *Multinational Companies, Knowledge and Technology Transfer*, Contributions to Management Science, DOI 10.1007/978-3-319-02033-4,

- What kinds of collaboration exist with these partners? (R&D, production, production process, design etc.)
- Do you carry out innovation and technological activities related to product/ production process/organizational improvement or development? If yes, please specify.
- Are any sorts of knowledge or technology related to product - production process - business management transferred by your partners? If yes, please specify how and in what way.

D. **Relationship with Suppliers**

- How many direct suppliers does your company have?
  - Who are they?
  - Where are they located? (domestic, abroad)
  - Is it important for your company that suppliers are close by?
- How are these suppliers selected by your firm?
- What are the most important criteria to be supplier of your firm?
- What kinds of collaborations are carried out with these selected suppliers? What kind of benefits and problems do you experience while collaborating? Please specify.
- What are the most important expectations and demands of your company from these suppliers? (Quality control, production, JIT etc.)
- Do you supervise and follow these suppliers regularly?
- Do you have any evaluation and reward system?
- Does your company transfer any kind of knowledge and technology to these suppliers related to product / production process / business management? If yes, please specify.
- Does your company carry out training programs or activities to these suppliers? If yes, please specify.
- Do you send your own expert staff for a certain period of time to suppliers to help various issues? If yes, please specify.
- Does your company provide any other support to these suppliers? If, yes please specify.

E. **Views, Future Projections and Government Policies related to Industry**

- What do you think about the future of Automotive Industry and your company?
  - What are the challenges and opportunities related to sector?
  - How will Turkey Automotive Industry be affected from these?
- What do you think about the local suppliers' capabilities and competitive levels in Turkey?

  - What are their strengths and weaknesses compared to China, South Korea, Malaysia, Thailand, Germany, France and Italy?
  - What kind of suggestions do you have to increase their capabilities and competitiveness?

- What are the future strategies of your company about technology and production in terms of global competition?
- What do you think about the MNCs role in the development of Turkey automotive industry?
- What do you think about the government role in the development and upgrading of local suppliers' technology capabilities? What kinds of policies are applied by government?
- Do you have any suggestions

  - to enhance KTTs from MNCs to local supplier firms?
  - to increase technological capabilities of local supplier firms?

Appendix J: Summary Statistics for Key Indicators in the Survey according to Firm Descriptions

	Local Firms (LF)					Foreign Firms (FF)				
	Obs.	**Mean**	**St. Dev.**	**Min**	**Max**	**Obs.**	**Mean**	**St. Dev.**	**Min**	**Max**
Age (year)	120	28.84	13.78	4	73	45	18.56	13.85	2	53
Turnover (TL)	105	49272931	68056943	800000	488000000	38	97316793	112000000	4000000	400000000
Employment	119	255.33	280.48	15	2275	42	404.36	551.20	16	3011
Engineer	119	16.38	21.25	1	170	42	27.50	26.93	2	106
White-Collar	119	36.92	33.84	2	195	42	67.24	103.87	3	642
Blue-Collar	118	204.01	245.70	5	2100	41	318.27	501.69	10	2774
Foreign Share (%)	120	0	0	0	0	45	76.00	29.10	0	100
Export Intensity (%)	120	34.53	27.50	0	100	45	44.00	33.20	0	100
R&D Expenditures (TL)	100	913846	3628707	0	36000000	33	2363398	4332883	0	20000000
R&D Intensity (%)	104	2.55	4.83	0	31	35	2.91	3.50	0	15
NPAT	114	3.75	12.59	0	122	40	1.62	4.12	0	22
NOPAT	114	1.56	4.63	0	42	40	0.82	2.02	0	10
NAPAT	112	2.23	8.15	0	80	39	0.82	2.22	0	12
Worktogether (year)	118	18.12	10.29	2	55	45	15.80	11.36	1	55
Subcontracting Agr. (%)	113	13.17	27.55	0	100	42	9.00	24.00	0	100

A. Sönmez, *Multinational Companies, Knowledge and Technology Transfer*,
Contributions to Management Science, DOI 10.1007/978-3-319-02033-4,

	All Firms (AF)					Direct Supplier Firms (DSF)				
	Obs.	Mean	St. Dev.	Min	Max	Obs.	Mean	St. Dev.	Min	Max
Age (year)	165	26.04	14.50	2	73	132	26.38	14.31	2	73
Turnover (TL)	143	62039832	84353133	800000	488000000	114	66568299	82750000	800000	400000000
Employment	161	294.20	374.41	15	3011	129	325.08	403.91	16	3011
Engineer	161	19.42	23.39	1	170	130	22.24	25.06	1	170
White-Collar	161	44.70	61.16	2	642	129	49.05	65.42	2	642
Blue-Collar	159	233.47	333.07	5	2774	127	258.24	361.75	10	2774
Foreign Share (%)	165	20.68	37.07	0	100	132	22.00	38.00	0	100
Export Intensity (%)	165	36.83	29.41	0	100	133	35.00	28.30	0	100
R&D Expenditures (TL)	133	1323629	3865176	0	36000000	107	1540915	4263426	0	36000000
R&D Intensity (%)	139	2.64	4.52	0	31	111	2.77	4.87	0	31
NPAT	154	3.20	11.05	0	122	125	3.58	12.12	0	122
NOPAT	154	1.37	4.12	0	42	125	1.51	4.48	0	42
NAPAT	151	1.87	7.13	0	80	123	2.10	7.82	0	80
Worktogether (year)	163	17.48	10.61	1	55	131	18.59	10.85	1	55
Subcontracting Agr. (%)	155	12.16	26.61	0	100	125	12.00	27.30	0	100

Source: Author calculations based on survey results
Legend: NPAT (Number of Owned and Applied Patents); NOPAT (Number of Owned Patents); NAPAT (Number of Applied Patents); Worktogether (number of years worked for the most important customer); Subcontracting Agr. (subcontracting agreements or contracts)

Appendix K: Description of the Explanatory Variables used in the Econometric Analysis

Variable	Description
F1RMAGE	Natural logarithm of firm age in years
FIRMSIZE	Natural logarithm of the total employees in 2008
EXPINT	Export intensity in percentage in 2008 (exports divided by total sales)
FOREIGN	A dummy variable taking the value of 1 if the share of firm's capital owned by foreigners equals at least to 10%, 0 otherwise
DIRECTSUPP	A dummy variable taking the value of 1 if the firm is a direct supplier of at least one automotive main manufacturer (AMM) in Turkey, 0 otherwise
LOCALGROUP	A dummy variable taking the value of 1 if the firm is part of a larger group, a parent company or holding in Turkey, 0 otherwise
CLIENTMNC	A dummy variable taking the value of 1 if the firm has a multinational company (MNC) among its customers, 0 otherwise
RDDEPT	A dummy variable taking the value of 1 if the firm has a seperated R&D department, 0 otherwise
PRODUCTINNO	A dummy variable taking the value of 1 if the firm carries out product innovation activities , 0 otherwise
SENGPER	Share of engineer personnel in total number of employees in percentage
HIGHTECHPROD	A dummy variable taking the value of 1 if the firm produces high technology products, 0 otherwise
HIGHDESCAP	A dummy variable taking the value of 1 if the design capability of the firm is high, 0 otherwise
ONESUPP	A dummy variable taking the value of 1 if the firm stated that its most important customer prefers to study with only one supplier per item purchased, 0 otherwise
SSUBCONT	The share of subcontracting agreements in total contracts in percentage
RDSUB	A dummy variable taking the value of 1 if the firm gets a financial subsidy from government or various institutions related to R&D (KOSGEB – TUBITAK – TEYDEB – TTGV – EU etc.), 0 otherwise

A. Sönmez, *Multinational Companies, Knowledge and Technology Transfer*,
Contributions to Management Science, DOI 10.1007/978-3-319-02033-4,

Appendix L: Descriptive Statistics for the Explanatory Variables used in the Econometric Analysis

Variable	*Variable Type*	*Obs.*	*Mean*	*Std. Dev.*	*Min*	*Max*
FIRMAGE(ln)	Continuous	165	3.133	0.736	0	4.32
FIRMSIZE(ln)	Continuous	161	5.201	0.987	2.71	8.01
EXPINT (%)	Continuous	165	36.832	29.412	0	100
FOREIGN	Dummy	165	0.273	0.447	0	1
DIRECTSUPP	Dummy	165	0.800	0.401	0	1
LOCALGROUP	Dummy	164	0.366	0.483	0	1
CLIENTMNC	Dummy	163	0.859	0.349	0	1
RDDEPT	Dummy	165	0.600	0.491	0	1
PRODUCTINNO	Dummy	164	0.732	0.444	0	1
SENGPER (%)	Continuous	161	8.035	8.548	0	66.20
HIGHTECHPROD	Dummy	164	0.518	0.501	0	1
HIGHDESCAP	Dummy	165	0.418	0.495	0	1
ONESUPP	Dummy	162	0.414	0.494	0	1
SSUBCONT (%)	Continuous	155	12.161	26.607	0	100
RDSUB	Dummy	165	0.509	0.501	0	1

Source: Author calculations based on survey results

A. Sönmez, *Multinational Companies, Knowledge and Technology Transfer*,
Contributions to Management Science, DOI 10.1007/978-3-319-02033-4,

Appendix M: Correlation Matrix

A. Sönmez, *Multinational Companies, Knowledge and Technology Transfer*,
Contributions to Management Science, DOI 10.1007/978-3-319-02033-4,

Variables	(1)	(2)	(3)	(4)	(5)	(6)	(7)	(8)	(9)	(10)	(11)	(12)	(13)	(14)	(15)
(l) FIRMAGE	1														
(2) FIRMSIZE	0.315 ***	1													
(3) EXPINT	−0.068	0.187 **	1												
(4) FOREIGN	−0.387 ***	0.094	0.141 *	1											
(5) DIRECTSUPP	0.120	0.243 ***	−0.120	0.102	1										
(6) LOCALGROUP	0.060	0.298 ***	−0.150 *	0.026	0.223 ***	1									
(7) CLIENTMNC	0.032	0.136 *	0.118	0.127	0.066	0.051	1								
(8) RDDEPT	0.106	0.184 **	0.127	−0.000	0.056	0.176 **	0.102	1							
(9) PRODUCTINNO	0.099	0.177 **	0.116	0.033	0.153 *	0.082	0.053	0.373 ***	1						
(10) SENGPER	−0.253 ***	−0.207 ***	−0.016	0.238 ***	0.104	−0.038	0.105	0.162 **	−0.072	1					
(11) HIGHTECHPROD	−0.061	0.071	0.098	0.210 ***	0.155 **	0.044	−0.043	0.092	0.060	0.136 *	1				
(12) HIGHDESCAP	−0.026	0.018	−0.073	0.005	−0.006	0.003	0.026	0.191 **	0.091	0.067	0.055	1			
(13) ONESUPP	−0.067	0.027	−0.053	0.095	0.102	0.023	−0.055	−0.080	−0.088	0.186 **	−0.009	−0.018	1		
(14) SSUBCONT	0.047	−0.024	−0.201 **	−0.062	0.000	0.040	0.034	−0.116	−0.093	0.121	−0.003	−0.064	−0.030	1	
(15) RDSUB	0.258 ***	0.178 **	0.093	−0.079	0.176 **	0.133 *	0.101	0.436 ***	0.200 **	0.083	0.060	0.021	−0.058	0.047	1

Source: Author calculations based on survey results

Notes: ***, **, * denote significance level of a correlation coefficient that is significant at 1%, 5% and 10%, respectively (two-tailed test)

References

Aitken, B. J., Hanson, G. H., & Harrison, A. E. (1997). Spillovers, foreign investment, and export behavior. *Journal of International Economics, 43*(1–2), 103–132.

Aitken, B. J., & Harrison, A. E. (1991). *Are there spillovers from foreign direct investment? Evidence from panel data for Venezuela*. Cambridge, MA: MIT.

Aitken, B. J., & Harrison, A. E. (1997). *Do domestic firms benefit from foreign direct investment? Evidence from Panel Data*. New York, NY: Mimeo, Columbia University.

Aitken, B. J., & Harrison, A. E. (1999). Do domestic firms benefit from direct foreign investment? Evidence from Venezuela. *American Economic Review, 89*(3), 605–618.

Aitken, B. J., Harrison, A. E., & Lipsey, R. E. (1996). Wages and foreign ownership: A comparative study of Mexico, Venezuela, and the United States. *Journal of International Economics, 40*, 345–371.

Akarsoy, T. A. (2005). Otomotiv Ana Sanayii Teknolojik Durum Değerlendirme Çalışması: 2002 Haziran'ından Günümüze Kaydedilen Gelişmeler (in Turkish). Bilim Teknoloji İnovasyon Politikaları Tartışma Platformu.

Akpınar, A. (2001). *Turkey's experience with bilateral investment treaties*. Paper presented at the OECD Investment Compact Regional Roundtable, May 28–29, Dubrovnik, Croatia.

Aksoy, Y. O. (2008). *Technology spillovers through foreign direct investment in Turkish manufacturing industry*. Unpublished MSc thesis, METU, Ankara, Turkey.

Alıcı, A. A., & Ucal, Ş. M. (2003). *Foreign direct investment, exports and output growth of Turkey: Causality analysis*. Presented at the European Trade Study Group, 11–13 September, Madrid, Spain.

Almeida, R., & Fernandes, A. M. (2008). Openness and technological innovations in developing countries: Evidence from firm-level surveys. *Journal of Development Studies, 44*, 701–727.

Alvaro, L., Chanda, A., Kalemli-Ozcan, S., & Sayek, S. (2004). FDI and economic growth: The role of local financial markets. *Journal of International Economics, 64*(1), 89–112.

Amemiya, T. (1981). Qualitative response models: A survey. *Journal of Economic Literature, 19*, 1483–1533.

Anfara, V. A., & Mertz, N. T. (2006). *Theoretical frameworks in qualitative research*. Thousand Oaks, CA: Sage Publications.

Aslanoğlu, E. (2000). Spillover effects of foreign direct investments on Turkish manufacturing industry. *Journal of International Development, 12*, 1111–1130.

Audretsch, D. B. (1998). Agglomeration and the location of innovative activity. *Oxford Review of Economic Policy, 14*, 18–29.

Audretsch, D. B., & Feldman, M. (1996). Knowledge spillovers and the geography of innovation and production. *American Economic Review, 86*(3), 630–640.

A. Sönmez, *Multinational Companies, Knowledge and Technology Transfer*, Contributions to Management Science, DOI 10.1007/978-3-319-02033-4,

Ayvaz, Y., Baldemir, E., & Ürüt, S. (2006). Yabancı Sermaye Yatırımlarının Verimlilik ve Kalkınmaya Etkilerinin Ekonometrik İncelemesi (in Turkish). *Yönetim ve Ekonomi, 13*, 177–186.

Bailey, K. D. (1987). *Methods of social research* (3rd ed.). London, England: Collier Macmillan.

Banga, R. (2003). *The differential impact of Japanese and U.S. foreign direct investments on exports of Indian manufacturing* (Working Paper). Delhi, India: Delhi School of Economics.

Barrios, S., Görg, H., & Strobl, E. (2003). Explaining firms' export behavior: R&D, spillovers and the destination market. *Oxford Bulletin of Economics and Statistics, 65*(4), 475–496.

Barrios, S., & Strobl, E. (2002). Foreign direct investment and productivity spillovers: Evidence from the Spanish experience. *Weltwirtschaftliches Archiv, 138*, 459–481.

Barro, R., & Sala-i-Martin, X. (1995). *Economic growth.* Cambridge, MA: McGraw-Hill.

Barry, F., Görg, H., & Strobl, E. (2005). Foreign direct investment and wages in domestic firms in Ireland: Productivity spillovers versus labour-market crowding out. *International Journal of the Economics of Business, 12*, 67–84.

Bartlett, M. S. (1954). A note on the multiplying factors for various chi square approximations. *Journal of the Royal Statistical Society, 16*(series B), 296–298.

Basant, R. (1997). Technology strategies of large enterprises in Indian industry, some explorations. *World Development, 25*, 1683–1700.

Bedir, A. (1999). *Gelişmiş Otomotiv Sanayilerinde Ana-Yan Sanayi İlişkileri ve Türkiye'de Otomotiv Yan Sanayinin Geleceği (in Turkish)*. SPO Thesis of Expertise, No. 2495, Ankara, Turkey.

Bell, M., & Pavitt, K. (1997). Technological accumulation and industrial growth: Contrasts between developed and developing countries. In A. Daniele & J. Michie (Eds.), *Technology, globalisation and economic performance*. Cambridge, England: Cambridge University Press.

Bennett, D. (2002). *Innovative technology transfer framework linked to trade for UNIDO action.* Vienna, Austria: UNIDO.

Berger, M. (2005). *Upgrading the system of innovation in late-industrialising countries—The role of transnational corporations in Thailand's manufacturing sector*. Unpublished PhD thesis, Christian-Albrechts-Universität, Mathematisch-Naturwissenschaftlichen Fakultät, Kiel, Germany.

Bergman, M. M. (2008). *Advances in mixed methods research: Theories and applications.* London, England: Sage Publications Ltd.

Bernard, A. B., & Jones, C. I. (1996). Technology and convergence. *Economic Journal, 106*, 1037–1044.

Bernard, M., & Ravenhill, J. (1995). Beyond product cycles and flying geese: Regionalization, hierarchy, and the industrialization of east Asia. *World Politics, 47*, 187–237.

Bertinelli, L., Pamukçu, M. T., & Strobl, E. (2006). *Technology spillovers due to foreign ownership and productivity of domestic firms: Analyzing the evidence.* Selected papers from the 13th international conference of ERF, Economic Research Forum (ERF), Cairo, Egypt (pp. 187–230).

Biber, S. N. H. (2010). *Mixed methods research: Merging theory with practice.* New York, NY: Guilford Press.

Bigsten, A., Collier, P., Dercon, S., Fafchamps, M., Gauthier, B., Gunning, J.W., Isaksson, A., Oduro, A., Oostendorp, R., Pattillo, C., Soderbom, M., Teal, E., & Zeufack, A. (2002). *Do African Manufacturing Firms Learn from Exporting*? Centre for the Study of African Economies Working Paper Series, 2002-09. Oxford, England: Oxford University.

Blalock, G., & Gertler, P. (2003). *Technology from foreign direct investment and welfare gains through the supply chain* (Working Paper). Department of Applied Economics and Management, Cornell University, Ithaca, NY.

Blalock, G., & Gertler, P. (2004). *Firm capabilities and technology adoption: Evidence from foreign direct investment in Indonesia* (Working Paper). Department of Applied Economics and Management, Cornell University, Ithaca, NY.

Blomström, M. (1986). Foreign investment and productive efficiency: The case of Mexico. *Journal of Industrial Economics, 35*(1), 97–110.

Blomström, M., Globerman, S., & Kokko, A. (2001). The determinants of host country spillovers from foreign direct investment: Review and synthesis of the literature. In N. Pain (Ed.), *Inward investment, technological change and growth* (pp. 34–66). London, England: Palgrave.

Blomström, M., & Kokko, A. (1997). *How foreign investment affects host countries* (Policy Research Working Paper, No. 1745). Washington, DC: The World Bank.

Blomström, M., & Kokko, A. (1998). Multinational corporations and spillovers. *Journal of Economic Surveys, 12*(2), 247–277.

Blomström, M., & Kokko, A. (1999). *Foreign direct investment and technology transfer: A survey*. International Conference on Asian-Europe on the Eve of the 21st Century. Bangkok, Thailand: Chulalongkorn University.

Blomström, M., Lipsey, R. E., & Zejan, M. (1994). *What explains developing country growth*? (NBER Working Paper, No. 4132). Cambridge, MA: National Bureau of Economic Research.

Blomström, M., & Persson, H. (1983). Foreign investment and spillover efficiency in an underdeveloped economy: Evidence from the Mexican manufacturing industry. *World Development, 11*(6), 493–501.

Blomström, M., & Sjöholm, F. (1999). Technology transfer and spillovers: Does local participation with multinationals matter? *European Economic Review, 43*(4–6), 915–923.

Blomström, M., & Wolff, E. N. (1994). Multinational corporations and productive convergence in Mexico. In W. J. Baumol, R. R. Nelson, & E. N. Wolff (Eds.), *Convergence of productivity: Cross national studies and historical evidence*. Oxford, England: Oxford University Press.

Borensztein, E., Gregorio, J., & Lee, J.-W. (1998). How does foreign direct investment affect economic growth? *Journal of International Economics, 45*, 115–135.

Bosco, M. G. (2001). Does FDI contribute to technological spillovers and growth? A panel data analysis of Hungarian firms. *Transnational Corporations, 10*(1), 43–68.

Braga, H., & Willmore, L. (1991). Technological imports and technological effort: An analysis of their determinants in Brazilian firms. *Journal of Industrial Economics, 39*, 421–432.

Briggs, S. R., & Cheek, J. M. (1986). The role of factor analysis in the development and evaluation of personality scales. *Journal of Personality, 54*, 106–148.

Cantwell, J. (1989). *Technological innovation and multinational corporations*. Oxford, England: Blackwell.

Cantwell, J. (1993). Corporate technological specialization in international industries. In M. C. Casson & J. Creedy (Eds.), *Industrial concentration and economic inequality*. Cheltenham, England: Edward Elgar.

Carson, D., Gilmore, A., Perry, C., & Gronhaug, K. (2001). *Qualitative marketing research*. London, England: Sage Publications Ltd.

Castellani, D., & Zanfei, A. (2002). *Multinational companies and productivity spillovers: Is there a specification error*? (Working Paper). Urbino, Italy: University of Urbino.

Caves, R. E. (1971). International corporations: The industrial economics of foreign investment. *Economica, 38*(149), 1–27.

Caves, R. E. (1974). Multinational firms, competition, and productivity in host-country markets. *Economica, 41*(162), 176–193.

Caves, R. E. (1996). *Multinational enterprise and economic analysis* (2nd ed.). Cambridge, MA: Cambridge University Press.

Çelikel, T. A. (2009). *Factors affecting research and development (R&D) collaboration of multinational enterprises (MNEs) and their local partner firms: A case study of Turkish automotive industry*. Unpublished PhD thesis, Department of Contemporary Management, Işık University, Istanbul, Turkey.

Çetin, C. (2009). *Determinants of technology transfer in developing economies: The case of Turkish manufacturing industries*. Unpublished MSc thesis, METU, Ankara, Turkey.

Chuang, Y., & Lin, C. (1999). Foreign direct investment, R&D and spillover efficiency: Evidence from Taiwan's manufacturing firms. *Journal of Development Studies, 35*(4), 117–137.

Cohen, W. M. (2010). Fifty years of empirical studies of innovative activity and performance. In B. H. Hall & N. Rosenberg (Eds.), *Economics of innovation* (Vol. 1, pp. 129–210). Amsterdam, Netherlands: North Holland.

Cohen, W. M., & Levinthal, D. A. (1990). Absorptive capacity: A new perspective on learning and innovation. *Administrative Science Quarterly, 35*(1), 128–152.

Collins, H. M. (2001). Tacit knowledge, trust and the Q of sapphire. *Social Studies of Science, 31* (1), 71–85.

Corden, W. M. (1967). Protection and foreign investment. *Economic Record, 43*, 209–232.

Crespo, N., & Fontoura, M. P. (2007). Determinant factors of FDI spillovers—What do we really know. *World Development, 35*(3), 410–425.

Creswell, J. W. (2009). *Research design: Qualitative, quantitative, and mixed methods approaches*. Thousand Oaks, CA: Sage Publications.

Creswell, J. W. (2010). *Designing and conducting mixed methods research* (2nd ed.). Thousand Oaks, CA: Sage Publications.

Cronbach, L. J. (1951). Coefficient alpha and the internal structure of tests. *Psychometrika, 16*(3), 297–334.

Damijan, J. P., Majcen, B., Knell, M., & Rojec, M. (2001). *The role of FDI, absorptive capacity and trade in transferring technology to transition countries: Evidence from firm panel data for eight transition countries*. Geneva, Switzerland: UN Economic Commission for Europe.

Das, S. (1987). Externalities, and technology transfer through multinational corporations: A theoretical analysis. *Journal of International Economics, 22*, 171–182.

Dayar, E., & Pamukçu, T. (2011). *Analysis of foreign ownership, R&D and spillovers in developing countries: Evidence from Turkey*. Paper presented at 17th International conference of Economic Research Forum, 20–22 March, 2011, Antalya, Turkey.

De Negri, J. A., & Turchi, L. M. (Eds.). (2007). *Technological innovation in Brazilian and Argentine firms*. Rio de Janeiro, Brazil: Institute for Applied Economic Research.

Devellis, R. F. (2003). *Scale development: Theory and applications* (2nd ed.). Thousand Oaks, CA: Sage Publications.

Dicken, P. (2007). *Global shift: Mapping the changing contours of the world economy* (5th ed.). London, England: Sage Publications Ltd.

Djankov, S., & Hoekman, B. (1998). *Avenue of technology transfer: Foreign investment and productivity change in the Czech Republic* (CEPR Discussion Paper No. 1883). London, England: Centre for Economic Policy Research.

Djankov, S., & Hoekman, B. (2000). Foreign investment and productivity growth in Czech enterprises. *World Bank Economic Review, 14*(1), 49–64.

Dosi, G. (1988). Sources, procedures and microeconomic effects of innovation. *Journal of Economic Literature, 26*(3), 1120–1171.

Drever, E. (2003). *Using semi-structured interviews in small-scale research: A teacher's guide*. Edinburgh, Scotland: Scottish Council for Research in Education, The SCRE Centre.

Driffield, N. (2001). The impact on domestic productivity of inward investment in the UK. *Manchester School, 69*(1), 103–119.

Driffield, N., Munday, M., & Roberts, A. (2002). Foreign direct investment, transactions, linkages, and the performance of the domestic sector. *International Journal of the Economics of Business, 9*(3), 335–351.

Dunning, J. (1958). *American investment in British manufacturing*. London, England: George Allen and Unwin.

Dunning, J. (1979). Explaining patterns of international production: In defence of the eclectic theory. *Oxford Bulletin of Economics and Statistics, 41*(2), 269–295.

Dunning, J. (1993). *Multinational enterprises and the global economy*. Wokingham, England: Addison-Wesley Publishing Company.

Dutz, M., Us, M., & Yılmaz, K. (2005). *Turkey's foreign direct investment challenges: Competition, the rule of law, and EU accession*. London, England: World Bank and the Centre for Economic Policy Research.

Eden, L., Levitas, E., & Martinez, R. J. (1996). The production, transfer and spillover of technology: Comparing large and small multinationals as technology producers. *Small Business Economics, 9*, 53–66.

Ekmekçi, U. (2009). *Determinants of knowledge transfer from foreign direct investments to local supplier firms: The case of Turkish automotive industry*. Unpublished PhD thesis, Department of Management Engineering, Institute of Science and Technology, Istanbul Technical University, Istanbul, Turkey.

Ellet, W. (2007). *The case study handbook, how to read, discuss, and write persuasively cases*. Boston, MA: Harvard Business School Press.

Erdil, E., Pamukçu, T. M., Erden, Y., Göksidan, H. T., & Kepenek, E. (2011). *Türkiye ekonomisinde yabancı sermayeli firmaların Ar-Ge etkinliklerinin analizi* [Analysis of research and development (R&D) activities of foreign firms in the Turkish Economy] (in Turkish, p. 350). Türkiye Uluslararası Yatırımcılar Derneği (YASED) için hazırlanan rapor (report prepared for YASED), Istanbul, Turkey.

Erdilek, A. (1986). Turkey's new open-door policy of direct foreign investment: A critical analysis of problems and prospects. *METU Studies in Development, 13*, 171–191.

Erdilek, A. (2003). A comparative analysis of inward and outward FDI in Turkey. United Nations conference on trade and development. *Transnational Corporations, 12*(3), 79–105.

Ernst, D., & Kim, L. (2002). Global production networks, knowledge diffusion, and local capability formation. *Research Policy, 31*, 1417–1429.

Evenson, R. E., & Westphal, L. E. (1995). Technological change and technology strategy. In J. Behrman & T. N. Srinivasan (Eds.), *Handbook of development economics* (Vol. 3, pp. 2211–2299). Amsterdam, Netherlands: Elsevier Science.

Findlay, R. (1978). Relative backwardness, direct foreign investment, and the transfer of technology: A simple dynamic model. *Quarterly Journal of Economics, 92*, 1–16.

Froman, R. D. (2001). Elements to consider in planning the use of factor analysis. *Southern Online Journal of Nursing Research, 2*(5), 1–22.

Gabriel, P. P. (1967). *The international transfer of corporate skills: Manager contracts in less developed countries*. Cambridge, MA: Harvard University Press.

George, A. L., & Bennett, A. (2005). *Case studies and theory development in the social sciences*. London, England: MIT Press.

George, D., & Mallery, P. (2003). *SPSS for windows step by step: A simple guide and reference* (4th ed.). Boston, MA: Allyn and Bacon.

Germidis, D. (1977). *Transfer of technology by multinational corporations*. Paris, France: Development Centre of the OECD.

Gerring, J. (2007). *Case study research, principles and practices*. New York, NY: Cambridge University Press.

Ghosh, S. (2002). *Characteristics and determinants of technology transfer to developing countries*. Unpublished PhD thesis, Boston University, Graduate School of Arts and Sciences, Boston, MA.

Girma, S., Görg, H., & Pisu, M. (2003). *The role of exports and foreign linkages for FDI productivity spillovers* (Working Paper). Nottingham, England: University of Nottingham.

Girma, S., Greenaway, D., & Wakelin, K. (2001). Who benefits from foreign direct investment in the UK? *Scottish Journal of Political Economy, 48*(2), 119–133.

Girma, S., & Wakelin, K. (2001). *Regional underdevelopment: Is FDI the solution? A semi-parametric analysis* (GEP research paper 2001/11). Nottingham, England: University of Nottingham.

Giroud, A. (2003). *Transnational corporations, technology, and economic development: Backward linkages and knowledge transfer in South-East Asia*. Northhampton, MA: Edward Elgar.

Glass, A. J., & Saggi, K. (1998). International technology transfer and the technology gap. *Journal of Development Economics, 55*, 369–398.

Glass, A. J., & Saggi, K. (2002). Multinational firms and technology transfer. *Scandinavian Journal of Economics, 104*(4), 495–513.

Globerman, S. (1979). Foreign direct investment and 'Spillover' efficiency benefits in Canadian manufacturing industries. *Canadian Journal of Economics, 12*(1), 42–56.

Goffin, K., & Koners, U. (2011). Tacit knowledge, lessons learnt and new product development. *Journal of Product Innovation Management, 28*(2), 300–318.

Gomm, R., Hammersley, M., & Foster, P. (2000). *Case study method: key issues, key texts.* London, England: Sage Publications.

Görg, H., & Greenaway, D. (2001). *Foreign direct investment and intra-industry spillovers: A review of literature.* Globalisation and Labor Market Programme (Resarch Paper, 2001/37), Nottingham, England.

Görg, H., & Greenaway, D. (2004). Much ado about nothing? Do domestic firms really benefit from foreign direct investment? *World Bank Research Observer, 19*(2), 171–197.

Görg, H., & Strobl, E. (2001). Multinational companies and productivity spillovers: A meta-analysis. *Economic Journal, 111*(475), 723–739.

Görg, H., & Strobl, E. (2002). *Spillovers from foreign firms through worker mobility: An empirical investigation* (GEP Research Paper 02/13). Nottingham, England: University of Nottingham.

Görg, H., & Strobl, E. (2003). Multinational companies, technology spillovers and plant survival. *Scandinavian Journal of Economics, 105*(4), 581–595.

Görg, H., Strobl, E., & Walsh, F. (2003). *Why do foreign firms pay more? The role of on-the-job training* (GEP Research Paper 02/15). Nottingham, England: University of Nottingham.

Greenaway, D., Sousa, N., & Wakelin, K. (2004). Do domestic firms learn to export from multinationals? *European Journal of Political Economy, 20*(4), 1027–1043.

Greene, J. C. (2007). *Mixed methods in social inquiry (research methods for the social sciences).* San Francisco, CA: Jossey-Bass, A Wiley Imprint.

Greene, W. H. (2008). *Econometric analysis* (6th ed.). New York, NY: Prentice Hall.

Griffith, R. (1999). Using the ADR establishment level data to look at the foreign ownership and productivity in the United Kingdom. *Economic Journal, 109*, 416–441.

Grosse, R. (1996). International technology transfer in services. *Journal of International Business Studies, 27*, 781–800.

Grossman, G. M., & Helpman, E. (1991). *Innovation and growth in the global economy.* Cambridge, MA: MIT Press.

Haddad, M., & Harrison, A. (1993). Are there positive spillovers from direct foreign investment? Evidence from panel data for Morocco. *Journal of Development Economics, 42*(1), 51–74.

Hair, J. F., Jr., Babin, B., Money, A. H., & Samouel, P. (2003). *Essential of business research methods.* Hoboken, NJ: John Wiley and Sons.

Hair, J. F., Jr., Black, W. C., Babin, B. J., Anderson, R. E., & Tatham, R. L. (2006). *Multivariate data analysis* (6th ed.). Upper Saddle River, NJ: Pearson Prentice Hall.

Hamel, J., Dufour, S., & Fortin, D. (1993). *Case study methods (qualitative research methods).* Thousand Oaks, CA: Sage Publications.

Hancock, D. R., & Algozzine, B. (2006). *Doing case study research: A practical guide for beginning researchers.* New York, NY: Teachers College Press, Columbia University.

Hanson, G. H. (2001). *Should countries promote foreign direct investment?* (G-24 Discussion, Paper 9). Geneva, Switzerland: UNCTAD.

Harris, R., & Robinson, C. (2003). Foreign ownership and productivity in the United Kingdom estimates for UK manufacturing using the ARD. *Review of Industrial Organization, 22*, 207–223.

Harris, R., & Robinson, C. (2004). Productivity impacts and spillovers from foreign ownership in the United Kingdom. *National Institute Economic Review, 187*, 58–75.

Haskel, J. E., Pereira, S. C., & Slaughter, M. J. (2002). *Does inward foreign direct investment boost the productivity of domestic firms?* (NBER Working Paper, No. 8724). Cambridge, MA: National Bureau of Economic Research.

Hines, J. R. (1995). Taxes, technology transfer, and R&D by multinational firms. In M. Feldstein, R. James, J. R. Hines, & R. G. Hubbard (Eds.), *Taxing multinational corporations* (pp. 51–62). Chicago, IL: University of Chicago Press.

Hobday, M. (1995). *Innovation in East Asia: The challenge to Japan*. Cheltenham, England: Edward Elgar.
Humphrey, J., & Memedovic, O. (2003). *The global automotive industry value chain: What prospects for upgrading by developing countries. Sectoral studies series*. Vienna, Austria: UNIDO.
Hymer, S. (1960). *The international operations of national firms: A study of direct investment*. PhD dissertation, Massachusetts Institute of Technology, Cambridge, MA.
Javorcik, B. S. (2004). Does foreign direct investment increase the productivity of domestic firms? Search of spillovers through backward linkages. *American Economic Review, 94*(3), 605–627.
Kaiser, H. (1970). A second generation little jiffy. *Psychometrika, 35*, 401–415.
Kaiser, H. (1974). An index of factorial simplicity. *Psychometrika, 39*, 31–36.
Kathuria, V. (1996). *Spillover effects of technology transfer to india: An econometric study*. Bombay, India: Indira Gandhi Institute of Development Research.
Kathuria, V. (2000). Productivity spillovers from technology transfer to Indian manufacturing firms. *Journal of International Development, 12*(2), 343–369.
Katrak, H. (1989). Imported technologies and R&D in a newly industrializing country. *Journal of Development Economics, 31*, 123–139.
Katrak, H. (1990). Imports of technology and the technological effort of Indian enterprises. *World Development, 18*, 371–381.
Katrak, H. (1997). Developing countries' imports of technology, in-house technological capabilities and efforts: An analysis of the Indian experience. *Journal of Development Economics, 53*, 67–83.
Keller, W. (1994). *Human capital and technology: Characteristics and interaction*. New Haven, CT: Mimeo, Yale University.
Keller, W. (1996). Absorptive capacity: On the creation and acquisition of technology in development. *Journal of Development Economics, 49*, 199–227.
Keller, W. (2004). International technology diffusion. *Journal of Economic Literature, 42*, 752–782.
Keller, W., & Yeaple, S. (2003). *Multinational enterprises, international trade, and productivity growth: Firm-level evidence from the United States* (NBER Working Paper 9504). Cambridge, MA: National Bureau of Economic Research.
Kepenek, Y., & Yentürk, N. (2003). *Türkiye ekonomisi (in Turkish)* (14th ed.). Istanbul, Turkey: Remzi Kitabevi.
Kim, L. (1997). *Imitation to innovation: The dynamics of Korea's technological learning*. Boston, MA: Harvard Business School Press.
Kim, L. (1998). Technology policies and strategies for developing countries: Lessons from the Korean experience. *Technology Analysis and Strategic Management, 10*(3), 311–323.
Kim, L. (2001). *The dynamics of technological learning in industrialization* (ISSJ 168/2001, UNESCO 2001, pp. 297–308). Malden, MA: Blackwell.
King, N., & Horrocks, C. (2010). *Interviews in qualitative research*. London, England: Sage Publications Ltd.
Kinoshita, Y. (1999). *Technology spillovers through foreign direct investment* (Working Paper, 221). Ann Arbor, MI: University of Michigan William Davidson Institute.
Kinoshita, Y. (2001). *R&D and technology spillovers through FDI: Innovation and absorptive capacity* (CEPR Discussion Paper DP2775). London, England: Centre for Economic Policy Research.
Kokko, A. (1992). *Foreign direct investment, host country characteristics, and spillovers*. Stockholm, Sweden: The Economic Research Institute.
Kokko, A. (1994). Technology, market characteristics, and spillovers. *Journal of Development Economics, 43*(2), 279–293.
Kokko, A. (1996). Productivity spillovers from competition between local firms and foreign affiliates. *Journal of International Development, 8*(4), 517–530.

Kokko, A. (2003). *Globalization and FDI policies, in the development dimension of FDI: Policy and rule-making perspectives* (pp. 29–39). New York, NY: UNCTAD.

Kokko, A., Tansini, R., & Zejan, M. C. (1996). Local technological capability and productivity spillovers from FDI in the Uruguayan manufacturing sector. *Journal of Development Studies, 32*, 602–611.

Kokko, A., Zejan, M., & Tansini, R. (2001). Trade regimes and spillover effects of FDI: Evidence from Uruguay. *Weltwirtschaftliches Archiv, 137*, 124–149.

Konings, J. (2001). The effects of foreign direct investment on domestic firms: Evidence from firm level panel data in emerging economies. *Economics of Transition, 9*(3), 619–633.

Köymen, S., & Sayek, S. (2010). *The role of human capital in productivity spillovers from FDI: An empirical analysis on Turkish manufacturing firms* (Discussion Papers, 10-03). Ankara, Turkey: Department of Economics, Bilkent University.

Krugman, P. (1979). A model of innovation, technology transfer, and the world distribution of income. *The Journal of Political Economy, 87*, 253–266.

Kugler, M. (2001). *The diffusion of externalities from foreign direct investment: The sectoral pattern of technological spillovers* (Working Paper). Southampton, England: University of Southampton.

Kuroda, A. (2001). *Technology transfer in Asia: A case study of auto parts and electrical parts industries in Thailand.* Bunkyo, Japan: Maruzen Planet.

Kuznets, S. (1966). *Modern economic growth: Rate, structure spread.* New Haven, CT: Yale University Press.

Kvale, S., & Brinkmann, S. (2008). *Interviews: Learning the craft of qualitative research interviewing* (2nd ed.). Thousand Oaks, CA: Sage Publications.

Lall, S. (1980). Vertical inter-Frm linkages in LDCs: An empirical study. *Oxford Bulletin of Economics and Statistics, 42*(3), 203–226.

Lall, S. (1992). Technological capabilities and industrialization. *World Development, 20*(2), 165–186.

Lall, S. (1996). *Learning from the Asian tigers.* London, England: MacMillan Press.

Lall, S. (1997). East Asia. In J. H. Dunning (Ed.), *Governments, globalization, and international business* (pp. 407–430). Oxford, England: Oxford University Press.

Lall, S., & Narula, R. (2004). Foreign direct investment and its role in economic development: Do we need a new agenda? *The European Journal of Development Research, 16*(3), 447–464.

Larrain, F. B., Lopez-Calva, L. F., & Rodriquez-Clare, A. (2000). *Intel: A case study of foreign direct investment in Central America* (Working Paper, 58). Cambridge, MA: Center for International Development, Harvard University.

Lavrakas, P. J. (2008). *Encyclopedia of survey research methods.* Thousand Oaks, CA: Sage Publications.

Lee, C. (2008). *Innovation, productivity and exports: Firm-level evidence from Malaysia* (Working Paper Series, March, Vol. 2008-06). Nottingham, England: Nottingham University Business School University of Nottingham Malaysia Campus.

Lenger, A. (2004). *Technological capability and economic growth: A study on the manufacturing industries in Turkey.* Unpublished PhD thesis, METU, Ankara, Turkey.

Lenger, A., & Taymaz, E. (2006). To innovate or to transfer? A study on spillovers and foreign firms in Turkey. *Journal of Evolutionary Economics, 16*, 137–153.

Levinsohn, J., & Petrin, A. (2003). Estimating production functions using inputs to control for unobservables. *Review of Economic Studies, 70*, 317–342.

Li, X., Liu, X., & Parker, D. (2001). Foreign direct investment and productivity spillovers in the Chinese manufacturing sector. *Economic Systems, 25*(4), 305–321.

Liao, T. F. (1994). *Interpreting probability models: Logit, probit, and other generalized linear models. Quantitative applications in the social sciences. Sage University Papers Series.* Thousand Oaks, CA: Sage Publications.

Lin, P., & Saggi, K. (2004). *Multinational firms and backward linkages: A survey and a simple model.* Hong Kong, China: Mimeo, Lingnan University and Southern Methodist University.

Lipsey, R. E., & Sjöholm, F. (2001). *Foreign direct investment and wages in Indonesian manufacturing* (NBER Working Paper, 8299). Cambridge, MA: National Bureau of Economic Research.

Lipsey, R. E., & Sjöholm, F. (2004). Foreign direct investment, education and wages in Indonesian manufacturing. *Journal of Development Economics, 73*, 415–422.

Liu, X., Siler, P., Wang, C., & Wei, Y. (2000). Productivity spillovers from foreign direct investment: Evidence from UK industry level panel data. *Journal of International Business Studies, 31*(3), 407–425.

Long, J. S. (1997). *Regression models for categorical and limited dependent variables*. Thousand Oaks, CA: Sage Publications.

Long, J. S. (2001). *Regression models for categorical and limited dependent variables. Advanced quantitative techniques in the social science series*. Thousand Oaks, CA: Sage Publications.

Lopez-Cordova, J. E. (2002). *NAFTA and Mexico's manufacturing productivity: An empirical investigation using micro-level data* (Working Paper). Washington, DC: Inter-American Development Bank.

Lorentzen, J. (2005). The absorptive capacities of South African automotive component suppliers. *World Development, 33*(7), 1153–1182.

Lorentzen, J., Mollgaard, P., & Rojec, M. (2003). Host-country absorption of technology: Evidence from automotive supply networks in eastern Europe. *Industry and Innovation, 10*(4), 415–432.

MacDougall, G. D. A. (1960). The benefits and costs of private investment from abroad: A theoretical approach. *Economic Record, 36*, 13–35.

Maddala, G. S. (1983). *Limited-dependent and qualitative variables in econometrics*. New York, NY: Econometric Society Monographs, Cambridge University Press.

Mansfield, E., & Romeo, A. (1980). Technology transfer to overseas subsidiaries by U.S.-based firms. *Quarterly Journal of Economics, 95*(4), 737–750.

Markusen, J. R. (1995). The boundaries of multinational enterprises and the theory of international trade. *Journal of Economic Perspectives, 9*(2), 169–189.

Markusen, J. R., & Venables, A. (1999). Foreign direct investment as a catalyst for industrial development. *European Economic Review, 43*, 335–356.

Marshall, C., & Rossman, G. B. (2010). *Designing qualitative research* (5th ed.). Thousand Oaks, CA: Sage Publications.

McDonald, I. M., & Solow, R. (1981). Wage bargaining and employment. *American Economic Review, 71*, 896–908.

McFadden, D. (1984). Econometric analysis of qualitative respond models. In Z. Griliches & M. Intriligator (Eds.), *Handbook of econometrics* (Vol. 2). Amsterdam, Netherlands: North-Holland.

McKelvey, M. (1998). Evolutionary innovations: Learning, entrepreneurship and the dynamics of the firm. *Journal of Evolutionary Economics, 8*, 157–175.

Mead, D. C. (1984). Of contracts and subcontracts: Small firms in vertically disintegrated production/distribution systems in LDC. *World Development, 12*(11/12), 1095–1106.

Merriam, S. B. (2009). *Qualitative research: A guide to design and implementation*. San Francisco, CA: John Wiley and Sons.

Meyers, L. S., Gamst, G., & Guarino, A. J. (2006). *Applied multivariate research: Design and interpretation*. Thousand Oaks, CA: Sage Publications.

Mills, A. J., Durepos, G., & Wiebe, E. (Eds.). (2009). *Encyclopedia of case study research*. Thousand Oaks, CA: Sage Publications.

MITI. (2010). *Otomotiv Sektörü Raporu* (*Automotive sector report*), Sektörel Raporlar ve Analizler Serisi (in Turkish, 2010/3). Ankara, Turkey: Ministry of Industry and Trade of Turkey.

MITI. (2011). *Otomotiv Sektörü Raporu* (*Automotive sector report*), Sektörel Raporlar ve Analizler Serisi (in Turkish, 2011/1). Ankara, Turkey: Ministry of Industry and Trade of Turkey.

Mody, A., Sanders, J., Suri, R., Rao, C., & Contreras, F. (1991). *International competition in the bicycle industry*. Washington, DC: World Bank.

Mollgaard, P. H., & Lorentzen, J. (2004). Exclusive safeguards and technology transfer: Subcontracting agreements in eastern Europe's car component industry. *European Journal of Law and Economics, 17*, 41–71.

Moran, T. (2001). *Parental supervision: The new paradigm for foreign direct investment and development*. Washington, DC: Institute for International Economic.

Nadiri, M. I. (1991). *Innovations and technological spillovers*. New York, NY: Mimeo, NBER, New York University.

Narula, R. (2004). Understanding absorptive capacities in an innovation systems context: Consequences for economic and employment growth. *Merit Research Memorandum, 2004–003*.

Narula, R., & Driffield, N. (2012). Does FDI cause development? The ambiguity of the evidence and why it matters. *European Journal of Development Research, 24*, 1–7.

Narula, R., & Dunning, J. H. (2000). Industrial development, globalization and multinational enterprises: New realities for developing countries. *Oxford Development Studies, 28*(2), 141–167.

Narula, R., & Marin, A. (2003). FDI spillovers, absorptive capacities and human capital development: Evidence from Argentina. *MERIT-Infonomics Research Memorandum Series*, 2003-016. Maastricht, Netherlands: MERIT, Maastricht Economic Research Institute on Innovation and Technology.

Navaretti, G. B., & Venables, A. J. (2004). *Multinational firms in the world economy*. Oxford, England: Princeton University Press.

Nelson, R. R., & Winter, S. (1982). *Evolutionary theory of economic change*. Cambridge, MA: Harvard University Press.

Neuman, W. L. (2006). *Social research methods: Qualitative and quantitative approach* (6th ed.). Boston, MA: Pearson International Edition.

Nunnally, J. C. (1978). *Psychometric theory* (2nd ed.). New York, NY: McGraw-Hill.

Nunnally, J. C., & Bernstein, I. H. (1994). *Psychometric theory* (3rd ed.). New York, NY: McGraw-Hill.

OECD. (1997). *Oslo manual. The measurement of scientific and technological activities: Proposed guidelines for collecting and interpreting technological innovation data. OECD and eurostat, European Commission* (p. 92). Paris, France: OECD Publishing.

OECD. (2002). *Frascati manual. The measurement of scientific and technological activities: Proposed standard practice for surveys on research and experimental development. OECD and eurostat, European Commission* (p. 256). Paris, France: OECD Publishing.

OECD. (2005). *Oslo manual. The measurement of scientific and technological activities: Guidelines for collecting and interpreting innovation data. OECD and eurostat, European Commission* (3rd ed., p. 164). Paris, France: OECD Publishing.

OECD. (2009). *Main science and technology indicators database* (Vol. 2009, No. 1). Paris, France: OECD Publication.

Okamoto, Y. (1999). Multinationals, production efficiency, and spillover effects: The case of the US auto parts industry. *Weltwirtschaftliches Archiv, 135*, 241–260.

Olley, S., & Pakes, A. (1996). The dynamics of productivity in the telecommunications equipment industry. *Econometrica, 64*(6), 1263–1297.

Önis, Z. (1994). Liberalization, transnational corporations and foreign direct investment in Turkey: The experience of the 1980s. In F. Senses (Ed.), *Recent industrialization experience of Turkey in a global context* (pp. 91–109). Westport, CT: Greenwood Press.

Pack, H. (1992). *Technology gaps between industrial and developing countries: Are there dividends for latecomers*? Proceedings of the World Bank Annual Conference on development economics, supplement to the world bank economic review and world bank research observer (pp. 283–302), World Bank, Washington, DC.

Pack, H. (2005). Econometric versus case study approaches to technology transfer. In B. Hoekman & B. S. Javorcik (Eds.), *Global integration and technology transfer* (pp. 29–50). Washington, DC: Palgrave Macmillan and the World Bank.

Pack, H., & Saggi, K. (1997). Inflows of foreign technology and indigenous development. *Review of Development Economics, 1*(1), 81–98.

Pack, H., & Saggi, K. (2001). Vertical technology transfer via international outsourcing. *Journal of Development Economics, 65*, 389–415.

Pack, H., & Westphal, L. E. (1986). Industrial strategy and technological change: Theory versus reality. *Journal of Development Economics, 22*(1), 87–128.

Pallant, J. (2007). *SPSS survival manual, a step by step guide to data analysis using SPSS for windows* (3rd ed.). New York, NY: McGraw Hill.

Pamukçu, T. (2003). Trade liberalization and innovation decisions of firms: Lessons from post-1980 Turkey. *World Development, 31*, 1443–1458.

Pamukçu, T., & Taymaz, E. (2009). *Spillovers in the MENA region: The case of Turkey* (Working Paper Series 463). Cairo, Egypt: Economic Research Forum for the Arab Countries (ERF).

Perez, T. (1997). Multinational enterprises and technological spillovers: An evolutionary model. *Journal of Evolutionary Economics, 7*, 169–192.

Polanyi, M. (1962). *Personal knowledge: Towards a post-critical philosophy*. Chicago, IL: University of Chicago Press.

Portelli, B., & Narula, R. (2004). Foreign direct investment through acquisitions and implications for technological upgrading, case evidence from Tanzania. *Merit Research Memorandum*. Maastricht, Netherlands: MERIT, Maastricht Economic Research Institute on Innovation and Technology.

Radosevic, S. (1999). *International technology transfer and catch-up in economic development*. Cheltenham, England: Edward Elgar Publishing Limited.

Ramachandran, V. (1993). Technology transfer, ownership, and investment in human capital. *Review of Economics and Statistics, 75*, 664–670.

Rath, A. (1994). Technology transfer and diffusion. In J.-J. Salomon, F. Sagasti, & C. Sachs-Jeantet (Eds.), *The uncertain quest: Science, technology and development*. New York, NY: United Nations University.

Rhee, Y. (1990). The catalyst model of development: Lessons from Bangladesh's success with garment exports. *World Development, 18*(2), 333–346.

Rhee, J. W., & Belot, T. (1989). *Export catalysts in low-income countries* (Working Paper). Washington, DC: World Bank.

Rodriquez-Clare, A. (1996). Multinationals, linkages, and economic development. *American Economic Review, 86*(4), 851–873.

Romer, P. (1986). Increasing returns and long-run growth. *Journal of Political Economy, 94*(5), 1002–1037.

Romer, P. (1990). Endogeneous technological change. *Journal of Political Economy, 98*(5), 71–102.

Rubin, H. J., & Rubin, I. S. (2004). *Qualitative interviewing: The art of hearing data* (2nd ed.). Thousand Oaks, CA: Sage Publications.

Safarian, A. E. (1966). *Foreign ownership of Canadian industry*. Toronto, Ontario, Canada: Mcgraw-Hill.

Saggi, K. (2002). Trade, foreign direct investment, and international technology transfer: A survey. *The World Bank Research Observer, 17*(2), 191–235.

Schumpeter, J. A. (1934). *The theory of economic development: An inquiry into profits, capital, credit, interest and the business cycle*. Cambridge, MA: Harvard University Press.

Seidman, I. (2006). *Interviewing as qualitative research: A guide for researchers in education and the social sciences* (3rd ed.). New York, NY: Teachers College Press.

Sekaran, U. (2003). *Research method for business: A skill-building approach* (4th ed.). New York, NY: John Wiley and Sons.

Sekaran, U., & Bougie, R. (2009). *Research methods for business: A skill building approach* (5th ed.). West Sussex, England: John Wiley and Sons Ltd.

Şenses, F., & Taymaz, E. (2003). Unutulan Bir Toplumsal Amaç: Sanayileşme Ne Oluyor? Ne Olmalı? [A Social Objective Forgotten: Industrialization] (in Turkish). In A. H. Köse, F. Şenses, & E. Yeldan (Eds.), *İktisadi Kalkınma Kriz ve İstikrar*. Istanbul, Turkey: İletişim Yayınları.

Sinani, E., & Meyer, K. (2004). Spillovers of technology transfer from FDI: The case of Estonia. *Journal of Comparative Economics, 32*, 445–466.

Sjöholm, F. (1999a). Technology gap, competition and spillovers from direct foreign investment: Evidence from establishment data. *Journal of Development Studies, 36*(1), 53–73.

Sjöholm, F. (1999b). Productivity growth in Indonesia: The role of regional characteristics and direct foreign investment. *Economic Development and Cultural Change, 47*(3), 559–584.

Smeets, R. (2008). Collecting the pieces of the FDI knowledge spillovers puzzle. *The World Bank Research Observer, 23*, 107–138.

Solow, R. (1956). A contribution to the theory of economic growth. *Quarterly Journal of Economics, 70*, 65–94.

Sönmez, A., & Pamukçu, M. T. (2012). Foreign direct investment and technology spillovers in the Turkish manufacturing industry. In I. H. Yetkiner, E. Erdil, & M. T. Pamukçu (Eds.), *Dynamics, innovation policy and economic growth through technological advancements*. Hershey, PA: IGI Global.

SPO. (2007). *Ninth plan for economic development (2007-2013)*. Report of the special expert committee on automotive industry (in Turkish, No: 2736-ÖİK: 687), Ankara: State Planning Organization, Turkey.

Stevens, J. P. (2002). *Applied multivariate statistics for social sciences* (4th ed.). Hillsdale, NJ: Erlbaum.

Tabachnick, B. G., & Fidell, L. S. (2007). *Using multivariate statistics* (5th ed.). Boston, MA: Pearson Education.

Tanaka, H., Iwaisako, T., & Futagami, K. (2007). Dynamic analysis of innovation and international transfer of technology through licensing. *Journal of International Economics, 73*, 189–212.

Tashakkori, A., & Teddlie, C. (1998). *Mixed methodology: Combining qualitative and quantitative approaches*. Thousand Oaks, CA: Sage Publications.

Tashakkori, A., & Teddlie, C. (2010). *SAGE handbook of mixed methods in social and behavioral research* (2nd ed.). Thousand Oaks, CA: Sage Publications.

Taymaz, E., & Lenger, A. (2004). *Multinational corporations as a vehicle for productivity spillovers in Turkey*. Winter PhD Conference (Working Paper, No 04-09). Aalborg, Denmark: Danish Research Unit for Industrial Dynamics (DRUID).

Taymaz, E., & Yılmaz, K. (2007). Productivity and trade orientation: Turkish manufacturing industry before and after the customs union. *Journal of International Trade and Diplomacy, 1* (1), 127–154.

Taymaz, E., & Yılmaz, K. (2008a). *Foreign direct investment and productivity spillovers: Identifying linkages through product-based measures*. Paper presented to the 35th EARIE Conference, Toulouse, France.

Taymaz, E., & Yılmaz, K. (2008b). *Integration with the global economy: The case of Turkish automobile and consumer electronics industries* (Working Paper No: 37). The World Bank, the Commission on Growth and Development. Istanbul, Turkey: Koc University-TUSIAD Economic Research Forum.

Techakanont, K. (2002). *A study on inter-firm technology transfer in the Thai automobile industry*. Unpublished PhD thesis, Graduate School for International Development and Cooperation, Hiroshima University, Higashihiroshima and Hiroshima, Japan.

Techakanont, K., & Terdudomtham, T. (2004). Evolution of inter-firm technology transfer and technological capability formation of local parts firms in the Thai automobile industry. *Journal of Technology Innovation, 12*, 2–20.

Teddlie, C., & Tashakkori, A. (2008). *Foundations of mixed methods research: Integrating quantitative and qualitative approaches in the social and behavioral sciences*. Thousand Oaks, CA: Sage Publications.

Teece, D. J. (1977). Technology transfer by multinational firms: The resource cost of transferring technological know-how. *Economic Journal, 87*, 242–261.

Temenggung, D. (2006). *Productivity spillovers from foreign direct investment: Indonesian manufacturing industry's experience 1975-2000*. Canberra, Australian Capital Territory, Australia: Economics Division, Research School of Pacific and Asian Studies, ANU.

Thomas, R. M. (2003). *Blending qualitative & quantitative research methods in theses and dissertations*. Thousand Oaks, CA: Sage Publications Company.

Travers, M. (2001). *Qualitative research through case studies*. London, England: Sage Publications Ltd.

TUBITAK. *Report for 2023 vision*. Otomotiv Yan Sanayii Parça Sektörü (in Turkish). Draft report, appendix-6A, Ankara, Turkey: TUBITAK. Retrieved from http://www.tubitak.gov.tr/tubitak_content_files/vizyon2023/mm/Ek6a.pdf.

Tung, R. L. (1994). Human resource issues and technology transfer. *International Journal of Human Resource Management, 5*, 808–825.

Ullrich, D. (1998, Winter). Intellectual capital = competence × commitment. *Sloan Management Review*, 15–26.

UN. (1999). *World investment report 1999*. Geneva, Switzerland: United Nations.

UNCTAD. (1997). *World investment report 1997: Transnational corporations, market structure and competition policy*. New York, NY: United Nations.

UNCTAD. (1998). *World investment report 1998: Trends and determinants*. New York, NY: United Nations.

UNCTAD. (2000). *Cross-border mergers and acquisitions and development*. World Investment Report 2000. New York, NY: United Nations.

UNCTAD. (2003). *World investment report 2003: FDI policies for development: National and international perspectives*. New York, NY: United Nations.

UNCTAD. (2005). *World investment report 2005: Transnational corporations and the internationalization of R&D*. New York, NY: United Nations.

UNCTAD. (2011). *World investment report 2011: Non-equity modes of international production and development*. New York, NY: United Nations.

UNCTC. (1994). Costs and conditions of technology transfer through transnational corporations: A regional perspective. In J. H. Dunning (Ed.), *Technology transfers to developing countries, United Nations library on transnational corporations* (Vol. 18). New York, NY: Routledge.

Verbeek, M. (2008). *A guide to modern econometrics* (3rd ed.). New York, NY: John Wiley and Sons.

Wagner, R. K., & Sternberg, R. J. (1985). Practical intelligence in real-world pursuits: The role of tacit knowledge. *Journal of Personality and Social Psychology, 49*(2), 436–458.

Wang, J. (1990). Growth, technology transfer, and the long-run theory of international capital movements. *Journal of International Economics, 29*, 255–271.

Wang, J., & Blomström, M. (1992). Foreign investment and technology transfer: A simple model. *European Economic Review, 36*(1), 137–155.

Wasti, S. N., Kozan, M. K., & Çınar, A. K. (2009). Ana Sanayi Firmalarının Yan Sanayiye Yaptıkları İlişkiye Has Yatırımın Öncülleri: Türk Otomotiv Sanayinde bir Araştırma (in Turkish). *ODTÜ Gelişme Dergisi, 35*, 315–340.

Wasti, N. S., Kozan, M. K., & Kuman, A. (2006). Buyer-supplier relationships in the turkish automotive industry. *International Journal of Operations and Production Management, 26*, 947–970.

Wengraf, T. (2001). *Qualitative research interviewing: Semi-structured, biographical and narrative methods*. London, England: Sage Publications Ltd.

Wong, P. K. (1991). *Technological development through subcontracting linkages*. Tokyo, Japan: Asia Productivity Organization.

Xu, B. (2000). Multinational enterprises, technology diffusion, and host country productivity growth. *Journal of Development Economics, 62*, 477–493.

Yamashita, S. (1991). *Transfer of Japanese technology and management to the ASEAN countries*. Tokyo, Japan: University of Tokyo Press.

Yılmaz, K. (2006). *Türkiye için Doğrudan Yabancı Yatırım Stratejisi'ne Doğru* (in Turkish). Istanbul, Turkey: YASED

Yılmaz, B., & Barbaros, A. (2005). *Development and determinants of foreign direct investment in Turkey: A comparison with Ireland and Spain*. Prepared for the Conference on Elites and EU Enlargement, Germany.

Yin, R. K. (1989). *Case study research: Design and methods* (Rev. ed.). Newbury Park, CA: Sage Publications.

Yin, R. K. (2003). *Applications of case study research* (2nd ed.). Thousand Oaks, CA: Sage Publications.

Zukowska-Gagelmann, K. (2000). Productivity spillovers from foreign direct investment in Poland. *Economic Systems, 24*(3), 223–256.

About the Author

Alper Sönmez was born in 1980 in Istanbul, Turkey. He received his B.Sc. in Economics in 2002 with the First-Class Honours Degree in the Faculty of Economic and Administrative Sciences from Yıldız Technical University (YTU), Turkey. After graduation, he was accepted at *Ph.D. on BS program* in Economics at Middle East Technical University (METU), Turkey. He worked as a teaching & research assistant, and continued his Ph.D. Studies in the Department of Economics at METU from 2002 to 2012. He also spent one year at Cornell University, Charles H. Dyson School of Applied Economics and Management (AEM), as a *Visiting Fellow* in 2011. He received his Ph.D. Degree in Economics in 2012. He also received *METU Thesis of the Year Award* granted by *METU Prof. Dr. Mustafa N. Parlar Education and Research Foundation* in 2012. In addition, he was granted a third award in *The Fourth Contest for Academic Studies* organized by the Central Bank of the Republic of Turkey (CBRT) in 2013. Since February 2013, he has been affiliated with the Department of Economics at Selcuk University, Turkey. His research interests include foreign direct investment, economies of technology and innovation, knowledge and technology spillovers, automotive industry, MNCs, global production networks, and R&D.

Contact: alpersonmez80@gmail.com

A. Sönmez, *Multinational Companies, Knowledge and Technology Transfer*, Contributions to Management Science, DOI 10.1007/978-3-319-02033-4,

MIX
Papier aus verantwortungsvollen Quellen
Paper from responsible sources
FSC® C105338

If you have any concerns about our products,
you can contact us on
ProductSafety@springernature.com

In case Publisher is established outside the EU,
the EU authorized representative is:
Springer Nature Customer Service Center GmbH
Europaplatz 3, 69115 Heidelberg, Germany

Printed by Libri Plureos GmbH
in Hamburg, Germany